DIAMOND & DIAMOND-LIKE FILM APPLICATIONS

HOW TO ORDER THIS BOOK

Diamond & Diamond-Like Film Applications

Nucleation and Growth
Modeling and Phase Equilibria
Properties and Characterization
Diamond-Like Carbon
Wide Bandgap Nitrides and Carbides

PROCEEDINGS
Third International Symposium on Diamond Films
St. Petersburg, Russia
June 16–19, 1996

EDITED BY

Peter J. Gielisse
V. I. Ivanov-Omskii
G. Popovici
M. Prelas

TECHNOMIC
PUBLISHING CO., INC.
LANCASTER · BASEL

Diamond and Diamond-Like Film Applications

a TECHNOMIC® publication

Published in the Western Hemisphere by
Technomic Publishing Company, Inc.
851 New Holland Avenue, Box 3535
Lancaster, Pennsylvania 17604 U.S.A.

Distributed in the Rest of the World by
Technomic Publishing AG
Missionsstrasse 44
CH-4055 Basel, Switzerland

Printed in the United States of America
10 9 8 7 6 5 4 3 2 1

Main entry under title:
  Diamond and Diamond-Like Film Applications

A Technomic Publishing Company book
Bibliography: p.
Includes index p. 495

Library of Congress Catalog Card No. 98-60401
ISBN No. 1-56676-655-9

Table of Contents

v

SESSION 2: NUCLEATION AND GROWTH

SESSION 5: DIAMOND-LIKE CARBON AND OTHER MATERIAL

Foreword

The main objective of this conference was to bring together scientists and business people active in the fields of diamond, diamond films, diamond-like carbon, nanodiamond, cubic boron nitride, carbon nitride and related materials, for the purpose of scientific information exchange and to discuss potential technical collaborations and business alliances.

Symposium topics included current applications of diamond materials as in electronics, coated tools, heat spreaders and medical technology, as well as developing areas of interest such as coated detectors, cold cathode emitters for electronics, flat panel displays and UV detectors. Papers on the fundamental aspects of materials growth and characterization, and contributions relating to higher growth rates, larger specimen areas, the lowering of defect densities and the development of processing reactors, were sought as well.

The symposium was attended by over 200 participants. The peer reviewed papers selected from the invited, delivered and poster presented papers, are presented in these proceedings.

The organizational committee would like to thank the following organizations which through their sponsorship, made the symposium possible: Russian Ministry of Science and Scientific Policy; Ioffe Physico-Technical Institute, Russian Academy of Sciences; Institute of Physical Chemistry, Russian Academy of Sciences; Office of Naval Research, Europe; Florida State University, USA; Florida Agricultural and Mechanical University, USA; and University of Missouri, Columbia, USA.

Keynote Speakers

N. Bulienkov (Institute of Physical Chemistry, Russia)
"Atomic Construction of DLC"
V. Chelnokov (Ioffe Institute, Russia)
"Electronics of Wide Band-Gap Semiconductors"
J. Davidson (Vanderbilt University, USA)
"Diamond as an Active Sensor Material"
A. Gicquel (Paris Nord University, France)
"Temperature and Hydrogen Atom Concentrations in Microwave
Plasmas Used for Diamond Depositions"
P. Gielisse (Florida State University, USA)
"Mechanical Properties of Diamond, Diamond-Like Films and
Related Materials"
T. Izumi (Tokai University, Japan)
"ECR Defects in As-Growth and Ion-Implanted Diamond Films"
M. Kamo (NIRIM, Japan)
"Characterization of Phosphorus-Doped Diamond Prepared by
Microwave Plasma CVD"
S. Khasawinah (Boston University, USA)
"Radiation Effects on Diamond Films"
S. Mitura (Technical University of Lodz, Poland)
"Applications of Amorphous Diamond Films and Coatings"
S. Pimenov (General Physics Institute, Russia)
"Ultrathin Diamond Films: Synthesis, Properties and Applications"
Yu. Pleskov (Institute of Electrochemistry, Russia)
"Photochemistry of Synthetic Diamond Electrodes"
G. Popovici (University of Missouri, USA)
"Lithium in Diamond"

B. Spitsyn (Institute of Physical Chemistry, Russia)
"Hydrogen in Diamond Films"
N. Suetin (Moscow State University, Russia)
"Modeling of CVD Reactor"
E. Terukov (Ioffe Institute, Russia)
"Photoluminescent Properties of Nanodiamond and a-C:H Films"

CONFERENCE ORGANIZING COMMITTEE

B. Spitsyn (co-chair)
Institute of Physical Chemistry, Russia

J. Davidson
Vanderbilt University, USA

M. Prelas (co-chair)
University of Missouri, USA

V. I. Ivanov-Omskii
Ioffe Physico-Technical Institute, Russia

P. Gielisse (co-chair)
Florida State University, USA

S. Mitura
Technical University of Lodz, Poland

G. Popovici (co-chair)
University of Missouri, USA

E. Terukov
Ioffe Physico-Technical Institute, Russia

V. Chelnokov
Ioffe Physico-Technical Institute, Russia

TECHNICAL PROGRAM COMMITTEE

E. Terukov (co-chair)
Ioffe Physico-Technical Institute, Russia

M. Prelas
University of Missouri, USA

A. Ya. Vol' (co-chair)
Ioffe Physico-Technical Institute, Russia

V. Ralchenko
General Physics Institute, Russia

P. Gielisse (co-chair)
Florida State University, USA

H. Schmidt
SI Diamond, USA

K. Bigelow
Norton Diamond Films, USA

T. Stoica
Institute of Physics and Technology of
    Materials, Romania

T. Izumi
Tokai University, Japan

N. Suetin
Moscow State University, Russia

A. Gicquel
Paris Nord University, France

R. Sussmann
De-Beers, UK

A. Gippius
Lebedev Physical Institute, Russia

J. Szmidt
Technical University of Warsaw, Poland

K. Gleason
MIT, USA

V. Varichenko
Belarus State University, Belarus

W. Hsu
Sandia, USA

R. Wilson
Hughes Aircraft, USA

M. Kamo
NIRIM, Japan

R. Zachai
Daimler Benz, Germany

V. Mironov
Almazy Rossii-Sakha Co. Ltd., Russia

INTERNATIONAL ADVISTORY COMMITTEE

A. Badzian
Pennsylvania State University, USA

P. Koidl
Fraunhofer Institut, Freiburg, Germany

V. Chelnokov
Ioffe Physico-Technical Institute, Russia

H. Okushi
Electrotechnical Laboratory, Tsukuba, Japan

L. Bouilov
Institute of Physical Chemistry, Russia

A. Sokolowska
Technical University of Warsaw, Poland

N. Fujimori
Sumitomo Electric Co., Japan

V. Vavilov
Lebedev Physical Institute, Russia

A. Hiraki
Osaka University, Japan

V. Varnin
Institute of Physical Chemistry, Russia

LOCAL ARRANGEMENTS COMMITTEE

E. Terukov (co-chair)
Ioffe Physico-Technical Institute, Russia

I. Kotina
St. Petersburg Nuclear Physics Institute, Russia

A. Vul' (co-chair)
Ioffe Physico-Technical Institute, Russia

S. Gordeev
Central Sci. Research Inst. Materials, Russia

TECHNICAL AND INDUSTRIAL EXHIBITS COMMITTEE

E. Terukov
Ioffe Physico-Technical Institute, Russia

N. Niculesco
FAMU/FSU, College of Engineering, USA

S. Gordeev
Central Science Research Institute of
   Materials, Russia

CONFERENCE SECRETARIAT

I. Trapeznikova and O. Konkov

DIAMOND AND DIAMOND-LIKE
FILM APPLICATIONS

Diamond Based p-i-n Transistor

A. A. MELNIKOV* AND V. S. VARICHENKO
Belarussian State University
220050 Minsk
Belarus

A. M. ZAITSEV**
University of Hagen
58084 Hagen
Germany

and

A. S. SHULENKOV
Institute RM
220115 Minsk
Belarus

1. INTRODUCTION

Diamond based semiconductor structure with an active insulating zone is a very perspective component of diamond electronics. The work of such device is based on space charge limited current (SCLC) or electron-hole recombination and space charge limited double injection current in insulator. The main advantage of the structure is, first of all, a lower concentration of defects in the dielectric

---

*Author to whom correspondence should be addressed. Dr. Alexander A. Melnikov, Belarussian State University, Fr. Skorina 4, 220050 Minsk, Belarus. E-mail: MELNIKOV@ HEII.BSU.MINSK.BY.
**Also at Belarussian State University 220080 Minsk Belarus.

active zone than in doped one, that allows to receive high mobility of the carriers in diamond, and second, in elimination of restrictions, connected with deep acceptor level of B in diamond. P-i-M and p-i-n diodes, p-i-p structures, p-i-p field effect transistor (FET) with a Schottky barrier gate and p-i-p FET with a SiO_2 gate insulator have been already developed and described [1–5]. Use of i-diamond as working area leads to the improvement of working parameters of active diamond devices. In p-i-p FET the holes are injected from p-diamond and transported by SCLC mechanism in the i-region. However currents of double injection significantly exceed SCLC [6]. One of the variants of high power insulated gate p-i-n transistor is described [7].

In this report, diamond p-i-n FET with a Schottky barrier gate was fabricated for the first time.

2. EXPERIMENTAL

The cross section of the p-i-n FET with a Schottky barrier gate is shown in Figure 1. The transistor is a planar p-i-n structure with i-zone covered by Al. The sizes p- and of n-areas made 50×240 mcm and length of i layer was 20 mcm. The transistor was produced on a polished plate of a natural crystal of a type IIa using standard photolithography and ion implantation. p-Type region was manufactured by boron implantation with distributed energy 25–150 KeV and total doze 2×10^{16} cm^{-2} and with subsequent thermal annealing at 1400°C in vacuum. n-Type region—by lithium implantation with energy 50 KeV and doze 4×10^{16} cm^{-2}. Graphitized layer after annealing was removed by etching in a hot solution of H_2SO_4: CrO_3.

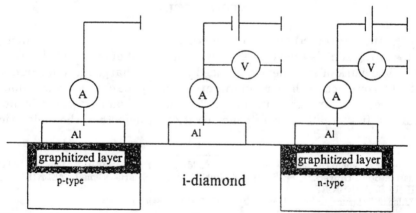

Figure 1. Cross section of diamond based p-i-n transistor with a Schottky barrier gate.

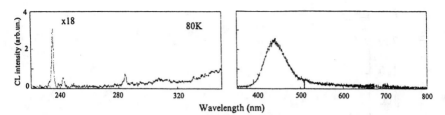

Figure 2. CL spectrum of natural diamond substrate on which has been fabricated planar p-i-n transistor.

Selection of diamond substrates (IIa type natural diamond) has been performed using cathodoluminescence technique. Intensity of free exciton line was chosen as an indicator of crystalline quality of substrates. The choice of this parameter is done because the current of double injection grows with increase of lifetime of injected electrons and holes [6]. As the intensity of free exciton directly reflects lifetime of nonequilibrium charge carriers, in crystals of diamond with intensive radiation of free exciton it will be possible to observe high enough current of double injection. A p-i-n transistor was made on diamond, spectrum cathodoluminescence of which is presented in Figure 2. It is characterized by presence of intensive free exciton and A-band, which points at rather low concentration of levels recombination in a substrate.

3. RESULTS AND DISCUSSION

Earlier, as it was shown [2,5], the operation of such p-i-n diodes on dia-

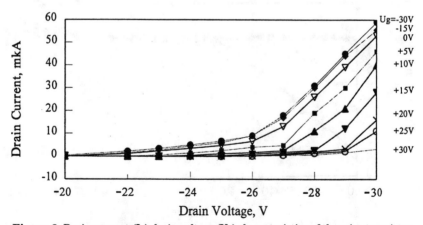

Figure 3. Drain current (I_d)-drain voltage (V_d) characteristics of the p-i-n transistor.

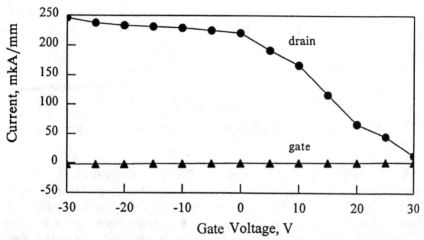

Figure 4. Drain current (I_d)-gate voltage (V_g) and gate current (I_g)-gate voltage (V_g) characteristics of the p-i-n transistor at V_d = 30 V.

mond is based on double carrier injection in insulating diamond with a limitation of the current by space charge and by electron-hole recombination. Drain current-drain voltage (I_d-V_d) and drain current-gate voltage (I_d-V_g) characteristics of the p-i-n transistor were plotted in Figure 3 and Figure 4, respectively.

It is possible to operate I_d by V_g in a depletion mode with the transconductance ~8 mcS/mm at positive V_g and with the transconductance ~0.8 mcS/mm at negative V_g.

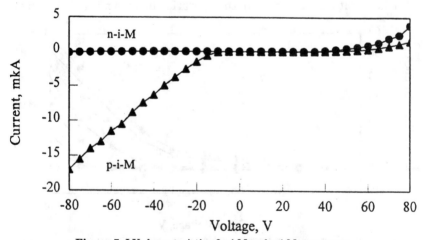

Figure 5. I-V characteristic of p-i-M and n-i-M structures.

In made p-i-n transistor an opportunity of control of a current of double injection is connected with the different height of the potential barrier of the n-diamond–i-diamond for electrons and p-diamond–i-diamond for holes. The parameters of a potential barrier s-diamond–i-diamond depend mainly on concentration and distribution of traps in i-area and concentration of charge carrier [6,8,9]. Distinction of potential barriers p–i-diamond and n–i-diamond specifies a various kind I-V characteristics of p-i-p and n-i-n structures. In case p-i-p structure on the given crystal its opening occurs at voltage 10–30 V and output in a mode SCLC 20–40 V [1,9], in case n-i-n structure injection current does not occur before voltage 100 V. I-V characteristics p-i-M and n-i-M structures, shown in Figure 5, significantly differ. The direct currents in a p-i-M structure much exceed currents in n-i-M structure. It specifies the fact that height of a barrier n–i-diamond for electrons is higher than barrier p–i-diamond for holes.

While applying a positive voltage to the gate of p-i-n transistor, the height of p–i-diamond barrier increases and results in the reduction of a holes flow through i area. The quantity of the holes, reaching of n-area, significantly decreases; that causes reduction of injection of electrons and fall of double injection currents. While applying a negative voltage to the gate, the increase of a holes flow, reaching of n-area, influences a high barrier for electrons for n–i-diamond in a smaller degree. It results in the reduction of the transconductance of the transistor. As p-i-n diodes work up to high temperatures [2], the opportunity of control of currents of double injection in the diamond p-i-n transistor opens a new way for high power high temperature diamond transistors.

4. CONCLUSION

Diamond p-i-n FET with a Schottky barrier gate was fabricated for the first time. The change of voltage applied to gate of the structure leads to the control of double injection current passing through i-region. Diamond p-i-n FET with a Schottky barrier gate operates in depletion modes.

ACKNOWLEDGEMENT

The authors would like to thank V. Dravin for ion implantation. This work was partially supported by DAAD and INTAS project No. 94-1982.

REFERENCES

1. Denisenko, A. V., A. A. Melnikov, A. M. Zaitsev, V. I. Kurganskii, A. J. Shilov, J. P. Gorban and V. S. Varichenko. 1992. *Mat. Sci. and Eng.*, B11:273.

2. Melnikov, A. A., A. V. Denisenko, A. M. Zaitsev, W. R. Fahrner and V. S. Varichenko. *Proc. Applied Diamond Conference 1995 Applications of Diamond Films and Related Materials: Third International Conference*, Gaithersburg, MD, USA, August 21–24 1995, A. Feldman, W. A. Yarbrough, M. Murakawa, Y. Tzeng and M. Y. Yoshikawa, eds., *NIST Spec. Publ.* 885, 639.

3. Pan, L. S. and D. R. Kania, eds. 1995. Diamond: *Electronic Properties and Applications*. Kluwer Academic Publishers, 472 p.

4. Miyata, K., K. Nishimura and K. Kobashi. 1995. *IEEE Trans. Electron. Dev.*, 42(11):2010.

5. Guseva, V. B., E. A. Konorova, Y. A. Kuznetsov and V. F. Sergienko. 1978. *Sov. Phys. Semicond.*, 12:290.

6. Lampert, M. A. and D. Mark. 1970. *Current Injection in Solids*. New York: Academic Press.

7. Huang, A. Q. 1995. *IEEE Electron Dev. Lett.*, 16(9):408–410.

8. Denisenko, A. V., A. A. Melnikov, A. M. Zaitsev, B. Burchard and W. R. Fahrner. 1995. *Diam. Films & Tech.*, 5.

9. Rusetski, M. S., A. V. Denisenko, A. A. Melnikov and A. M. Zaitsev. 1996. *Proc. ISDF3*, 16–19 May 1996, S. Peterburg, Russia. In print.

Diamond as an Active Sensor Material

J. L. DAVIDSON* AND W. P. KANG

AES Department and ECE Department
Vanderbilt University
Nashville, TN 37235

ABSTRACT: Diamond has attractive properties as an advanced electronic material. Its combination of high carrier mobility, electric breakdown, and thermal conductivity results in the largest calculated figures of merit for speed and power of any material. Recently, the discovery and development of what might be called sophisticated secondary effects in diamond, are opening interesting sensor applications. Boron doped diamond will change electrical resistance with strain (piezoresistance), meaning it can be used as a strain gauge on rugged electronic microsensors for pressure and acceleration sensing. This paper will present some critical issues of diamond for microelectromechanical sensing applications such as its rupture stress and edge stress of diamond diaphragms and the high temperature responses of a diamond pressure sensor. We will describe an all diamond pressure microsensor that measures pressure at more than 300°C. Also, we have observed that layered diamond films can behave as chemical sensors measuring hydrogen, oxygen and many other chemicals' concentrations. For example, a diamond-based chemical gas sensor using Pd/i-diamond/p^+-diamond metal-insulator-semiconductor diode structure has been made and the behavior of a new diamond-based chemical gas sensor has been studied. Also, the creation and properties of diamond microtips as field emitters are discussed.

1. INTRODUCTION

Following is an introductory overview of the diamond examples that are the subject of this paper.

---

*Author to whom correspondence should be addressed.

Microelectromechanical Piezoresistance (PZR) Based Effects and Microdevices. Boron doped diamond will change electrical resistance with strain (piezoresistance), meaning it can be used as a strain gauge on rugged electronic microsensors for pressure and acceleration sensing. We have studied the PZR effect in *p*-doped diamond [1] and built an all diamond pressure microsensor [2] that measures pressure at more than 300°C. Diamond deposition processing and silicon photolithographic and etching techniques were used to create undoped diamond diaphragms a few millimeters in diameter and 5–10 microns thick. Delineated and electrically isolated doped diamond resistors nominally 100 microns wide by 500 microns long by 4 microns thick, with thick film silver based interconnect are fabricated on top of the diaphragm. Zero strain values of the resistors are nominally a few hundred Kohms. Isolation ratio is greater than 10^3. As the membrane is flexed by pressure, the $\Delta R/R$ piezoresistance (PZR) of the diamond resistors was measured. Various PZR configurations and temperature behaviors are being examined.

Thin Film Diamond Diode Chemical/Gas Sensors. We have observed that layered diamond films can behave as chemical sensors measuring hydrogen, oxygen and many other chemicals' concentrations. For example, a diamond-based chemical gas sensor using Pd/*i*-diamond/p^+-diamond metal-insulator-semiconductor diode structure has been made and the behavior of a new diamond-based chemical gas sensor is being studied. Hydrogen sensing characteristics of the device have been investigated as a function of hydrogen partial pressure and temperature. The hydrogen sensitivity is found to be large, repeatable and reproducible. The adsorption/desorption of hydrogen gas by the sensor causes large changes in its current-voltage (I-V) and capacitance-voltage-frequency (C-V-F) characteristics. The operating principle and the sensing mechanism are under investigation. Analysis of the steady state reaction kinetics confirm that the sensing mechanism is attributable to barrier height modulation effects upon hydrogen adsorption/desorption in the diamond Schottky diode. The finding opens the way to practical implementation of diamond- based gas detectors with better performance and temperature tolerance than those based on conventional semiconductor technology.

Micro-Patterned Diamond Field Emitter Vacuum Diode Arrays. Electron field emission from arrays of patterned pyramids of polycrystalline diamond for vacuum diode applications is being investigated. High current emission from the patterned diamond microtip arrays was obtained at low electric fields. An emission current from the diamond microtips of 0.1 mA was observed for a field of <10 V/μm. Field emission for these diamond microtips exhibits significant enhancement in total emission current compared to silicon emitters. Moreover, field emission from patterned polycrystalline diamond pyramidal tip arrays is unique in that the applied field is found to be lower compared to that required for emission from Si, Ge, GaAs, and metal surfaces. The fabrication process utilizes selective deposition and molding of polycrystalline diamond film in a silicon cavity mold and subsequent creation of free standing polycrystalline diamond diaphragm with diamond pyramidal microtip array. The effect of doping, hydrogen plasma tip sharpening and high temperature annealing on the current emission characteristics are currently under investigation.

2. DISCUSSION AND RESULTS

As this paper is a collection of examples, the descriptions following will be per example.

2.1 Microelectromechanical Piezoresistance (PZR) Based Effects and Microdevices

Pressure Sensors. PZR based pressure sensors can be derived from micromechanical diaphragm structures. This configuration can be achieved in diamond because polycrystalline diamond films (PDF) piezoresistors can be directly patterned on undoped diamond diaphragms. There are two advantages in using PDF as diaphragm material. First, PDF is chemically inert and can serve as an etch-stop with large etching time tolerance. Second, PDF can be deposited with control by adjusting deposition time and other growth parameters resulting in selectability of diaphragm thickness. Therefore, the full-scale pressure range can be varied with PDF as diaphragm material. The sensitivity of diaphragm sensors depends on a variety of parameters, such as the diaphragm size, the impurity doping concentration and the gauge location. Resistor placement on a diaphragm can be chosen to achieve maximum strain and optimize the PZR effect.

The stress pattern due to pressure difference across a diaphragm is known [3]. The radial strain at radius r in the diaphragm is

$$\varepsilon_{rr} = \frac{3P(1-v^2)}{8Et^2}(R_o^2 - 3r^2) \tag{1}$$

where R_o is the radius of the diaphragm, P is the pressure difference, t is thickness, E is Young's modulus, v is Poisson's ratio which is 1 to 2 for diamond. The tangential strain can be expressed as

$$\varepsilon_{\theta\theta} = \frac{3P(1-v^2)}{8Et^2}(R_o^2 - r^2) \tag{2}$$

In our previous work the PZR effect in *free-standing* PDF resistors was examined by determining gauge factors as a function of applied strain on samples with different doping concentrations. Variations in doping concentrations were achieved by in-situ solid source doping with different concentration boron compound wafers [4]. By varying dopant concentration, resistivity values from 0.01–200 ohm-cm at 300 K were observed. Resistance measurements over a temperature range of 25°C to 225°C were performed for various doping concentrations. Figure 1 shows the relative change in resistance of the resistors over the strain investigated. Gauge factors between 10 to 100 were observed, depending on doping concentration and PDF grain size.

More recently [5], and noted here, we have examined the dependency of the gauge factor (as determined by calibrated flexing of attached free-standing PDF) on the boron doping concentration, using SIMS analysis to determine the boron dopant concentration. Figure 2 shows those results. The gauge factor, and hence PZR, apparently has a maximum at moderate doping levels at 25°C.

Pressure Sensor Fabrication, Measurement. Polycrystalline diamond films for the pressure sensor were grown using an ASTeX (from ASTeX, Inc., Woburn, MA) microwave plasma assisted chemical vapor deposition diamond deposition system. Typical process parameters for the growth of these films were: pressure, 40 torr; substrate temperature, 850°C; gas flow rate, 500 sccm hydrogen and 5 sccm methane; microwave power, 1500 watts. The average growth rate was nominally 0.5 microns per hour. Coherent, pinhole free, diamond membranes may be produced by utilizing the above process in as little as ten hours deposition time to achieve

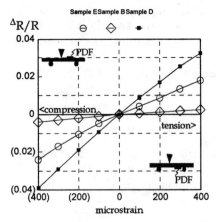

Figure 1. Resistance change vs. strain^2.

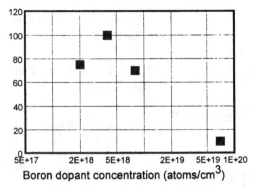

Figure 2. The longitudinal gauge factor of boron doped PDF as a function of the dopant concentration.

thickness of 5 microns on a 2 cm × 2 cm silicon substrate which had been pre-deposition polished with 0.1 micron diamond powder. Films were annealed for one minute at 850°C in argon to remove adsorbed or incorporated hydrogen [6].

The p-type boron doped PDF piezoresistors were fabricated on undoped diamond films via in-situ boron solid source doping method mentioned above. A lift-off mask composed of a printed dielectric (ESL, Inc. type 4771-G) was used to achieve selective deposition of the doped diamond resistors. The solid wafer boron compound source was placed under the substrate during resistor deposition. Boron out-gases at the processing temperatures and is incoporated into the growing diamond film [4]. The dielectric mask was then lifted off (etched away) by an HF etch. Terminal metallization was achieved by using silver based thick film conductor paste (ESL, Inc. type 590) printed and fired at 500°C. The diamond membrane was accomplished with an isotropic silicon etch of the back side of the substrate. The process flow to achieve the diamond pressure microsensor is shown in Figure 3. The layout of the first circular diaphragm design of the sensor is shown in Figure 4. The dashed circle represents the edge of the diaphragm. Four resistors (width = 0.5 mm) are positioned on top of the diaphragm. Two resistors (R1, R2) have their length oriented radially to examine longitudinal PZR, and the other two resistors (R3, R4) have their length orthogonal to the radius to examine transverse PZR. The conductor pads are off the diaphragm to avoid interference. Two other resistors are placed off diaphragm, on opposite corners, for reference purposes. The doped diamond resistors directly deposited on the diamond diaphragm, exhibited excellent adherence. The diaphragm thickness was 5 microns and diameter 0.92 cm. The dia-

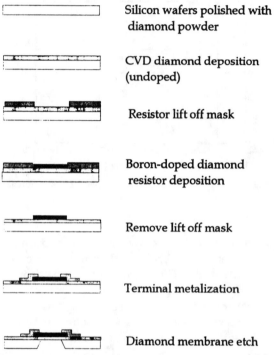

Silicon wafers polished with diamond powder

CVD diamond deposition (undoped)

Resistor lift off mask

Boron-doped diamond resistor deposition

Remove lift off mask

Terminal metalization

Diamond membrane etch

Figure 3. Diamond pressure sensor process flow overview.

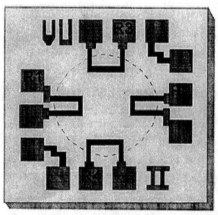

Figure 4. Example design pattern for diamond pressure sensor.

phragm was highly insulative (>1 GΩ) as determined by I-V measurements at room temperature. Pressure differential was produced on a specially designed fixture to examine the PZR of these diamond resistors. Pressure was applied to the membrane in the range of 0 to 250 mmHg. The resistances of the four resistors were taken individually at varying pressure. Figure 5 shows the resistance change versus pressure of these resistors from a typical sample. It can be seen that the resistance change of the two longitudinal PZR, identified as R1 and R2, is larger than the other two transverse resistors, R3 and R4. R1 has piezoresistance of 0.1% per 100 mmHg and R2 has a slightly lower response. This is due to slight asymmetry of the alignment of this particular device, where more length of R1 is on the membrane than R2. R3 and R4 have a diminished PZR, as they are in the transverse configuration. The resistivity of these resistors was 4 to 5 ohm-cm and the gauge factor was estimated to be 7. [Membrane rupture stress (in compression) was observed to be greater than 350 mmHg (our gauge limit) in one sample membrane.] To further examine the rupture stress of diamond diaphragms, several more samples were tested. The rupture stress of the diamond diaphragms ranged from $1.84 \pm 0.42 \times 10^{11}$ to $5.20 \pm 1.2 \times 10^{11}$ dyne/cm^2 with an average of $3.40 \pm 0.79 \times 10^{11}$ dyne/cm^2. The difference in rupture stress among these samples was because of imperfections in the diamond diaphragms and differences in thickness and boundary clamping. Compared to the rupture stress of silicon, 6.24×10^8 dyne/cm^2 [12], diamond film has a rupture stress nominally three orders of magnitude larger. Hence, the diamond diaphragm is more rugged than silicon, can withstand much

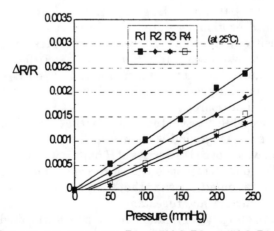

Figure 5. $\Delta R/R$ vs. pressure, no pressure, R1 = 173 kΩ, R2 = 145 kΩ, R3 = 171 kΩ, R4 =

higher pressure conditions and a thinner diaphragm could be achieved using diamond.

The scaling effects of microelectronics can be applied to achieve smaller structure [7]. Other designs intended to better utilize radial and transverse strain in a resistor bridge configuration have been fabricated and described elsewhere [8]

2.2 Thin Film Diamond Diode Chemical/Gas Sensors

The diamond-based chemical gas sensor was fabricated in a layered Pd/i-diamond/p-diamond Schottky diode configuration on a tungsten substrate as shown in insert of Figure 6. Heavily doped p-type polycrystalline diamond films, doped in-situ from a boron solid source, 5–10 μm thick were deposited on the surface of the tungsten substrate using PECVD. A thin (0.1 μm) undoped diamond interfacial layer was then deposited on the p-diamond layer. The films were deposited using a mixture of 99% hydrogen and 1% methane gas, and a substrate temperature of 850°C. Films were then annealed for one minute at 850°C in argon. Raman spectroscopy of these diamond films shows the 1332 cm^{-1} peak indicating characteristic sp^3 bonding and a slight broad graphite peak at 1580 cm^{-1}. SIMS studies of the p-diamond layer indicate a boron doping concentration of 10^{18} cm^{-3}. Finally, a palladium electrode was thermally evaporated on the i-diamond surface to complete the device.

Typical I-V characteristics of the diamond-based hydrogen sensor, before and after exposure to H$_2$ are shown in Figure 6, for the device operating at 27°C. The figure shows a large change in the I-V characteristics of the device, shifted from near ohmic to rectifying behavior, upon the device's exposure to 0.01 torr of H$_2$ gas. The effect is reproducible and repeatable upon adsorption/desorption of hydrogen gas. Figure 7 shows the detection sensitivity, ΔI, versus hydrogen partial pressure for the device operating at a constant temperature under a fixed (-0.4 V) foward bias voltage. The figure shows a rapid increase in ΔI upon H$_2$ adsorption at low hydrogen partial pressures, followed by a trend to saturation at higher H$_2$ partial pressures.

A typical repeatability test of the sensor for the detection of H$_2$ gas in an open air (1 atm.) background is shown in Figure 8. The device, maintained at 85°C, was exposed to a series of H$_2$ pulses with a fixed flow rate of 10 ml/min. The device responds with a recovery time in seconds upon turning the H$_2$ gas on/off, showing that the detection of hydrogen is fast, sensitive, reproducible, and repeatable.

The current-voltage (I-V) relationship of a Schottky diode is given by [9]

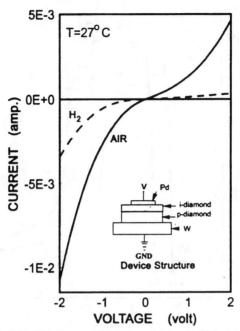

Figure 6. *I-V* of Pd-diamond H_2 sensor: (a) in air, (b) in 0.01 torr H_2.

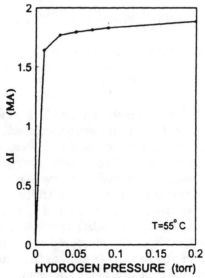

Figure 7. Change in diode current, ΔI, with hydrogen partial pressure at 55°C.

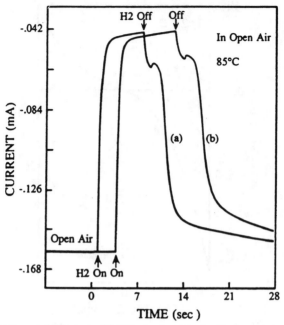

Figure 8. Repeatability test of Pd-diamond sensor for H_2 detection in open air.

$$I = I_0 \exp(qV/nkT)\{1 - \exp(-qV/kT)\} \tag{3}$$

where the saturation current density I_0 is given by

$$I_0 = AA^{**}T^2 \exp(-q\phi_b/kT) \tag{4}$$

Here ϕ_b is the Schottky barrier height, n is an ideality factor, A^{**} is the effective Richardson constant, T is the measurement temperature in $°K$, and k is Boltzmann's constant. Figure 9 is log of $I/\{1 - \exp(-qV/kT)\}$ against V, for the sensor operating at 27°C in air and hydrogen environments, showing a linear relation over the entire reverse bias region, indicating a barrier dominated conduction mechanism controlled by thermionic field emission. Therefore, accurate determination of I_0 and can be extracted from the reverse I-V characteristics to the zero-voltage intercept and the gradient of the plot, respectively. The measured values of I_0 and calculated values of ϕ_b are given in Figure 9. As seen from the figure, ϕ_b changes by 0.05 eV after exposure to H_2. Ideality factors determined from the straight line sections of the plots were 1.15 for the diode examined and they remain unchanged, but the barrier heights obtained change from 0.55 eV in air to

0.60 eV in hydrogen. In forward bias the curve departs from linearity due to series resistance effects or space-charge-limited conduction. These results are discussed further elsewhere [10].

The observed hydrogen-induced barrier height modulation effect can be related to the hydrogen reaction kintics that occur at Pd/i-diamond interface of the Schotty diode [10].

$$H_{2(g)} \overset{c_1}{\underset{d_1}{\rightleftharpoons}} 2H_a \overset{c_2}{\underset{d_2}{\rightleftharpoons}} 2H_{ai} \tag{5}$$

where g and a stand for gaseous and adsorbed species, respectively. The adsorption rate constants are denoted as c_1 and c_2 and the desorption rate constants are denoted as d_1 and d_2. Under the stady state condition in Equation (5), the hydrogen coverage, θ, at the interface can be expressed as:

$$\theta/(1-\theta) = (c_1/d_1)^{1/2} (P_{H_2})^{1/2} \tag{6}$$

If the change in barrier height, $\Delta\phi_b$, induced by hydrogen adsorption [10] is proportional to the hydrogen coverage, θ, i.e., $\Delta\phi_b = \Delta\phi_{max} \theta$, then Equation (6) can be reduced to

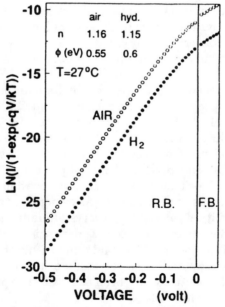

Figure 9. Plot of $\ln (I/(1 - \exp(-qV/KT)))$ vs. voltage.

$$(1/\Delta\phi_b) - (1/\Delta\phi_{max}) = (1/\Delta\phi_{max})(d_1/c_1 P_{H_2})^{1/2} \qquad (7)$$

where $\Delta\phi_{max}$ is the maximum barrier change. Furthermore, the barrier height of a Schottky barrier diode is related to the I-V characteristics. Thus, one finally obtains

$$1/(\ln(I_0/I_{0g})) = 1/(\ln(I_0/I_{0g(max)})) + [1/(I_0/\ln(I_{0g(max)}))](d_1/c_1)P_{H_2})^{1/2} \qquad (8)$$

Equation (8) defines a linear plot of $1/\ln (I_0/I_{0g})$ versus $(1/P_{H_2})^{1/2}$ and plots of $1/\ln (I_0/I_{0g})$ versus $(1/P_{H_2})^{1/2}$ of the data for the diamond sensor diode are indeed linear, indicating hydrogen reaction kinetics in the Pd-diamond diode structure.

C-V measurements of the Pd/i-diamond/p^+-diamond device for several signal frequencies before and after exposure to H_2 gas were made [10] but space is not available to exhibit here. A large decrease in capacitance upon H_2 absorption was evident.

2.3 Micro-Patterned Diamond Field Emitter Vacuum Diode Arrays

A new field emission device using sharp tips of well patterned pyramids of polycrystalline diamond for the development of vacuum field emitter diode arrays is described. The use of local field enhancement at sharp points constructed by molding and micromachining techniques in diamond material as reported here is a new development in this area.

The diamond films were characterized by Raman spectroscopy showing the 1332 cm^{-1} peak indicating a good diamond film, along with a low broad graphite peak at 1580 cm^{-1}. The microstructure of the free standing diamond diaphragm with patterned microtips was examined by scanning electron microscopy (SEM). Figure 10 shows a patterned polycrystalline diamond pyramidal microtip on a polycrystalline diamond membrane. The diamond pyramid has a base dimension of 3μm $\times$ 3μm and a tip radius of about 200 A. As evident by these SEM micrographs, the capability of patterning polycrystalline diamond microtips on diamond film is demonstrated. Selective deposition and molding of undoped or doped polycrystalline diamond film in a silicon mold and subsequent creation of free standing polycrystalline diamond diaphragm with diamond pyramidal microtip have been achieved. The process technique is compatible with IC technology.

The electrical properties of the microtip arrays were characterized for

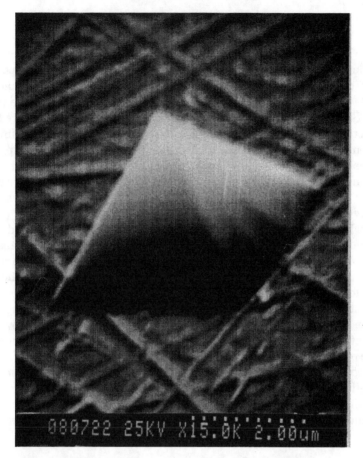

Figure 10. Diamond pyramidal microtip on a diamond membrane.

electron emission. The current emission characteristics of the emitter array in diode configuration were measured under vacuum in forward and reverse bias.

Device operation in the field emission process is controlled by the Fowler-Nordheim tunneling process [11]. The emission current density is given by:

$$J = K_1(E^2/\Phi)\exp[-(K_2\Phi^{3/2})/E] \qquad (9)$$

where K_1 and K_2 are constants, Φ is the work function of the emitting surface. The electric field E is defined as:

$$E = (\beta V)/d \qquad (10)$$

where V is the anode-cathode voltage, and β is the field enhancement factor mainly controlled by the cathode radius of curvature at the point of emission, and d is the spacing between the cathode tip and the anode.

The field at the apex of the tip is inversely proportional to the tip radius. Sharp needle tips of diamond-based pyramidal structures are fabricated to enhance high electric field at the apex which yields high emission current. The relative change in emission current of the microtip array versus applied field is shown in Figure 11. The data indicate that a high current (0.1 mA) in forward bias can be obtained under a low electric field of less than 10 V/μm at a pressure of 10^{-6} torr. No emission current was observed in the reverse bias regime.

Fowler-Nordheim field emission behavior of the diamond-based pyramids was analyzed and shown in Figure 12. Linearity indicates that the I-V characteristics follow the Fowler-Nordheim relation. The results indicate good field emission behavior in two distinct regimes. The positive slope of the linear plot at lower electric field may indicate electron emission due to negative electron affinity which can occur in diamond. At higher electric field regime, electron emission attributed to positive electron affinity dominates the emission curve as indicated by the negative slope of the Fowler-Nordheim plot.

A comparison of the emission characteristics between diamond tips and silicon tips is shown in Figure 13. It can be seen that electron field emission for the diamond microtips, in forward bias, exhibits significant enhancement both in total emission current and stability compared to

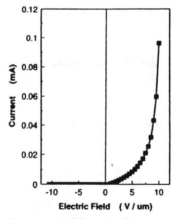

Figure 11. Emission current of the microtip array versus applied field.

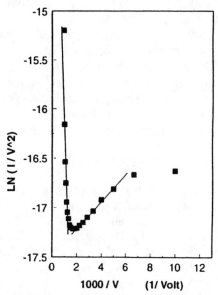

Figure 12. Fowler-Nordheim field emission behavior of diamond-based pyramids.

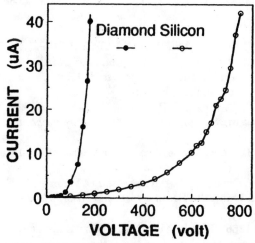

Figure 13. Comparison of emission characteristics between diamond tips and silicon tips.

pure silicon emitters. Moreover, field emission from patterned polycrys-talline diamond pyramidal tip arrays is unique in that the applied electric field is found to be considerably lower compared to that required for emission from Si, Ge, GaAs, and metal surfaces.

Controlled structures of polycrystalline diamond-based pyramidal arrays on diamond membranes have been fabricated for vacuum microelectronics applications. High electron emissions under low electric field have been obtained. This electron emission behavior suggests many practical applications of such diamond-based vacuum microdiodes. The effect of doping, hydrogen plasma tip sharpening and high temperature annealing on the current emission characteristics are currently under investigation [12].

REFERENCES

1. Wur, D. R., J. L. Davidson, W. P. Kang and D. L. Kinser. 1993. *Third International Symposium on Diamond Materials, 183rd Meeting of the Electrochemical Society,* Honolulu, Hawaii.
2. Wur, D. R., J. L. Davidson and W. P. Kang. 1993. "A Polycrystalline Diamond Film Piezoresistive Microsensor," *Transducers 1993,* Yokohama, Japan.
3. 1968. "Design Considerations for Diaphragm Pressure Transducers," *Micro-Measurements Tech. Note,* p. 129.
4. Edwards, L. M. and J. L. Davidson. 1992. *Proceedings of 3rd European Conference on Diamond, Diamond-Like and Related Coatings,* Heidelberg, Germany.
5. Wur, D. R., J. L. Davidson, J. P. Wang and Y. C. Ling. 1994. "Dopant Concentration Dependence of Piezoresistive Effect in Boron Doped Polycrystalline Diamond Films," *Fifth Annual Diamond Workshop,* Wayne State University, Troy, MI.
6. Frenklach, M. 1989. *J. Appl. Phys.* 65(12):5142–5149.
7. Davidson, J. L., J. S. Shor, D. R. Wur and A. D. Kurtz. 1993. *Third International Symposium on Diamond Materials, 183rd Meeting of the Electrochemical Society,* Honolulu, Hawaii.
8. Wur, D. R., J. L. Davidson, W. P. Kang and D. L. Kinser. 1995. "Polycrystalline Diamond Pressure Sensor," *IEEE J. Microelectromechanical Systems,* 4, No. 1.
9. Sze, S. M. 1981. *Physics of Semiconductor Devices,* Wiley-Internscience, New York.
10. Kang, W. P., Y. Gurbuz, J. L. Davidson and D. V. Kerns. 1995. "A Polycrystalline Diamond Thin-Film-Based Hydrogen Sensor," *Sensors and Actuators B,* 24–25:421–425.
11. Spindt, C. A. 1966. "Research into Micron-Size Emission Tubes," *IEEE Conf. on Tube Techniques.*
12. Kang, W. P. and J. L. Davidson. (submitted). "Patterned Diamond Field Emitter Arrays," *J. Vac. Sci. Tech.*

Field Electron Emission from Diamond-Like Carbon Films

A. V. KARABUTOV, V. I. KONOV, V. G. RALCHENKO,*
E. D. OBRAZTSOVA, V. D. FROLOV,
S. A. UGLOV AND M. S. NUNUPAROV
General Physics Institute
Russian Academy of Sciences
ul. Vavilova 38
Moscow 117942, Russia

H.-J. SCHEIBE
Fraunhofer-Institut
Dresden, Germany

V. E. STRELNITSKIJ
Kharkov Institute of Physics and Technology
Kharkov, Ukraine

V. I. POLYAKOV AND N. M. ROSSUKANYI
Institute of Radio Engineering and Electronics
Moscow, Russia

ABSTRACT: A comparative study of field electron emission properties of different diamond-like carbon films is reported. The films were produced by ion beam, laser-arc, r.f.-plasma depositions and magnetron sputtering on silicon and metallic substrates. Electron emission at fields as low as 4–8 V/μm was observed for some of the samples. The emission current versus applied field behavior, current time stability and surface homogeneity of emission properties were studied in dependence of film structure, composition, thickness and conductivity. Correlations between emission properties and film hardness, optical band gap, sp^3-bonded carbon content were found, and factors which improve the field

---

*Author to whom correspondence should be addressed.

electron emission from DLC films were discussed. Raman spectroscopy, deep-level transient charge spectroscopy, photoelectrical and electroconductivity methods were used to characterize surface and bulk electronic properties of samples.

KEY WORDS: low field electron emission, diamond-like carbon film, sp³-bonded carbon.

INTRODUCTION

$\mathbf{S}$ome diamond and diamond-like carbon (DLC) films possess excel-lent and stable field electron emission properties with high current densities at electric fields as low as 3 V/μm [1,2]. However those proper-ties are strongly influenced by film microstructure, so the fields of 20–60 V/μm are often needed to obtain electron emission [3–5]. The study of the field electron emission of diamond materials also was stimulated by an existence of the negative electron affinity on some diamond surfaces [6,7], making these materials candidates for vacuum microelectronics applications including cold cathode flat panel displays. On the other hand, it is not easy to obtain homogeneous and low-cost polycrystalline diamond films at large substrate area, which is necessary for displays. For this reason, emission properties of DLC films are of interest due to their high chemical stability similar to diamond, low-cost and low-temperature growth on large area substrates, very low film roughness, and the possibility of changing the film electroconductivity over a broad range.

We report the results of characterization of field electron emission properties of DLC films produced by various techniques onto Si and me-tallic substrates.

RESULTS

The following DLC and amorphous carbon (a-C) film series were stud-ied:

1. DLC films prepared by pulsed laser-arc onto Si substrates [8]
2. a-C films produced by ion beam deposition on Ti and Cu substrates [9]
3. a-C:H films deposited from C_6H_6 in r.f. glow discharge plasma on Si substrates [10]
4. DLC films produced by magnetron sputtering of glassy carbon tar-gets onto Si [11]

The emission measurements were performed in an oil-free vacuum of 10^{-7} Torr. Four tungsten probes with tip radius of about 20 μm were placed above the sample surface, the gap between tips and film surface being fixed in 10–100 μm range. A high DC voltage (up to 8 kV) was applied between the sample (cathode) and each probe (anode) to measure the field electron emission current. The values of the tip radius, the tip-sample distance and applied voltage were used to calculate the electric field, E, at the sample surface under the probes; a tip current of 1 μA corresponds to current density on the surface of $\approx$1 mA/mm^2. Three field emission characteristics were investigated: the emission current versus applied electric field (I-E) behavior for selected surface area of minimum radius 10 μm, current time stability, and surface homogeneity of the emission.

The typical emission (I-E) curves are shown in Figure 1 for a 300 nm thick DLC film produced by magnetron sputtering in nitrogen atmosphere. For that sample the lowest electron emission fields of 4–10 V/μm were observed. The measurements were performed in a cyclic manner by gradually increasing the voltage, and then by decreasing it until the emission current deceases. The following features in I-E characteristics common for all the films studied were noted. The "switch-off" emission process always shows a smooth exponential decrease of the current to zero at field E_{off} (see decreasing branches in Figure 1). For "turn-on" emission process a more complicated behavior was observed. First, the emission current jumps from "zero" value (undetectable level) to a few μA at a threshold field E_{th}. Second, in rare cases, the emission current reversibly increases from zero above a turn-on field E_{on} without the jump. Also a combined process often occurs, i.e., at first the emission increases smoothly, then the current jumps up at $E_{th} > E_{on}$ (see Figure 1). Thus, as a rule, an essential hysteresis of I-E curves is observed for the first cycle, but the repetition in the measuring cycles leads to the shift of increasing branch of I-E curve to lower fields, so the hysteresis is reduced or almost vanishes. Note that the evolution of I-E curves from cycle to cycle was reproduced for a selected spot on the film surface, if the field was not applied for several hours between measurements [compare Figures 1(a) and 1(b)].

The I-E emission characteristics are well approximated by Fowler-Nordheim (FN) plots for low field sections of both increasing and decreasing I-E branches (Figure 2). The electron work function formally calculated from FN-plots (neglecting possible field inhomogeneities) was found to be $\Phi_{FN} = 0.15$ eV for first increasing I-E branch and $\Phi_{FN} = 0.067$ eV for decreasing one. These very low work function values could be ascribed either to specific electronic properties of the carbon material itself, or to strong electric field inhomogeneities (for instance, field enhancement on sharp protrusions), and we will discuss this later. The dif-

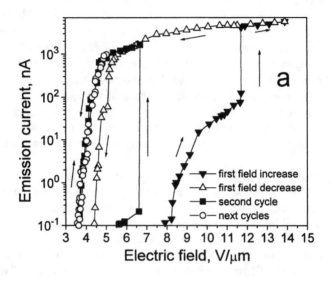

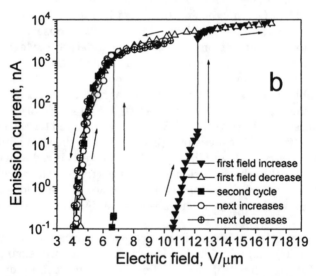

Figure 1. Electron emission current versus electric field for a DLC film deposited by magnetron sputtering: for a few initial measuring cycles (a) and cycling after 3 hours current off (b). The behavior of *I-E* plots is typical for all types of DLC films studied.

28

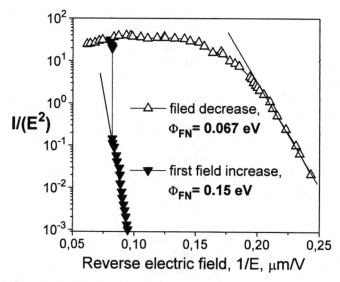

Figure 2. The Fowler-Nordheim representation of field emission plots of DLC film shown in Figure 1(b).

ferences in work function determined for the two branches may indicate that reversible changes in film surface structure occurred under a field/current action at $E > E_{th}$ (e.g., water sorption/desorption on a film surface). Inspection of the surface with an optical microscope ($2\,\mu$m resolution) showed no visible craters or other traces of a film modification/damage after many measuring cycles, even if turn-on current jumps were recorded. Nevertheless, an occurrence of submicron damages is not excluded. If the threshold field is not exceeded the emission properties are reversible and homogeneous (as follows from the straight slope of the increasing branch in Figure 2). But after a current jump the deviation from the FN-plot for decreasing branch may indicate a rise of surface inhomogeneity of emission. Another reason may be a limiting influence of a low conductive sub-surface layer, since there is an ohmic constant slope of I-E plot in linear coordinates in the high-current region (1–$10\,\mu$A).

The time stability of the emission current for some of the DLC films studied is good: the I-E curves are reproducible after many measuring cycles, or after many hours without field applied (Figure 1). The emission current is constant at different fixed fields during at least one hour (Figure 3). Note that, the relative value of the low frequency current noise decreases with electric field. We compared the DLC samples with glassy carbon and highly oriented pyrolitic graphite, which have poor stability and good spatial homogeneity, and for which the electron emission occurs at higher fields (>30–60 V/μm). The emission properties of our DLC films often had an essential surface inhomogeneity: low field emission spots, and even "no emission" spots.

To study the influence of the growth process parameters (substrate temperature, ion energy, doping gases, hydrogen incorporation) and the post-growth annealing on the field electron emission, DLC films prepared by various techniques were investigated. Their growth parameters and properties are shown in Tables 1–3.

Samples F1 and F2 prepared by laser-arc technique at lower temperatures (see Table 1) displayed Raman spectra with one band centered at 1500 cm^{-1} corresponding to the amorphous carbon phase. As the deposition temperature increased, that single band gradually transformed into two bands corresponding to D and G peaks of microcrystalline graphite. They became clearly resolved for sample F4 [12,13]. For amorphous carbon samples F1 and F2, there is a high concentration, up to 80% of sp^3-carbon fraction, that provides high values of elastic modulus and optical band gap. With the rise of graphitization (samples F3 and F4) the sp^3-carbon content, elasticity and optical band gap decrease [8]. These film structure changes affect electronic properties, in particular, electrical resistivity is higher for more "diamond-like" samples F1 and F2 and lower for more "graphitic" samples F3 and F4. The interesting trend ob-

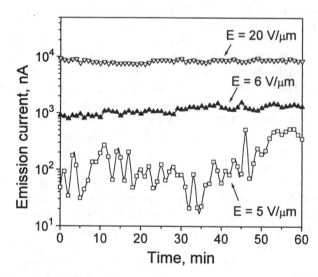

Figure 3. The typical field electron emission current time stability for DLC film shown in Figure 1.

Table 1. *Diamond-like carbon films prepared by pulsed laser-arc.*

Parameter/Sample No.	F1	F2	F3	F4
1. Film thickness, h, nm		70–80 nm for all films		
2. Deposition temperature, T_s, °C	23	80	150	350
3. Percentage of sp^3-carbon	>80%	60%	25%	20%
4. Optical band gap, E_g, eV	2.0	1.4	0.7	0.36
5. Young's modulus, Y, GPa	450	300	160	80
6. Resistivity, ρ_d, Ohm cm	2×10^7	7×10^8	800	600
7. Surface potential (photo-emf), meV	−160	−100	−60	−30
8. Lowest emission field, E_{off}, V/μm	>18	>20	>30	>60
9. Threshold emission field, E_{th}, V/μm	>60	>100	>80	>130
10. Spatial dispersion of E_{off}, %	+140	+100	+40	+20
11. FN-plot work function, Φ_{FN}, eV	0.14	0.15	0.21	0.29

Table 2. *a-C:H films deposited onto Si from C_6H_6 in r.f.-plasma [10].*

Parameter/Sample No.	H1	H2	H3	H4
1. Film thickness, h, nm	900	1050	1000	1000
2. Bias voltage, U, eV	1500	600	300	150
3. Microhardness, GPa	37	18	8	6.5
4. Film mass density, ρ, g/cm^3	2.4	2.0	1.85	1.7
5. Optical band gap, E_g, eV	1.2	1.45	1.52	1.6
6. sp^3/sp^2 carbon ratio	1.3	0.5	0.4	0.3
7. C-H sp^3/sp^2 bonding ratio	4	4	0.7	1.3
8. Hydrogen contents, atomic %	10	9	12	8
9. Resistivity, ρ_d, Ohm cm	10^8	10^9	5×10^{11}	10^{13}
10. Lowest emission field, E_{off}, V/μm	>20	>90	>100	>200
11. Threshold emission field, E_{th}, V/μm	>50	>130	>200	>200
12. Spatial dispersion of E_{off}, %	+50	+200	—	—

Table 3. a-C films deposited onto Cu (#C1–C4) and Ti (#T1–T5) substrates from ion beam.

Parameter/Sample No.	C1	C2	C3	C4	T1	T2*	T3	T4	T5
1. Film thickness, h, μm	0.05	0.6	5.1	3.0	36	36	32	51	8.2
2. Ion energy, E_i, eV	50	50	50	50	750	750	750	750	1500
3. Microhardness, GPa	>100	>100	50	80	54	54	28	20	26
4. Doping gas	no	no	N_2	O_2	no	no	N_2	H_2S	no
5. Resistivity, ρ_d, Ohm cm	6×10^6	5×10^9	2×10^6	7×10^7	10^6	3×10^3	16	8	—
6. Lowest emission field, E_{off}, V/μm	>20	>50	>30	>200	>200	>8	>20	>15	>40
7. Threshold emission field, E_{th}, V/μm	>80	>100	>90	>200	>200	>50	>50	>60	>80
8. Spatial dispersion of E_{off}, %	+200	+200	+300	—	—	+400	—	—	+300

*Post-growth annealing in vacuum for 12 hours at T = 590°C.

32

served is that emission field E_{off} decreases with an sp³-carbon fraction increase (Figure 4). The work function calculated from the FN-plot showed a similar tendency (Table 1). Besides, variation of emission from spot to spot (spatial dispersion of E_{off}) decreases for more "graphitic" samples.

Similar field emission dependences on the film structure were found for the hydrogenated carbon film series (Table 2). For most of the samples the wide Raman band centered near 1500–1550 cm⁻¹ corresponding to the amorphous carbon phase was observed [14]. The film hardness, density and sp³-bonded carbon content in the a-C:H films increase while optical band gap and resistivity decrease with a rise of bias voltage applied to the substrate during deposition. Again, the emission fields were decreased when hardness and sp³-carbon content were increased (Figure 4). However, the influence of the sp³-carbon-hydrogen bonds content in a-C:H films on the field emission was not found, though the negative electron affinity of diamond was reported just for hydrogen-terminated surfaces [7,15].

The non-hydrogenated 0.05–50 μm thick a-C films were deposited from ion beam onto Ti and Cu substrates (Table 3). Carbon ion energy was varied from 50 to 1500 eV, some samples were doped with N or S during film deposition. It was found that the following factors positively influence field emission:

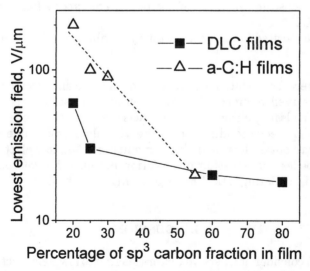

Figure 4. The influence of sp³-carbon fraction on emission field for DLC (laser-arc) and a-C:H films.

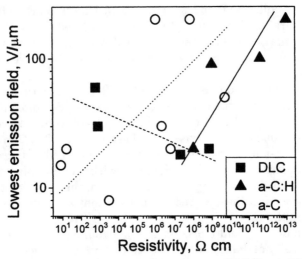

Figure 5. Field electron emission dependence on DLC film resistivity.

- nitrogen or sulphur doping (compare samples T3, T4 versus T1, and C3 versus C2)
- post-growth annealing resulting in film graphitization (sample T2 versus T1)
- nanometer-scale thickness of the poorly conductive film (sample C1 versus C2)
- higher deposited ion energy resulting in film graphitization (sample T5 versus T1)

No universal correlation between field emission and electroconductivity was observed, but it could be traced within each series (Figure 5). The emission fields may either increase or decrease with conductivity. In general, the high electroconductivity of the DLC film is preferable, but is not necessary to obtain low field electron emission. Yet, low electroconductivity impedes the emission observation because the voltage drop on some semi-insulating DLC films was as high as 400 V.

DISCUSSION

One can consider at least two factors leading to low field electron emission observed for diamond and DLC films. The first one is a strong electric field inhomogeneity (geometric field enhancement) at the film surface with the high (about 5 eV) work function. Hence, the field enhancement by 100–1000 times on surface microprotrusions or narrow conductive domains (channels) between insulating grains in the film is required to take place. Another approach implies the existence of a very low energy barrier to emit electrons into vacuum from the smooth film surface. In this case, the work function of the order of 0.1 eV has to be assumed possibly due to some mechanisms based on negative electron affinity (see, for example, Reference [16]). A combination of field enhancement and low work function may also be considered.

Some interesting issues about the field electron emission from DLC films are worth noting. First, the field emission parameters for DLC films referred to in Tables 1 and 2 improve with increase of sp^3-bonded carbon content, which is controlled by the film growth conditions (substrate temperature, ion energy). Such a trend looks logical if one considers the sp^3-carbon clusters to be disordered fragments of diamond lattice with a low electron affinity. Another possible explanation of the improved emission is an internal field enhancement on narrow conductive sp^2-bonded carbon channels into insulating sp^3 matrix. The rise of sp^3/sp^2 ratio will promote field enhancement (samples F1 and F2 versus F3 and F4). Because of a very low roughness of DLC films it is difficult to ascribe the field enhancement to surface microprotrusions, while it could be possible in the case of polycrystalline diamond films.

Another fact is the improvement of the field emission for nitrogen- or sulphur-doped films, as well as for post-growth annealed film (12 hours in vacuum at 590°C). Both doping and annealing lead to a strong decrease in resistivity, that may indicate an enrichment of the samples with graphitic component. Thus, to obtain good low field emission properties for DLC films the presence of both sp^3 and sp^2 carbon fractions seems needed. Perhaps the best emission corresponds to some optimum in the sp^3/sp^2 ratio for each growth method and is determined by the cluster structure of the DLC film.

Using deep level transient charge spectroscopy (Q-DLTS) [17] and photoelectrical characterization of the films, it was found that for good field emission films, the integral electron density of states is higher than for poor emission films. Besides, as a rule, the surface band bending for the former is more negative than for the latter (line 7 of Table 1), that evidences a reduction of the surface tunneling barrier for films with good emission. This may mean that high concentration of defect states at the surface of films promotes the negative surface band bending, and thus, leads to low field electron emission. Further study is needed to understand the correlations observed.

CONCLUSION

DLC films produced by four deposition methods (ion beam, laser-arc, r.f.-plasma depositions and magnetron sputtering) were studied for field electron emission at room temperature. For the best samples examined, a stable low field emission at fields of 4–10 V/μm was observed and a work function value as low as 0.067 eV from the formal Fowler-Nordheim analysis was calculated. It was found that the emission properties strongly depend on film structure and composition determined by deposition conditions, and for some DLC films, fields higher than 200 V/μm were required to obtain emission. It was concluded that the positive influence on the field emission is produced by the following factors:

- high concentration of the sp^3-carbon fraction in DLC film
- nitrogen or sulphur doping of the amorphous carbon film
- nanometer-scale thickness of a poorly conductive film
- post-growth annealing resulting in film graphitization

The emissions current versus applied field plots for DLC films show a hysteresis for branches corresponding to increasing and decreasing field with the presence of a turn-on current jump for increasing field. The repetition of the measuring cycles strongly reduces the hysteresis. A high conductivity of the DLC film is preferable, but not necessary to observe the low field electron emission.

ACKNOWLEDGEMENTS

This work was supported by Diagascrown Co. Ltd., Moscow, Russia.

REFERENCES

1. Wang, C., A. Garcia, D. C. Ingram, M. Lake and M. E. Kordesch. 1991. *Electron. Lett.*, 27:1459–1460.
2. Xu, N. S., Y. Tzeng and R. V. Latham. 1993. *J. Phys. D. Appl. Phys.*, 26:1776–1780.
3. Geis, M. W., J. C. Twichell, J. Macaulay and K. Okano. 1995. *Appl. Phys. Lett.*, 67:1328–1330.
4. Zhu, W., G. P. Kochanski, S. Jin and L. Seibles. 1995. *J. Appl. Phys.*, 78(4):2707–2711.
5. Karabutov, A. V., V. I. Konov, S. M. Pimenov, V. D. Frolov, E. D. Obraztsova, V. I. Polyakov and N. M. Rossukanyi. 1996. *J. de Phys. IV*, 6-C5:113–118.
6. Himpsel, F. J., J. A. Knapp, J. A. Van Vechten and D. E. Eastman. 1979. *Phys. Rev. B*, 20(2):624–627.

7. Van der Weide, J. Z. Zhang, P. K. Baumann, M. G. Wensell, J. Bernholc and R. J. Nemanich. 1994. *Phys. Rev. B,* 50(8):5803–5806.

8. Scheibe, H.-J. and B. Schultrich. 1994. *Thin Solid Films,* 246:92–102.

9. Aksenov, I. I. and V. E. Strelnitskij. 1991. *Surface and Coatings Technology,* 47:98–105.

10. Kononenko, T. V., V. G. Ralchenko, V. I. Konov and V. E. Strelnitskij. 1992. In *Diamond Optics V, Proc. SPIE,* 1759:106–114.

11. Uglov, S. A., V. E. Shoub, A. A. Beloglazov, E. N. Loubnin, E. D. Obraztsova, A. S. Chernikov and V. I. Konov. 1994. *Physica i Chimia Obrabotki Materialov,* 6:68–74 (in Russian).

12. Scheibe, H.-J., D. Drescher and P. Alers. 1995. *Fresenius J. Anal. Chem.,* 353:695–697.

13. Obraztsova, E. D., F. Fujii, S. Hayashi and H.-J. Scheibe. In preparation.

14. Nistor, L. C., J. Van Landuyt, V. G. Ralchenko, T. V. Kononenko, E. D. Obraztsova and V. E. Strelnitskij. 1994. *Appl. Phys. A.,* 58:137–144.

15. Bandis, C. and B. B. Pate. 1995. *Phys. Rev. Lett.,* 74(5):777–780.

16. Pan., L. S. 1995. *Mater. Res. Soc. Proc. Fall Meeting,* Boston, MA, Paper DD8.2.

17. Arora, B. M., S. Chakravarty, S. Subramanian, V. I. Polyakov, M. G. Ermakov, O. N. Ermakova and P. I. Perov. 1993. *J. Appl. Phys.,* 73):1802–1806.

Stable Low-Field Electron Emission from Diamond-Like Films

A. A. DADYKIN AND A. G. NAUMOVETS
Institute of Physics
Natl. Academy of Sciences of Ukraine
Prospekt Nauki 46,
Kiev 22, UA-252022 (Ukraine)

N. V. NOVIKOV, V. D. ANDREEV, T. A. NACHALNAYA
AND V. A. SEMENOVICH
Institute for Superhard Materials
Natl. Academy of Sciences of Ukraine
Avtozavodskaya St. 2
Kiev 74, UA-254074 (Ukraine)

ABSTRACT: We have obtained and studied stable low-field ($\sim 10^5$ V/cm) electron emission from diamond-like a-C and i-C films deposited on Ni and W tips and on flat Si, Ni and Cr substrates. The 10 to 500 nm thick films were fabricated in a RF discharge (CH_4 + H_2 mixture, 13.56 MHz). The Fowler-Nordheim plots of the emission current are broken lines similar in shape to the emission characteristics for other semiconducting films like those of ZnS, NiO and SiO_2.

1. INTRODUCTION

The great interest in electron emission from diamond was kindled by the work of Himpsel et al. [1] which disclosed that a (111) surface of diamond possesses a negative electron affinity (NEA). From that time on, hopes are pinned on NEA diamond materials as promising candidates for stable and efficient cold cathodes [2–5]. Many attempts have

38

been made to fabricate stable field emission cathodes from graphite, with relatively good results [6–9]. A few research groups have been investigating the so-called low-field electron emission (LFEE) from various materials such as semiconductors, metals and composite systems [10–16]. LFEE is conventionally called cold electron emission, observed in electric fields of about 10^5 V/cm (as calculated from macroscopic geometry of electrodes). LFEE is distributed very nonuniformly over the cathode surface. It has been found that the emitting sites contain usually an excessive concentration of carbon, alkali and other impurities [13,14,17] although neither stoichiometry nor their structural and chemical state have been documented. The mechanisms of LFEE have not been elucidated unequivocally, but it cannot be ruled out that its origin may be related, in some way, to a specific role played by carbon [14,16]. It must be admitted that the mechanism(s) of electron emission from diamond cathodes are not sufficiently clear [4]. All these considerations prompt further comparative investigations of electron emission from various carbon-containing materials, most of all from diamonds with their rich diversity of bulk and surface properties.

The objective of this work was to examine emission characteristics of a number of cathodes based on diamond-like and other semiconducting films and to gather new data on possible emission mechanisms.

2. EXPERIMENTAL

Glass, pyroceram and silicon plates, "brushes" of nickel tips as well as single crystal silicon and tungsten tips, were utilized as substrates for the deposition of diamond-like films. The resistivity and thickness of the films were varied in the ranges 10^2 to 10^{13} Ohm·cm and from 10 to 500 nm, respectively. Raman scattering (RS), infrared absorption spectroscopy (IAS) and electron spin resonance (ESR) were used to evaluate the phase composition as well as the concentration of impurities and other defects. Two types of diamond-like films, i-C and a-C, were synthesized. The RS spectra of the a-C films showed either a broad asymmetrical peak with a maximum at 1500 cm^{-1} or two peaks at 1350 and 1550 cm^{-1}. Contrary to this, the RS spectra for the i-C films contained two broad intense lines at 1350 and 1600 cm^{-1} [Figure 1(a)]. The ratio of their intensities mirrors the relative number of sp^2- and sp^3-bonds in a sample. This ratio could be intentionally varied in between broad limits (from 30 to 80%) by choosing different deposition regimes.

Several types of specimens were fabricated and tested to compare their emission characteristics:

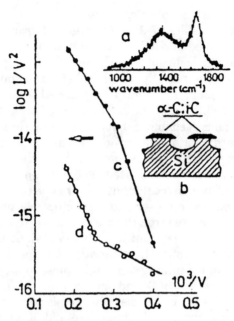

Figure 1. (a) Raman scattering spectrum for an i-C diamond-like film. (b) Mushroom-like configuration of emitters. (c) Typical FN-plot for an emission site on a-C/graphite/Ni emitter. (d) The same for a-C/Ni emitter. I is measured in Amperes, V in volts.

1. Single Si and W tips, clean and coated with i-C and a-C films
2. Diamond-like films deposited on silicon, glass and pyroceramic plates through a mesh mask having square holes sized from 50 to 200 μm
3. The same films with "under-etched" edges [a mushroom-like configuration, see Figure 1(b)]
4. "Brushes" of Ni tips uncoated and coated with diamond-like films. On such brushes, the film was located predominantly on the tip apexes.

The tip cathodes were tested in field emission microscopes (Müller's projectors). To visualize the emission current distribution over a tip surface and to simultaneously measure current-voltage (I-V) curves for small areas on the tip apex, two different types of microscope tubes were used. In one case a probe hole was made in the luminescent screen of the microscope, and electrons passing through it were collected by a Faraday cup. In another microscope tube, the luminescent screen was deposited onto a vacuum-tight optic-fiber window. By using a flexible optic-fiber light guide which collected the light emitted from a small area on the screen and transported it to a photomultiplier, or by recording the image with a videocamera, one could obtain local I-V curves [18].

Flat diamond cathodes were tested in parallel-plate diode cells. Their anodes represented either a flat luminescent screen (for visualization of the emission current distribution) or a flat massive metal anode. In the latter case only I-V plots could be recorded, but, on the other hand, one avoided the problem of contamination of the cathode by species sputtered from the luminescent screen.

The measurements were carried out in a vacuum of ~10^{-6} torr.

3. RESULTS AND DISCUSSION

All types of emitters examined in this work have shown rather similar I-V characteristics in the sense that, when plotted in the Fowler-Nordheim (FN) coordinates $\log(I/V^2)$ versus $1/V$, they represented broken lines. In the most general case the plots have a more or less extended "shelf" segment where the emission current is only weakly dependent on voltage (Figure 2, plot 1). The presence of the shelf segment and its extent depend on the type of deposition and anneal of a film, i.e., on the film phase composition as well as on the nature and density of impurities and

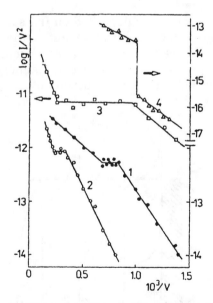

Figure 2. The Fowler-Nordheim plots. (1) typical for all diamond-like films investigated (this particular plot is for an a-C/graphite/Ni emitter). (2) NiO_x on Ni, (3) ZnS on Cr. (4) Typical for a-C and SiO_2 on Si and ZnS on W. I is measured in amperes, V in volts.

lattice defects. The spatial distribution of the emission current is generally highly nonuniform and reveals the existence of bright point-like emission sites and of rather extended emission regions with a lower current density (almost homogeneous over the surface). The FN plots with the shelf segment are typical of cathodes that give both point-like and extended emission, whereas the plots for point-like sites have usually no shelf segment [Figure 1(c) and (d)].

The two plots depicted in Figure 1 relate to a-C films, synthesized on different substrates. In one case (plot d) the film was deposited on Ni tips while in the other case (plot c) the Ni tips, prior to a-C deposition, were coated with an intermediate graphite layer, a few tens of a nanometer thick. In both cases the plots are broken lines, one of them concave and the other convex. This result shows that the substrate has an essential impact upon the emission characteristics of the film, either via the electron properties of the interface itself (e.g., via the presence of interface states) or by influencing the growth regime and structure of the film.

From the measurements with flat emitters the range of the fields providing appreciable emission currents is estimated at $(1-2) \times 10^5$ V/cm. This is about 2 to 3 orders of magnitude lower than the usual field emission level. The estimated current densities within localized emission sites can reach 10 A/cm^2.

When a diamond-like film is deposited on flat substrates through a mask, the emission sites are most abundant at the edges of the film "patches." The edges correspond to penumbral regions in the deposition process and are wedge-shaped. As revealed by scanning electron microscopy, the film has usually an island structure in these areas. It is likely

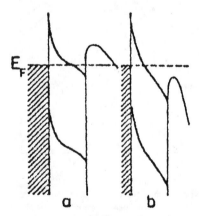

Figure 3. Hypothetical band schemes for emission in (a) low and (b) high electric fields.

that the most favorable conditions for emission will always be attained somewhere at the film edge. It seems also possible that interface stress, concentrated at the edge, can affect the growth process itself, the segregation of impurities, or the formation of conducting channels.

Xu, Tzeng and Latham [16] have recently detected a number of close similarities in the low field electron emission of diamond coated Mo and polished graphite surfaces. In the present work we have carried out comparative measurements of emission properties of diamond-like-carbon and some other semiconducting films deposited on the same substrates under identical conditions. As illustrated in Figure 2 (curves 2 and 3), thin films of NiO prepared on Ni by anodic oxidation and ZnS films evaporated on Cr show FN characteristics the shape of which is very similar to those of the diamond-like films (curve 1).

A number of specimens, such as a-C and SiO_2 films on Si and ZnS films on W, showed FN plots exemplified by curve 4 in Figure 2. One sees a sharp rise in current at some voltage, which signals a switching from one emission regime to another. Such an I-V characteristic is typical for field emission from n-semiconductors with acceptor surface states which cause an upward band bending at the surface [19]. In this case the sharp growth of the field emission current is due to compensation of the band bending by the external electric field when both the transparency of the potential barrier and the surface electron density are rapidly increasing. A similar behavior is predicted for emission from a dielectric with charge traps that are filled and limit the current at low fields, but which at some critical field, become ionized [20].

Comparative tests for long-term stability of emission characteristics have not yet been performed. We can only state that mean values of the emission current remained reproducible during at least a few days in a vacuum of $\sim 10^{-6}$ torr. Short-time current fluctuations $\Delta I/I$ did not exceed $\approx 5\%$ at voltages in the "shelf" region of the FN plots or above them. This indicates that the films are rather resistant with respect to residual gases, or ion sputtering.

We cannot say if the diamond-like surfaces under study (or at least on some parts thereof) had a natural negative electron affinity (NEA). However, it seems likely that there occurs a transition from positive electron affinity at low fields to an effective NEA at high fields due to field penetration (Figure 3). This could also explain the appearance of the two different slopes in the FN plots (Figures 1 and 2). The high electric field near the substrate that facilitates the injection of electrons into the film can be caused, for instance, by the existence of ionized (positively charged) traps within the film [21]. It is understood that the model depicted in Figure 3 is at present hypothetical and illustrative only. According to calculations made by Huang et al. [4] an appreciable low-field electron emission from diamond can only be expected if one assumes the presence of surface states and of subbands attributed to volume lattice defects.

4. CONCLUSION

These results indicate the potential for some diamond-like thin films to act as cold electron emitters. The similarity of their characteristics with those of other semiconducting films, prompts a broader research effort into possible common mechanisms underlying low-field electron emission.

REFERENCES

1. Himpsel, F. J., J. A. Knapp, J. A. Van Vechten and D. E. Eastman. 1979. *Phys. Rev. B., 20*:624.

2. Van Der Wiede, J. and R. J. Nemanich. 1994. *J. Vac. Sci. Technol. B., 12*:2475.

3. Xie, C., C. N. Potter, R. L. Fink, C. Hilbert, A. Krishan, D. Eichman, N. Kumar, H. K. Schmidt, M. N. Klark, A. Ross, B. Lin, L. Fredin, B. Baker, D. Patterson and W. Brookover. 1994. *Le Vide, Les Couches Minces*, Suppl. au N 271:229.

4. Huang, Z.-H., P. Cutler, N. Miskovsky and T. Sullivan. 1994. *Le Vide, Les Couches Minces*, Suppl. au N 271:92 Preprint.

5. Derbyshire, K. 1994. *Solid State Technology, 37*:55.

6. Baker, F. S., A. R. Osborn and J. Williams. 1974. *J. Phys. D., 7*:2105.

7. Hosoki, S., S. Yamamoto, M. Futamoto and S. Fukuhara. 1979. *Surface Sci., 86*:723.

8. Bondarenko, B. V., E. S. Baranova, A. Yu. Cherepanov and E. P. Sheshin. 1985. *Radiotekhnika i Elektronika, 30*:2234.

9. Latham, R. V. and M. A. Salim. 1987. *J. Phys. E., 20*:181.

10. Bibik, V. F., P. G. Borziak and A. F. Yatsenko. 1968. *Ukrainskii Fiz. Zhurnal, 13*:868.

11. Bayliss, K. H. and R. V. Latham. 1986. *Proc. Roy. Soc. (Lond.) A, 403*:285.

12. Latham, R. L. and M. S. Mousa. 1986. *J. Phys. D., 19*:699.

13. Niederman, Ph., N. Sankarraman and R. J. Noer. 1986. *J. Appl. Phys., 59*:892.

14. Yatsenko, A. F., G. G. Kulishova and L. N. Starovoitova. 1988. *Izv. AN SSSR, ser. fiz. 52*:1530; English Translation: *Bull. Acad. Sci. USSR Phys. Ser.* (USA), 52:64.

15. Latham, R. V. and N. S. Xu. 1992. *Le Vide, Les Couches Minces*, Suppl. au N 260:215.

16. Xu, N. S., Y. Tzeng and R. V. Latham. 1993. *J. Phys. D.*, 26:1776.
17. Bibik, V. F., P. G. Borziak, G. A. Zykov, N. G. Nakhodkin and A. F. Yatsenko. 1973. *Phys. Stat. Sol (a)*, 16:151.
18. Dadykin, A. A. and A. G. Naumovets. 1992. *Acta Phys. Polonica A.*, 81:131.
19. Stratton, R. 1962. *Phys. Rev.*, 125:67.
20. Bube, R. H. 1962. *J. Appl. Phys.*, 33:1733.
21. Elinson, M. I. and D. V. Zernov. 1957. *Radiotekhnika I Elektronika*, 2:75, (in Russian).

UV-Sensors from Diamond-Like Carbon Films

E. I. TOCHITSKY, N. M. BELIAVSKY, O. G. SVIRIDOVITCH,
S. D. LOPATIN AND I. V. DUDARTCHIK
Engineering Center "Plasmoteg"
Belarus Academy of Sciences
Minsk, Belarus

ABSTRACT: Metal-semiconductor-semiconductor (MSS) UV-cells (Al/C/In$_2$O$_3$: SnO$_2$) with diamond-like carbon base layers have been fabricated. The diamond-like carbon films on single crystal silicon (100) substrate were obtained by means of pulsed laser deposition (1060 nm, 25 ns long). The influence of substrate temperature on film microstructure was investigated by transmission electron microscopy and electron diffraction. The results of these investigations have been in turn used to optimize the parameters of diamond-like film growth process. The improvements of photocells quality can be achieved by use of high quality diamond-like films. By measuring the photosensitive characteristics of the MSS-cells we have found a photo-EMF of 15–20 mV in the wavelength range from 300 to 350 nm.

KEY WORDS: UV-sensors, diamond-like carbon films, laser deposition, microstructure of films

1. INTRODUCTION

The measurement, utilisation and conversion of ultraviolet (UV) irradiation is of a great practical interest. The traditional materials for such converters have been Si, CdS and GaAs. Sensors made from these materials have low radiation resistance and often need the use of cryogenics environments.

46

The combination of a large bandgap, high thermal conductivity and radiation resistance, makes diamond a valuable material for the manufacture of radiation sensors. Based on polycrystalline diamond-like carbon films, prepared from the gas phase, we have made UV (190–230 nm) sensors using a special electrode system. Future investigations will be directed at the design of vertical heterostructures and on device microminiaturisation. At high light absorption in the desired spectral range, it is difficult to measure the transverse photoconductivity of thick films, since the greatest part of the film volume is not light modulated due to near surface absorption. Therefore, the investigation of photoelectric phenomene in thin diamond-like carbon films (300–600 nm) is of much practical interest. One of the more desirable methods for thin diamond-like carbon film synthesis is laser deposition at low temperature, which may make photosensitive integral structure possible. Current optical metrology requires integration of a photoreceiver, a photosignal intensifier and a control scheme, all on one crystal. Laser deposition of such films allow broad possibilities for modification of their structure and morphology, as well as the optical and electrophysical properties.

The objectives of the present work are the realization of diamond-like film photosensitive heterostructures using laser technology and the investigation of their photoelectric properties in the visible and UV spectral regions.

2. EXPERIMENTAL

Metal-semiconductor-semiconductor heterostructures (MSS) of $Al/C/In_2O_3$: SnO_2 type were executed on silicon substrates. The use of aluminum-oxide technology precluded the emergence of undesirable relief on the upper surface layers: aluminum electrodes and Al_2O_3 insulation between the elements. Carbon films were prepared in vacuum of 10^{-4} Pa on the glass and single crystal silicon (100) substrates. The thin films of 100–1000 nm were deposited onto heated substrates (273–673 K) by pulsed laser sputtering. Using an aluminum ittrium garnet laser (1060 nm), pulse time 25 ns at 10 Hz, the power density on the target was changed between the limits of $q = 5.10^8$ and 10^9 W/cm^2. The target material was pure graphite (type PMG-6) or pressed ultradispersed diamond powder (grains of 4–5 nm).

3. RESULTS AND DISCUSSION

The microstructure of the prepared films was studied by transmission electron microscopy and electron diffraction. Sputtering with pressed ultradispersed diamond powder targets yielded amorphous films with electron diffraction patterns consisting of three diffusion rings, corresponding to lattice parameters of $d = 4.1-4.4$; $d = 2.3-2.1$; $d = 1.17-1.26$ A. With graphite target sputtering the electron diffraction pattern shows more diffraction rings and in some cases a spot pattern indicating wide structural variation. An increase of the substrate temperature during ultradispersed diamond powder sputtering, promotes the formation of simultaneous quasi amorphous and crystalline phases, yielding diffraction patterns corresponding to interplanar distances: 4.18; 3.74; 2.98; 2.49; 2.35; 2.22; 2.07; 1.85; 1.75; 1.56 A. In some cases the electron diffraction pattern shows a first diffuse ring (graphite) displacement. The MSS structures were fabricated with diamond-like films having a finely-dispersed polycrystalline structure (grain size 10–15 nm) which included randomly dispersed microcrystallites (50 nm) of surface density $10^8/cm^2$.

An actual structure $Al/C/In_2O_3:SnO_2$ and the set-up for photopotential measurement under UV-irradiation is given in Figure 1. For preparation of MSS-heterostructures we have used SiO_2 coated Si (100) substrates. The substrates were treated in a peroxide-ammonia solution and were rinsed in deionised water. We then magnetron sputtered layers of Ta and Al, 20 and 500 nm thick respectively.

Photolithography developed a topological pattern of upper electrodes. For pattern transfer into the Al layer and simultaneous formation of insulation between the different elements, we have made use of porous anodized aluminum in the windows of the photoresistive mask. Porous anodizing used a phosphoric acid electrolyte which does not attack the photoresist. Next, a diamond-like carbon film was deposited onto the aluminum surface, followed by deposition of a transparent layer of $In_2O_3:SnO_2$ onto the sample surface. Photolithography and etching realized the upper transparent electrodes.

For the investigation of photo-electric phenomena in MSS-structures we have used: a high pressure mercury source with high emission of radiation in the visible and UV parts of spectrum; a voltage booster for the signal source, with a high internal resistance (capacitor structure); a device having a low resistance input (oscilloscope) and a universal oscilloscope with memory and a universal electrometer. The MSS-structures manufactured with high-resistance diamond-like semiconductor films, 500 nm thick, have shown a resistance of $(0.8-1.3) \cdot 10^7$ ohm and a capacity of 0.2–9.0 nF, the range of values being dependent the nature of the surface of the upper transparent electrode.

Measurement of the transverse electromotive force (EMF) on fabricated MSS-structures was made with mercury irradiation entering from

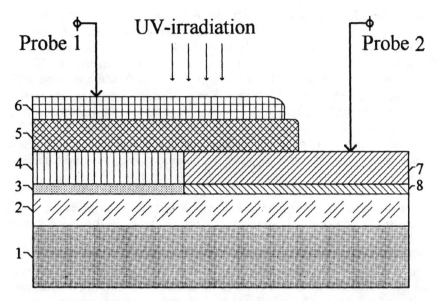

Figure 1. Cross section of a photosensitive MSS heterostructure. 1—Silicone, 2—SiO_2, 3—Ta_2O_5, 4—Al_2O_3, 5—photosensitive DLC layer, 6—In_2O_3:SnO_2 electroconductive layer, 7—Al, 8—Ta.

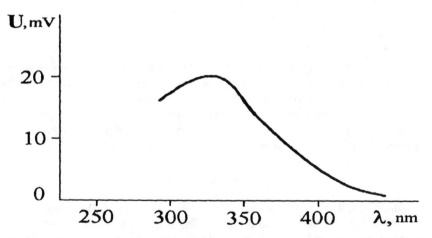

Figure 2. Spectral dependence of the photoelectromotive force of diamond-like film MSS structures.

the side of the transparent electrode. We have determined the spectral dependence of the photo potential of MSS-structures with interference light filters. These structures have photosensitivity in the visible from green to violet and beyond into the near ultraviolet. A maximum EMF of 15–20 mV is observed in the wavelength range 300–350 nm (Figure 2). The signal rise time was 15–20 ms. The results structure morphological influence on MSS-structures properties were determined and the conditions for diamond-like carbon film manufacturing were optimized. The resultant spectral dependence is good and in line with the photoelectric response of synthetic semiconductor diamond polycrystalline film electrodes.

The influence of diamond semiconductor properties on photoelectric emission are contributed to secondary photogeneration of carriers and their movement on the boundary with the electrode in the space charge field. Since the diamond bandgap is about 5.5 eV, carrier generation by direct interband transition is not possible under visible light or UV energy exitation. Evidently, photoexitation of electrons takes place from levels within the forbidden band to the conduction band (with concommitant formation of holes). Taking into consideration the photocurrent spectral dependence, (obtained by photochemical investigation [1]) and the photo potential spectral dependence of the MSS-heterostructures, fabricated by us, the interband levels lie at a depth of 2.0–3.5 eV from the zone conductivity bottom.

4. CONCLUSION

Diamond-like carbon film UV-sensor was fabricated and its photoelectric characteristics were studied. The results of these investigations made us consider diamond-like carbon films deposited with laser sputtering for photoconductivity in near UV-range of spectrum. Such films would allow the manufacture of photoelements in microelectronics with a great degree of integration.

REFERENCES

1. Pleskov, Y. V., A. Y. Sakhorova, A. E. Sevostianov and M. D. Krotova. 1991. "Films from Synthetic Semiconductor Diamond as Material of Electrode in Electrochemistry," *Proceedings, First International Symposium on Diamond Films,* ISDF-1, June 30–July 6.

Diamond-Like Film-Porous Silicon Systems: New Approach for Optoelectronic Devices Fabrication

YU. P. PIRYATINSKII
Institute of Physics
National Academy of Sciences of Ukraine
prospect Nauki, 46
252028, Kiev, Ukraine

V. A. SEMENOVICH
Institute for Superhard Materials
National Academy of Sciences of Ukraine
str. Avtozavodslaya, 2
254074, Kiev, Ukraine

N. I. KLYUI* AND A. G. ROZHIN
Institute of Semiconductor Physics
National Academy of Sciences of Ukraine
prospect Nauki, 45
252028, Kiev, Ukraine

ABSTRACT: A detailed experimental analysis is presented based on photoluminescence (PL) spectra from series of coated and uncoated porous silicon (PS) samples by diamond-like carbon (DLC) films using the time-resolved photoluminescence spectroscopy.

The PL spectra from uncoated PS samples has shown two main luminescence bands whose maximum intensities are located: around 450 nm with life time

---

*Author to whom correspondence should be addressed.

about some ns, and around 630 nm with life time about 20 μs. The PL spectra from coated PS samples has shown the shift of the band ~630 nm to the lower wavelength side (about 610 nm). The band ~450 nm drastically decreased and two new bands around 510 and 590 nm have arisen in this case. The DLC film-porous silicon systems survived durability tests. Environmental tests also revealed the protection imparted by DLC films to porous silicon devices.

Thus, PS + DLC systems appear to be a promising avenue for further investigation and may form the basis for the manufacture of new electronics and opto-electronics devices.

1. INTRODUCTION

Visible photoluminescence (PL) of porous silicon (PS) and the possibility of its optoelectronic applications have been extensively studied [1,2]. While there are many experimental studies going, they are not close to any application. This is due to the fact that there are at least two problems: how to synthesize porous silicon with uniform size distribution of pore walls and how to prevent deterioration of PS surface in atmospheric environments. Consequently, there is a need for a protective coating that would extend the lifetime of PS in uncontrolled environments and "tune" luminescence spectrum to arbitrary wavelengths through the structure and thickness of DLC films.

The "quality" of PS can be defined in several different ways: (1) size of pore walls; (2) structure of little crystallites of silicon; (3) surface morphology (namely periodic array of crystalline wires) and (4) defect density (residual complexes of SiO_2, SiH_2, etc.). These characteristics can be analysed by various characterization techniques [3]. PL can be used to analyse the PS quality and for relating this quality to the process parameters.

Two proposals explain the visible luminescence mechanism from PS based either on the Si quantum wire model (the 3-dimensional quantum confinement) or on a- Si [4,5]. Although it cannot be ruled out that its origin may be related to saturation of the PS surface with SiH_2, SiO_2, etc. complexes. All these considerations encourage further comparative investigations of visible PL from PS.

The present paper will focus on the application of PL to the study of the "quality" of PS and character changes due to deposition of DLC films different structure and thickness which may confirm the important role of the residual complexes in PL and to gather new data on a possible visible luminescence mechanism.

2. EXPERIMENTAL DETAILS

PS samples were formed from p,n-type (100) crystalline Si wafers with a resistivity of 4.5–40 $\Omega \cdot$cm. Aluminum was coated on the back side of wafers to obtain an ohmic contact. The front surface was etched in a HF electrolyte (HF:C_2H_5OH = 1:1 using a constant current density of 10–75 mA·cm^{-2} for 1.5–10 min.

DLC films were prepared in an rf (13.56 MHz) plasma discharge by decomposition of CH_4:H_2:N_2 gas mixture. The substrates for deposition were put directly on the cathode of diameter 200 mm which was cooled by water and capacitively connected to rf generator. The total pressure of the reaction chamber was varied from 0.1 to 0.8 torr. During the plasma decomposition experiments rf bias voltage was about 1900 volts. The thickness of the films was varied in the region between 100 and 10,000 nm.

The time-resolved photoluminescence spectroscopy (PLS) technique has been used to clarify the character changes due to the decomposition of DLC films on PS. The PL was excited by nitrogen laser at wavelength λ_e = 337.1 nm. The duration of the light pulse was 10 ns and the peak pulse power was 3 kW. Light intensity I_e was varied within the limits 10^{19} to 10^{23} photon/cm^2·s. Light emitted from the films was directed to an FEU-77 or ELU-FS photomultiplier through an SPM-2 monochromator. The experimental technique has previously been described in detail [6]. The experimental facility made it possible to measure PL kinetics and PL spectra at various moments with delay time t_d from the peak of the laser pulse. Time resolution was 0.1 ns or better. Life time for excited state τ_{F1} has been determined from deconvolution of decay curves. Instant and steady-state PL spectra have been compared. The latter represent superposition of fast (FPL) and slow (SPL) photoluminescence which manifest themselves in ns and μs time ranges respectively.

3. RESULTS AND DISCUSSION

Prior to study of the character changes in the porous silicon due to deposition DLC films, we had focused our attention on the results from PL of pristine DLC films. Analysis has shown that under band-band photoexciting, spectral position of the main maximum of PL spectra is stipulated by types of obtained films: for a-C:H:N films containing about 6 at·% nitrogen it corresponds to the spectral range of 2.0 eV (Figure 1). Also this spectrum shows a high energy features at 2.48 eV and 2.8 eV. Life time of excited state was about 5 ns. We suggest that both the excitation and the emission occur in strongly localized electron-hole pairs [7].

Figures 2 and 3 show a PL spectra from a coated and uncoated of PS made as described above from n-Si (100). Close examination of the PL spectra from uncoated PS samples has shown two main luminescence bands whose maximum intensities are located: around 450 nm with life

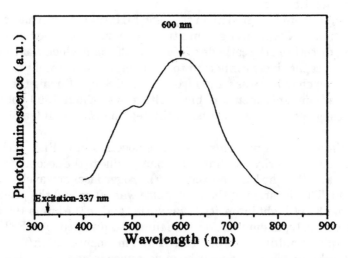

Figure 1. Room temperature PL spectrum from DLC films containing about 6 at · % nitrogen.

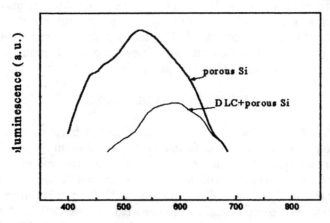

Figure 2. Room temperature PL spectra from PS and DLC + PS systems in the short wave range. Delay time $t_d = 1 \pm 0.1$ ns.

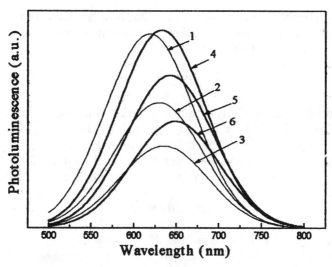

Figure 3. Room temperature PL spectra from PS (4, 5, 6) and DLC + PS systems (1, 2, 3). Delay time t_d: 0–1 μs (1 and 4); 5–6 μs (2 and 5); 10–11 μs (3 and 6).

time about some ns, and around 630 nm with life time about 20 μs. In this case the increase in the delay time from 20 to 30 μs in this wave range is accompanied by the shift of band luminescence from 630 to 650 nm. On the other hand, the PL spectra from coated PS samples has shown quite different results in these two wave ranges. First, we observed the shift of the band ~630 nm to the lower wavelength side (about 610 nm). In this case, the intensity of PL did not change. Second, in the short wave range the band ~450 nm drastically decreased and two new bands around 510 and 590 nm have arisen. An increase in the delay time from 1 to 20 ns leads to dramatic decreasing of the band around 450 nm and essentially increased the intensity of luminescence band around 590 nm. It should be noted that we have observed on the other samples of PS + DLC system not only shift of the main PL peak to short wavelengths but also very strong enhancement of the photoluminescence. Besides, the DLC film-porous silicon systems survived durability tests. Environmental tests also revealed the protection imparted by DLC films to porous silicon devices.

The observations can be interpreted in terms of the structural transformation that occurs in the neighborhood of residual complexes in the pores of PS, and that is promoted by absorption of deposited clusters of DLC which leads to bond breaking and results in smaller cluster sizes. The PL enhancement, therefore is caused by the increasing confinement of photogenerated electron-hole pairs. On the other hand the shift of PL spectra is a consequence of reaction between the decreased clusters and residual complexes of PS.

4. CONCLUSIONS

We have done comparative studies of the PL from coated and uncoated PS samples by DLC films. For samples with the DLC films deposited in the pores of porous Si two effects were observed: enhancement and shift of the photoluminescence. The increasing confinement of photogenerated electron-hole pairs was considered to be responsible for these effects.

REFERENCES

1. Cullis, A. G. and L. T. Canham. 1991. *Nature,* 353:335.
2. Halimaoni, A., C. Oules, G. Bomchil, A. Bsiesy, F. Gaspard, R. Herino, M. Ligeon and F. Muller. 1991. *Appl. Phys. Lett.,* 59:304.
3. Perez, J. M., J. Villalobos, P. McNeill, J. Prasad, R. Cheek, J. Kelber, J. P. Estrera, P. D. Stevens and R. Glosser. 1992. *Appl. Phys. Lett.,* 61:563.
4. Canham, L. T. 1990. *Appl. Phys. Lett.,* 57:1046.
5. Zheng, X. Q., C. E. Liu, X. M. Bao, F. Yan and H. Q. Yang. 1993. *Solid State Commun.,* 87:1005.
6. Piryatinskii, Yu. P. and M. V. Kurik. 1992. *Mol. Mat.,* 1:43.
7. Chernyshov, S. V., E. I. Terukov, V. A. Vassilyev and A. S. Volkov. 1994. *J. Non-Cryst. Solids,* 134:218.

Photovoltaic Cells with CVD Diamond Films: Correlation between DLTS Spectra and Photoresponse Kinetics

P. I. PEROV,* V. I. POLYAKOV, N. M. ROSSUKANYI,
A. I. RUKOVISHNIKOV, A. V. KHOMICH AND A. I. KRIKUNOV
*Institute of Radio Engineering & Electronics of RAS,
11 Mohovaya Street, Moscow, 103907, Russia*

V. P. VARNIN AND I. G. TEREMETSKAYA
*Institute of Physical Chemistry of RAS
Moscow, Russia*

S. M. PIMENOV AND V. I. KONOV
General Physics Institute of RAS, Moscow, Russia

ABSTRACT: Photovoltaic cells with polycrystalline and nanocrystalline CVD diamond films have been fabricated and investigated. The undoped and B-doped polycrystalline diamond films were grown on Si and W substrates by Hot Filament CVD technique, and nanocrystalline diamond films were grown on Si substrate by DC plasma CVD technique. Different after-growth treatments were applied to the films including annealing in air and etching in hydrogen or argon microwave plasma. Semitransparent Ti electrodes were deposited on top of the films, and the substrate was used as the second electrode. Interface band bending and recharging trapping center amount were shown to be strongly dependent on the specific treatment. Kinetics of the open circuit photovoltage and short circuit photoresponse at pulse illumination was measured and analyzed in comparison with DLTS spectra. The correlation between photoelectric signal decays and the DLTS spectra has

---

*Author to whom correspondence should be addressed.

been found. That lets one to ascribe some of the time constants of the photoelectric signal decays to the certain trap levels in the structures, and also to make some conclusions on localization of these traps.

KEY WORDS: diamond photovoltaics, CVD diamond, interface potential, trapping centers.

INTRODUCTION

Diamond film photovoltaics of interest for different applications including radiation and particle detectors and nuclear driven compact batteries [1-3]. The built-in electric field separates the charge carriers generated due to irradiation, and the carrier accumulation and escape processes in the structures with traps can strongly influence the kinetics of the response. The trap parameters in diamond film photovoltaic structures have been studied in this work with the charge-based DLTS [4,5], open circuit photovoltage kinetics at high illumination intensities, and short circuit photocurrent decay.

SAMPLES

Photovoltaic cells with polycrystalline and nanocrystalline diamond films have been fabricated. The undoped and B-doped polycrystalline diamond films were grown on Si and W substrates by Hot Filament CVD technique, and nanocrystalline diamond films were grown on Si substrate by DC plasma CVD technique. Different after-growth treatments were applied to the films including annealing in air at different temperatures and etching in hydrogen or argon microwave plasma. Semitransparent Ti or Ni electrodes were deposited on top of the films, and the substrate was used as the second electrode.

RESULTS AND DISCUSSION

Open circuit photovoltage kinetics under intense pulse illumination (sufficient to reach saturation of the photovoltage magnitude) allowed to find the band bending magnitude and direction (downward or upward) at the diamond-metal interface. The nanocrystalline diamond film treatment and preparation regimes had a strong effect [6] on band bending at interfaces in different structures, as well as on concentration of the recharging traps (which is roughly proportional to the maximum ΔQ value in Q-DLTS spectra), see Figure 1(a) and (b). Note that etching in hydrogen gas atmosphere results in a 20-fold decrease of the recharging trap concentration. The band bending at the diamond film/metal interface be-

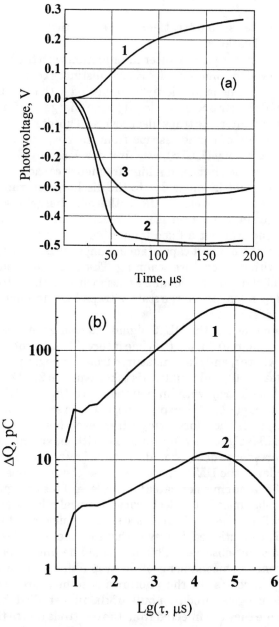

Figure 1. Open circuit photovoltage kinetics (a) and Q-DLTS spectra of the Ti/nanocrystalline diamond film/Si structures as dependent on the diamond film treatment type: 1—as grown (non-etched) diamond film; 2—etched in hydrogen gas atmosphere; 3—etched in ar-

came negative after such a treatment—the case that is desirable for the cold emitter applications.

Decays of the photoresponse after illumination with the high intensity light pulse showed in some samples clearly distinct exponential decay regions, Figure 2(a). The faster decay is connected with the pulse lamp intensity tail, and slower decay is related to the charge carrier escape rate from the bulk and interface traps (deep levels). The alternative technique for finding the rate of carrier escape from the deep levels, Deep Level Transient Spectroscopy (DLTS), was first developed by D. Lang in 1974 [7] making use of the change in a high frequency capacitance of the p-n junction or the Schottky diode due to the discharge of traps during the temperature scan. We developed the isothermal charge-based DLTS (Q-DLTS) technique [4,5], making use of direct measurements of the charge emitted from the traps at a fixed temperature with window rate scanning, instead of the indirect capacitance change measurements at a fixed rate window with temperature scanning. This technique can be used for wider range of structures, including those with the high frequency capacitance independent on charge state of traps, like Mott diodes or metal-insulator-metal structures.

In all the approaches, the DLTS signal is a change in measured value (capacitance or charge) during the time interval $t_2 - t_1$ of the whole relaxation process, where t_1 and t_2 are measured from the beginning of the emission process (at $t = 0$), and usually t_2 is taken equal to $2t_1$. For the case of a single discrete hole trap with an emission rate $1/\tau_p$, the DLTS signal is proportional to $\exp(-t_1/\tau_p) - \exp(-t_2/\tau_p)$, and this expression has a maximum at $t_{1m} = \tau_p \ln 2$. Therefore a single trap level should result in a single peak in the DLTS signal vs $\tau = t_1/\ln 2$ plot with maximum at $\tau_m = \tau_p$.

If the photoresponse decay would be defined by the carrier escape from a single deep level, the DLTS spectra and the photoresponse decay plots should give the same emission rates for this level if the charging processes do not differ substantially for electrical and optical filling the traps. Comparing Figure 2, (a) and (b), one can see such a coincidence for given sample. It should be mentioned, however, that for other samples such a good coincidence did not observed. This fact could be understood having in mind a difference in mechanisms of the trap filling with carriers during the charging processes. In the photoresponse measurement case, the photocarriers are generated in a bulk of the film studied, but in the Q-DLTS technique they are injected from the electrode due to the electrical charging pulse applied to the structure. Depending on the deep level localization in respect to the charge injecting electrode, the charge carrier transport to/from the traps can differ in a great extent from case to case. So, complex study of the photoresponse decay kinetics and DLTS spectra

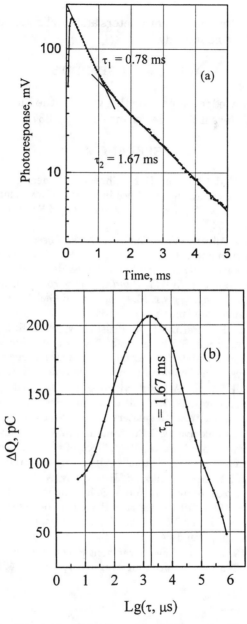

Figure 2. Semi-logarithmic plot of the photoresponse decay after illumination with a 0.2 ms light pulse (a) and Q-DLTS spectrum for the case of charging the sample with an 1V, 1 ms voltage pulse (b) for the Ti/diamond/Si structure based on the B-doped polycrystalline CVD

could be useful in finding the parameters and localization of the deep levels in the samples under study.

ACKNOWLEDGEMENTS

Authors acknowledge supporting this work by the Russian Foundation of Fundamental Research under grant #96-02-18373.

REFERENCES

1. Prelas, M. A., E. J. Charlson, E. M. Charlson, J. Meese, G. Popovici and T. Stacy. 1993. "Diamond Photovoltaics in Energy Conversion," *2nd International Conference on Application of Diamond and Related Materials*, MYU, Tokyo, pp. 329–334.
2. Perov, P. I., V. I. Polyakov, A. V. Khomich, N. M. Rossukanyi, A. I. Rukovishnikov, V. P. Varnin and I. G. Teremetskaya. 1995. "Diamond Photovoltaics: Characterization of CVD Diamond Film-Based Heterostructures for Light to Electricity Conversion," in *Wide Band Gap Electronic Materials, NATO ASI Series 3. High Technology*, v.1, M. Prelas and P. Gielisse, eds., Netherlands: Kluwer Academic Publisher, pp. 171–185.
3. Polyakov, V. I., P. I. Perov, N. M. Rossukanyi, A. I. Rukovishnikov, A. V. Khomich, M. A. Prelas, S. Khasawinah, T. Sung and G. Popovici. 1995. "Photovoltaic Effects in Metal/Semiconductor Barrier Structures with Boron Doped Polycrystalline Diamond Films," *Thin Solid Films*, 266:278–281.
4. Polyakov, V. I., P. I. Perov, A. I. Rukovishnikov, O. N. Ermakova, A. L. Aleksandrov and B. G. Ignatov. 1987. "Multifunctional System for Measurements of Semiconductor Structure Parameters," *Mikroelektronika*, 16:326–331.
5. Arora, B. M., S. Chakravarty, S. Subramanian, V. I. Polyakov, M. G. Ermakov, O. N. Ermakova and P. I. Perov. 1993. "Deep-Level Transient Charge Spectroscopy of Sn Donors in ALGaAs," *J. Appl. Phys.*, 73:1802–1806.
6. Polyakov, V. I., N. M. Rossukanyi, P. I. Perov, A. I. Rukovishnikov, A. V. Khomich, A. I. Krikunov, V. I. Konov, S. M. Pimenov, V. P. Varnin and I. G. Teremetskaya. 1995. "Photoelectrical Properties of Heterostructures Based on CVD Diamond Films," *Diamond Films '95 Conference*, Barcelona, Spain 10-15 September, paper 14.4.
7. Lang, D. V. 1974. "Deep Level Transient Spectroscopy: A New Method to Characterize Traps in Semiconductors," *J. Appl. Phys.*, 45(7):3023–3032.

The Effect of (a-C:H) Films on the Properties of the n-Si/p-(a-SiC:H) Solar Cells

L. S. AIVAZOVA
National Technical University of Ukraine
Kiev, 252056, Ukraine

N. V. NOVIKOV, S. I. KHANDOZHKO, A. G. GONTAR
AND A. M. KUTSAY
V. N. Bakul Institute for Superhard Materials
of the National Academy of Sciences of Ukraine
2, Avtozavodskaya St.
Kiev, 254074, Ukraine

INTRODUCTION

This paper deals with the study of the effect of a-C:H quarter-wave anti-reflection coatings on the spectral dependence characteristics of solar cells based on n-Si/p-(a-SiC:H)/a-C:H structures. These characteristics are shown to be affected by the a-C:H film thickness, which is responsible for the location of the reflection minimum of the film-solar cell system. Mechanical properties (nanohardness, adhesion and strength) have been studied for a-C:H films deposited on solar cells. The results obtained support the advisability of using a-C:H films not just as anti-reflection, but as protective coatings as well.

EXPERIMENTAL DETAILS

Solar cells based on n-Si/p-(1-SiC:H)/a-C:H structures were processed at the Kiev National Technical University. Cell operation characteristics were measured before and after the a-C:H film deposition. The a-C:H films have been deposited in a planar reactor with a plasma excited by a RF discharge of 13.56 MHz on to low temperature (350–550 K) substrates and at pressures between 1 and 5 Pa. The methane content of the methane-hydrogen varied between 20 and 100 vol. %. Methane concentration was one of the main factors defining the film growth rate. The highest growth rate, 50 nm/min, was achieved with film deposition from pure methane. Spectral characteristics of solar cells have been adjusted to account for the spectral sensitivity of the optical measuring system.

RESULTS AND DISCUSSION

Structural research (Raman spectroscopy, mass-spectroscopy of secondary ions, optical absorption edge) and studies of physico-mechanical properties (electrical conductivity, transmission, reflection coefficient, refractive index, nanohardness) of the deposited films support the presence of amorphous hydrogenated carbon.

The film optical gap, E_g, varies with methane concentration of the working mixture, at constant discharge power. With methane concentrations between 20 and 40%, the E_g-value is highest (1.4 eV). The film refractive index ranges from 1.9 to 2.1. At a low-current RF discharge and while increasing the methane content of the gas mixture from 20 to 40%, the nanohardness and Young modulus increased from 7 to 15.5 and from 79 to 127 GPa, respectively. At constant methane concentration, an increase in RF power resulted in a reduction of the film microhardness. The critical load on the hardness indenter at which the film starts to peel off the substrate was found to be at values as high as 200 g. This points to good adherence of the films to the silicon. Coatings of a-C:H films are dielectrics: at room temperature. Their electrical resistance falls between 10^{11}–10^{12} Ohm/cm^2.

The a-C:H quarter-wave anti-reflection coating allows the reflection coefficient of silicon to be reduced from 36 to 0.5% [1,2]. Integral electrical parameters of solar cells before and after applying a-C:H coatings while illuminated with a tungsten filament lamp (2450 K) are listed in Table 1. The data indicate the possibility of improving the solar cell efficiency ratio by 30–40%.

Figure 1 shows the spectral dependence of the solar cell short circuit current, I_{sc}, before and after applying a 70-nm thick quarter-wave anti-reflection coating. The location of its first interference reflection

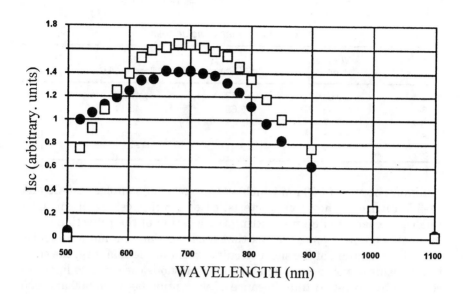

Figure 1. Spectral dependence of the solar cell short circuit current: ●—solar cell; □—solar cell and 70 nm thick film.

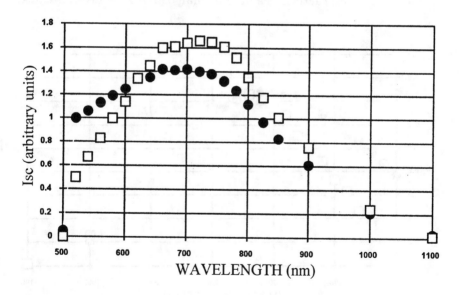

Figure 2. Spectral dependence of the solar cell short circuit current: ●—solar cell; □—solar cell and 85 nm thick film.

Table 1. Properties of solar cells.

Sample	P_s/P_{sa} mW	η_s Silicon	η_{sa} Silicon + DLC	η_{sa}/η_s
42–5–5	2.98/4.13	3.43	4.75	1.38
42–5–22	3.18/3.97	3.65	4.56	1.25
42–5–23	1.69/2.42	3.88	5.57	1.43
42–5–26	1.68/1.97	3.86	4.38	1.13
42–5–32	1.87/2.40	4.31	5.51	1.28

minimum occurs at 550 nm which was the case for samples 42-5-23 and 42-5-5 Figure 2 shows characteristics of a solar cell with an anti-reflection coating of 85 nm putting the location of the reflection minimum at 680 nm. Despite the fact that the reflection minimum coincides with the maximum of the short circuit current of the solar cell, the combined result is worse as compared to the previous case, due to light absorption by the a-C:H film. Because of absorption between 500 and 600 nm, the short circuit current of the solar cell becomes lower as compared to that of the cell without an anti-reflection coating (specimens 42-5-22 and 42-5-32). Characteristics of the solar cell with a 110 nm thick

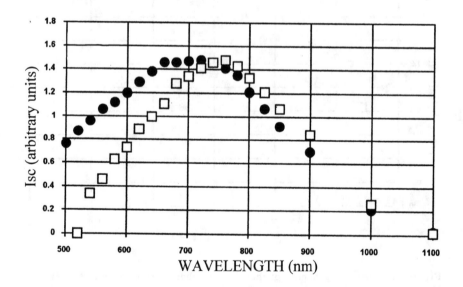

Figure 3. Spectral dependence of the solar cell short circuit current: ●—solar cell; □—solar cell and 110 nm thick film.

a-C:H anti-reflection protective film (specimen 42-5-26) are shown in Figure 3. In this case, the reflection minimum is at 880 nm. A considerable reduction of the short circuit current is observed in the 500–750 nm range due to the absorption which is caused by the increase in film thickness. As a result, the solar cell efficiency increases only slightly.

CONCLUSIONS

With optimally chosen process parameters for the a-C:H quarter-wave anti-reflection coating, solar cell efficiency can be increased by a factor of 1.4. The coating thickness in this case is 70 nm and the reflection minimum is at a wavelength of 550 nm. Nanohardness measurements of a-C:H films support the possibility for their use as protective coatings.

REFERENCES

1. Moravec, T. J. and J. C. Lee. 1982. *J. Vac. Sci. Technol.*, 20(3):338–400.
2. Gontar, A. G., A. A. Doroshenko, A. M. Kytsay and S. I. Khandozhko. 1995. *J. Chem. Vap. Dep.*, 4:15–21.

Simulation of Influence of Traps on the Operation p-i-p and p-i (M)-p Diamond Structures

M. RUSETSKY, A. DENISENKO,* A. MELNIKOV,**
V. VARICHENKO AND A. ZAITSEV*
Belarussian State University
220050 Minsk, Belarus

The **various kinds** of field transistors (FET) were created on diamond and diamond films [1–6]. However characteristic of FET, using p-type diamond as working areas, is limited by high activation energy of boron in diamond, 0.35 eV, and significant reduction of hole mobility owing to scattering on ionized impurity at high concentration of dopant. These restrictions can be removed in devices, using dielectric diamond as work- ing area. The work of such device is based on injection of charge carriers in i-diamond and their transport in a space-charge limited current mode. Effect of holes injection in i-diamond from p-area and SCLC mechanism already was observed in p-i-p structures in MIS-diodes [7–11]. A principle of work of p-i-p structure on diamond also was stated in [12], FET with Schottky gate was demonstrated on IIa type natural diamond [7]. The modeling of p-i-p transistor with SiO_2 gate insulator was spent in [13] neglecting traps in i-diamond. However presence of capture levels, which are always present in natural diamonds and diamond films, renders significant influence on SCLC [14].

---

*Also at University of Hagen, 58084 Hagen, Germany.
**Author to whom correspondence should be addressed.

In the present work with the help of 2D simulation the influence of traps on the electrical characteristics of p-i-p structures and FET on their basis was studied.

For modeling "ATLAS" program was used. The boron concentration in p^+ areas was 10^{18} cm^{-3}, activation energy 0.2 eV. As the program "ATLAS" does not allow to simulate continuous distribution of traps, discrete set from 15 donor-like levels was used. Their energy was counted from the top of valence band, the value of holes capture cross section was 10^{-14} cm^2. It is necessary to note, that the change of boron concentration within the limits of 10^{18}–10^{20} cm^{-3} does not render essential influence to results of modeling. The contacts to p^+ areas were simulated as Ohmic. Other parameters of a material are used as in [13]. In the simulation concentration and field dependencies of charge carriers mobility was taken into account. The results of modeling were compared with electrical characteristics of p-i-p structures and FET, made on a polished plate of a natural crystal of IIa type with the help ion implantation. p^+ areas were made with implantation of boron with distributed energy 25–150 KeV and total dose 2×10^{16} cm^{-2} with subsequent annealing in vacuum at temperature 1450°C during 60 min. Graphitized layer was removed by etching in a hot solution H_2SO_4:$CrCO_3$. The schematic image of structure is submitted on Figure 1.

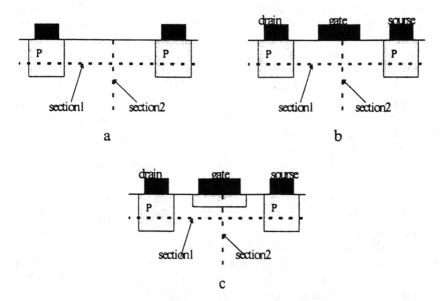

Figure 1. A cross-section view of p-i-p structure (a), FET (b) and FET with additional doping area (c).

As the character of traps distribution in diamond is rather diverse [10,15,16], two most probable distributions was simulated: Gaussian and exponential. In case of Gaussian distribution fitting parameters were total concentration of traps, width of distribution and location of the maximum. In case of exponential distribution-total concentration and the index of exponent. In Figure 2 simulated I-V curves of p-i-p structure with various distribution of traps are shown. The character of traps distribution have the most influence to initial area of I-V curve, where the current is determined by barrier emission. In the SCLC area the most impotent parameter is the total concentration of traps. With increase of traps concentration SCLC part of I-V curve is shifted in the area of high voltage.

In case of Gaussian distribution of density of trap states the heaviest influence on I-V curves renders a location of maximum of distribution (shift of a maximum of distribution from valence band towards conduction band results in increase slope of an initial region of I-V curve). Width change of distribution (at fixed total concentration of traps) results in insignificant change of a kind of I-V curve in its initial part. In exponential distribution case the slope of I-V curve is determined in index of exponent and its increase results in reduction of slope of initial area of I-V curve. In both cases increase of total concentration of traps results in increase of height of a potential barrier and shift of an initial region of I-V curve in the area of high voltage. Figure 3 shows the potential distribution in p-i-p structure on without traps and with exponential traps distribution. The I-V curves obtained from experiment and simulation for 4, 9 and 13 μm, are shown in Figure 4. As it is visible from the diagram both distributions with about identical accuracy describe experimental data. However in both cases at large length of i layer the modeling gives shift of initial part of the I-V curves to the high voltage region concerning experimental. Such divergence with experiment can be connected with unhomogeneous distribution of traps in the bulk of sample.

As both distributions with close accuracy describe experimental date for p-i-p structures at modeling of FET exponential distribution of traps was used. At modeling of FET with Schottky barrier gate the work function of the gate metal was 6 eV. The characteristics drain current-drain voltage (I_D-U_D) at $U_G = 0$ and I_D-U_G for FET with gate length 15 micron on pure diamond and diamond with traps are submitted on Figure 5 (curve 1 and 2 respectively). Presence of traps results in displacement of a range of working voltage of the transistor in high voltage area. Nevertheless the size I_D remains very small even at voltage, close to breakdown. Specific transconductance of such transistor does not exceed 2.5×10^{-8} S/μm for pure diamond and 1.2×10 S/μm for diamond with traps. It is

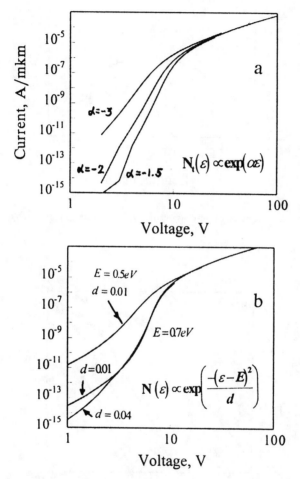

Figure 2. I-V curves of p-i-p structure with exponential (a) and Gaussian (b) traps distribution (total concentration of traps 1.8×10^{13} cm^{-3}).

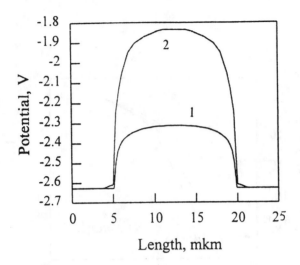

Figure 3. Potential distribution in p-i-p structure without traps (1) and with exponential traps distribution (2).

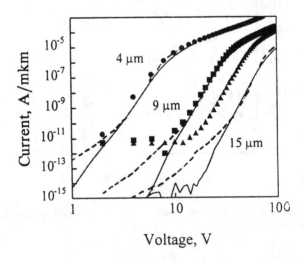

Figure 4. I-V curves of p-i-p structures. Solid lines—exponential traps distribution, dashed lines—Gaussian traps distribution, discrete points—experimental date.

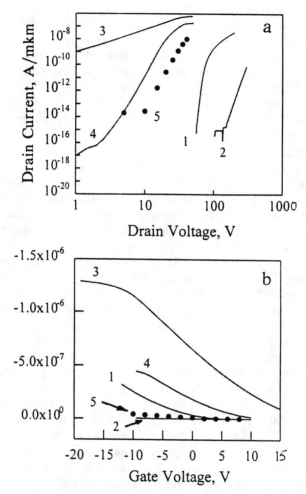

Figure 5. (a) I_D-U_D characteristics of FET $U_G = 0$; (b) I_D-U_G characteristics of FET (1,2-$U_D = -200$ V, 3,4-$U_D = -50$ V, 5-$U_D = -40$ V).

necessary to note, that the transistor with such parameters gate works only in depletion mode. As a gate voltage about -10 V junction gate—source opens, specific transconductance of the transistor decrease with increase of trap concentration.

It is necessary to note, that drain current of the transistor is much less than through p-i-p structure with appropriate width of i layer. It is connected to gate influence on the potential distribution of the transistor. In Figure 6(a) and 6(b) is shown the distribution of potential in p-i-p structure and FET at zero voltage on electrodes (curves 1 and 2 respectively). Potential barrier in the transistor is much higher, as results in significant reduction of I_D. It is, probably, connected to pinning of Fermi level in the middle of a band gap owing to interaction with metal of gate.

However Schottky barrier heights depends mainly not on metal of gate, but on processing of a surface and annealing [17,18], i.e., on density of surface states. The used version of a program "ATLAS" does not allow to simulate influence of surface states to Schottky barrier height. To investigate influence of surface states the layer under the gate thickness 0.1μm was doped with boron at concentration 10^{18} cm^{-3} [Figure 1(c)]. The role of doped area is similar to a role of surface states. In this case the Fermi level is pinned near the valence band and potential barrier height is thus fixed. The distribution of potential in such structure is represented in Figure 6 (curves 3). Potential barrier height in such structure significantly decreases. Reduction of a potential barrier results in significant shift of the I_D-U_D characteristics of the transistor in area of low voltage [Figure 5(a), curve 3—pure diamond, 4—diamond with traps]. The I_D-U_G characteristics are also shifted in the area of positive voltage, that allow to receive significant modulation of a drain current both in depletion and accumulation modes. Specific transconductance in such FET makes 4.3×10^{-8} S/μm for pure diamond and 2.8×10^{-8} S/μm at presence of traps. Increase of traps concentration, as in the previous case, results in reduction of working currents and transconductance of the transistor. Comparison of the experimental electrical characteristics of FET with Schottky barrier [Figure 5(a) and 5(b)] shows, that its characteristics are rather close to the characteristics of simulated FET in view of surface states in Schottky junction.

Thus, results of modeling show, that presence of traps in i-area in p-i-p structures and FET results in displacement of the electrical working characteristics in the high voltage area. The I_D-U_G and I_D-U_D characteristics of FET strongly depend on the surface states in Schottky junction. Controllable introduction of surface states allows to set potential barrier height and, thus shifts a working point of the transistor from the accumulation to the depletion mode.

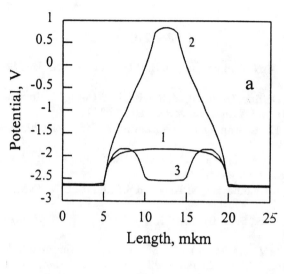

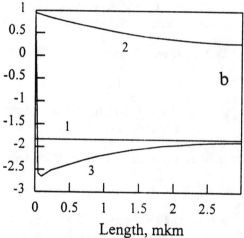

Figure 6. Potential distribution in p-i-p structure and FET (a—section 1; b—section 2). 1—p-i-p structure [Figure 1(a)], 2—FET [Figure 1(b)], 3—FET with addition doping area [Figure 1(c)].

This work was partially supported by DAAD and INTAS project No. 94-1982.

REFERENCES

1. Geis, M. W., N. N. Efremov and D. D. Rathman. 1988. *J. Vac. Sci. Technol.*, A6(3):1953.
2. Trai, W., M. Delfino, D. Hodul, M. Riaziat, L. Y. Ching, G. Reynolds and C. B. Cooper. 1991. *IEEE Electron Device Lett.*, 12:157.
3. Nishimura, K., K. Kumagai, R. Nacamura and K. Kobashi. 1994. *J. Appl. Phys.*, 76(12):8142.
4. Zeisse, C. R., C. A. Hewett, R. Nguyen, J. R. Zeidler and R. C. Wilson. 1993. *IEEE Electron Device Lett.*, 14:66.
5. Shiomi, H., Y. Nishibayshi, N. Toda and S. Shikata. 1995. *IEEE Electron Device Lett.*, 16:36.
6. Gildenblat, G. S., S. A. Grot and R. Bardzian. 1991. *IEEE Electron Device Lett.*, 12:37.
7. Denisenko, A. V., A. A. Melnikov, A. M. Zaitsev, V. I. Kurganskii, A. J. Shilov, J. P. Gorban and V. S. Varichenko. 1992. *Mat. Sci. and Eng.*, B11:273.
8. Miyata, K. and D. L. Dreifus. 1993. *J. Appl. Phys.*, 73:4448.
9. Ashok, S., K. Srikanth, A. Bardzian, T. Bardzian and R. Messier. 1987. *Appl. Phys. Lett.*, 50:763.
10. Matsumae, Y., Y. Akiba, Y. Hirose, T. Kurosu and M. Iida. 1994. *J. Appl. Phys.*, 33:L702.
11. Denisenko, A. V., A. A. Melnikov, A. M. Zaitsev, B. Burchard and W. R. Fahrner. 1995. *Diam. Films & Technology*, 5.
12. Burchard, B., A. V. Denisenko, A. A. Melnikov, A. M. Zaitsev, W. R. Fahrner, F. Peterman, K. Hoppe and B. Stanski. 1994. In *Advances in Diamond Science and Technology*, S. Saito et al. eds., Tokyo: MYU, 721.
13. Miyata, K., K. Nishimura and K. Kobashi. 1995. *IEEE Trans. Electron. Dev.*, 42:2010.
14. Lampert, M. A. and D. Nark. 1970. *Current Injection in Solids*. New York: Academic Press.
15. Mort, J., M. A., Machonkin and K. Okumura. 1991. *Appl. Phys. Lett.*, 59:455.
16. Maki, T. and T. Kobayashi. 1994. *Advances in New Diamond Sci. and Techn.* Tokyo: MYU, 733.
17. Das, K., V. Venkatesan, K. Miyata and D. L. Dreyfus. 1992. *The Sol. Films*, 212:19.
18. Geis, M. W., N. N. Efremov and J. A. von Windhein. 1993. *Appl. Phis. Lett.*, 63(7):952.

Surface Modification of Polymers for Medical Application by Means of Plasma-Ion Deposited Carbon Films

V. V. SLEPTSOV, V. M. ELINSON V. V. POTRAYSAY AND A. N. LAYMIN
Moscow State Aviation Technology University
27 Petrovka str.
103767 Moscow, Russia

YU. A. ROVENSKY
Cancer Center of Russian Academy of Sciences
24 Kashirskoye shosse
115478, Russia

ABSTRACT: Modification of a polymer material's surface is an excellent method to alter its properties and thereby expand its application.

Polymers present a class of materials widely used in medical technology where medical and biological properties are defined by surface characteristics: composition, charge and surface roughness.

In the present study, surface properties of polymer materials modified by carbon films are considered. The films have been formed by different plasma-ion methods. Electrostatic and electrokinetic potentials have been investigated in their dependence on method, deposition conditions gas mixtures, and *in vitro* adhesion to cells of connective tissue.

1. INTRODUCTION

Polymers present a broad class of materials frequently used in medicine [1]. They stand out because of their simplicity of treatment, chemical inertness and low cost. The most often used polymers, in

77

medical practice are teflon, polypropylene, polymethylmethacrylate (PMMA) and its derivatives, (e.g., hydroxyethylmethacrylate-HEMA), acrylamide and silicone. Disadvantages of polymers are their low mechanical strength, low biocompatibility, and the absence of longtime resistance to aggressive biological media. In many cases the growth of body-activity-products on the inner and outer surface of polymers has been observed. To solve the problem it would be useful to deposit carbon films, including diamond-like and carbine-containing ones, onto the surface of polymer products [2,3].

As has been shown before [2–4] carbon thin films can provide biocompatibility, in many cases increasing chemical resistance, mechanical properties of and antibacterial activity in a few cases [4]. Plasma-ion deposition also provides high adhesion between carbon films and polymer materials [2–6].

It should be mentioned that plasma-ion technology allows not only adjustment of film properties but also provides the possibility to change over a wide range, surface characteristics such as composition, charge, structure and surface roughness. The surface properties of the carbon films mainly define their medical and biological interface characteristics: adhesion to cells and proteins, cytotoxicity and the effect on surrounding media.

The aim of present study was twofold: the investigation of charge characteristics of carbon films formed by different plasma-ion methods on the surface of polymers, as a function of method, deposition conditions and composition of the gas media. Secondly, the establishment of the interrelation between charge properties and the nature of medical and biological applications.

2. EXPERIMENTAL DETAILS

Carbon films were obtained by deposition from directed plasma-ion flows of hydrocarbon vapours (method 1) [5,6] and by plasmachemical deposition. In method one treatment with argon ions was sometimes used. Wafers and sheets of PMMA, PET and HEMA were used as substrates. Cyclohexane and its mixtures with nitrogen were the working gases. Film thickness varied within the interval 20–200 mcm.

Adhesion to culturated cells (embryon mice fibroblasts) was investigated in the following way. The substrates were treated with 96% ethanol, washed in distilled water and after drying were placed into 2 cm² bottom square culture cups. The cups with the substrates were sterilised in UV radiation for 1 h. The cells were suspended in 199 cultural media. The volume of the cell suspension per a cup was 1.5 ml at a cell concentration 50.103 per 1 ml. The cells were cultured for 18–20 h, at 37°C in air with 5% CO_2. Then samples were fixed, colored with hemathoxylineosine according to the standard method and studied with a light microscope. Every sample was examined for 15–20 image fields (square of image was 0.6 mm^2).

Measurements of the electrostatic potential were carried out with the help of a voltmeter. Measurements of the electrokinetic potential were carried out by the flowing potential method described in [9].

3. RESULTS AND DISCUSSION

It should be pointed out that after carbon film deposition on polymers, any changes in charge characteristics, contact angle, relief and surface composition were noted. In the present paper only some of the results concerning charge property studies are given while an attempt has been made to find a correlation with adhesion on the connective tissue cultured cells. Figures 1 and 2 show results on measurements of electrostatic potential $|\varphi|$ of PET, PMMA, HEMA modified by film deposition in a mixture of C_6H_{12} + N_2 by two plasma-ion methods. The accuracy of the electrostatic potential measurements was 25–30%.

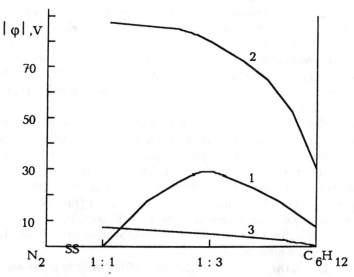

Figure 1. Electrostatic potential $|\varphi|$ of polymer materials modified by film deposit from directed plasma-ion flows of particles in C_6H_{12} + N_2 mixtures: 1-PMMA, 2-PET, 3-HEMA.

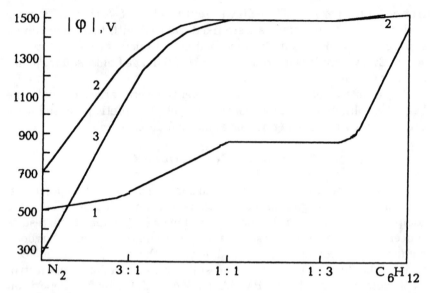

Figure 2. Electrostatic potential $|\varphi|$ of polymer material modified by plasma chemical deposition in $C_6H_{12} + N_2$ mixtures: 1-PMMA, 2-PET, 3-PET track membranes.

Several points should be noted while examining the curves: the absolute value of the electrostatic potential of carbon coated polymers made by method 2 is more than two orders of magnitude higher than for those in which method 1 was used; dependence of $|\varphi|$ on composition is very different for methods 1 and 2; PET track membranes (diameter of pore 0.4 mcm) demonstrate the same behavior as dense PET till 50% N_2 is reached in the working mixture. When the N_2 content exceeds 50% this dependence becomes steeper.

These effects could be explained if we take into account that mechanisms of film formation are different for methods 1 and 2. In the first case deposition is realised at a pressure of $(3-5)\ 10^{-4}$ torr, in the second case the pressure is $10^{-1} + 10^{-2}$ torr. In the first case the substrate and growing film are impacted by partially ionised beam from the autonomic ion source and the film forming process is occurring only on the surface of substrate. In the second case substrate and film are in contact with the plasma volume and it is necessary to take into account the collective interactions in the plasma.

The surface of the substrate and growing film, in spite of the same content of gas mixture, is treated by absolutely different particles in both cases. Differences in $|\varphi|$ value and types of dependences of $|\varphi|$ upon gas mixture composition could be explained on this basis.

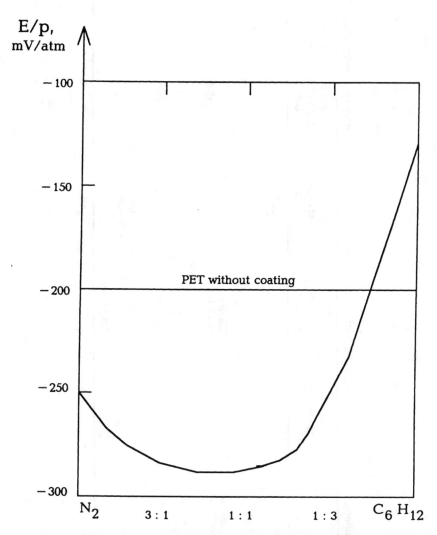

Figure 3. Electrokinetic potential of PET track membranes modified by plasma chemical deposition in $C_6H_{12} + N_2$ mixtures.

Table 1. Adhesion of connective tissue cells to carbon films deposited onto PET by different plasma-ion methods.

	Substrate	Depos. Method	Deposition Conditions	Average No. of Cells in 0.6 mm^2	Amount of Lightly Depos. Cells(%)	$\frac{\|\varphi\|}{B}$ ± 25%
1	Glass (control)	—	—	300.0	5	—
2	PET membrane	2	C_6H_{12}	11.0	51.6	1200
3	PET	2	C_6H_{12}	13.2	27.2	1300
			$d = 0.5 \times 10^{-6}$ m			
4	PET membrane	2	C_6H_{12}	15.2	75.9	1500
5	PET	1	$C_6H_{12} + N_2$ 3:1	34.6	13.6	30
6	PET	1	—	46.5	14.6	10
7	PET	1	C_6H_{12}	82.7	76.7	5
8	PET	1	$C_6H_{12} + N_2$ 1:1	92.4	29.6	0
9	PET	1	C_6H_{12}	189.4	35.3	5
10	PET	2	C_6H_{12}	196.7	12.2	~100
			$d > 3 \times 10^{-6}$ m			

Data gathered by IR-spectroscopy are in good agreement with such an explanation. The results show that in the first case, the film contains -NH$_2$ bonds, -NH bonds and at high N$_2$ content -CN bonds. In the second case the film contains only -NH$_2$ and -NH bonds; -CN bonds are absent.

Differences in φ behaviour for continuous and porous PET at high N$_2$ content in gas mixtures is connected with an increase of the shape and diameter of the pores because of N$_2$ etching during the plasma chemical treatment. In this case the mean free path of the particles is close to the pore diameter.

Measurement of the flow potential allowed determination of the sign of the charge for pure PET and the influence of film deposition on the charge characteristics (Figure 3). It could be established that film deposition from C$_6$H$_{12}$ decreases the charge of the membrane surface while an increase of N$_2$ content in gas mixture leads to an increase of charge on the surface. This could be connected with appearance of nitrogen-containing groups on the surface.

Data on the adhesion to embryo mice fibroblasts on continuous and porous PET after carbon film deposition, by different plasma-ion methods, are presented in Table 1. It gives the average number of cells and the content of weakly spread cells including cells with highly reduced cytoplasma.

The number of adsorbed cells (spread, weakly spread and unspread) characterizes the adhesive properties of the substrates—their capability to bond the cells percipitated from suspensions onto substrate surfaces. Content of cells achieved sufficiently exhibited spreading characterises complex of properties of this substrate; adhesion to cells, ability to absorb proteins, character of roughness, as well as the absence of cytotoxicity (excluding spreading cells promoting their further demise).

The table shows the general tendency: the lower the g value, the higher the adhesion to the cells. Nevertheless we must note the influence of the structure factor: samples 7 and 9 are different in their sp^3 and sp^2-hybridised states in accordance with the results of AES spectroscopy [5,6]. Sample 9 contains more sp^2-bonds. These films also possess different specific surface values. Surface roughness demonstrates the influence in samples 3 and 4. The content of weakly spread cells does not change proportional to the charge dependence. The main factors in this case are structure and surface composition.

4. CONCLUSION

The possibility for wide range monitoring of the charge characteristics of polymer surface coated with carbon films deposited by different plasma-ion methods has been demonstrated in this paper.

It was established that adhesion to connective tissue cells is proportional to the value of the electrostatic potential. The content of weakly spread cells, connected with cytotoxicity, is defined it seems to us, by structure, relief and surface composition.

REFERENCES

1. 1992. "Modern Approach to Development of Effective Stuture and Polymer Implants," *Proc. of 1st Intern. Conf.,* Moscow.

2. Thomson, L. A., F. C. Law, M. Ruston and F. Franks. 1991. *Biomaterials,* 12:37-39.

3. Mitura E., S. Mitura, P. Niedzelski, Z. Has, R. Wolviec, A. Jacubovski, I. Szmidt, A. Sokolovska, P. Louda, I. Marciniak and B. Koszu. 1993. "Diamonds Films 93," *4th Europ. Conf.,* ALbufeira, Portugal, p. 12,166.

4. Sleptsov, V. V., V. M. Elinson, N. V. Simakina, G. E. Khots and Yu. A. Riovensky. 1995. "C-BN and DIamond Crystallization Under Reduced Pressure," *Proceeding, Intern. Conf.,* Jablonna, Poland, p. 77.

5. Sleptsov, V. V., A. A. M. Kuzin, G. F. Ivanovsky, V. M. Elinson, S. S. Gerasimovich, A. M. Baranov and P. E. Kondrashov. 1991. *J. of Non-Cryst. Solid,* 136:53-59.

6. Ivanovsky, G. F., V. V. Sleptsov, V. M. Elinson and P. E Kondrashov. 1989. *Electronic Industry,* (in Russian), 12:26-31.

7. Sleptsov, V. V., V. M. Elinson, N. V. Simakina and A. M. Baranov. 1991. *Proc. of 179th Meeting of Electrochem. Soc., 2nd Int. Symp. on Diamond Materials,* Washington, DC, p. 126.

8. Sleptsov, V. V., V. M. ELinson, S. S. Gerasimovich and A. G. Vologirov. 1990. *Electronic Technique, (in Russian), 2(159):30–34.*

9. Sobolev, V. J., J. P. Sergeeva, N. V. Churaev and D. E. Ulberg. 1992. *J. Colloid and Interface Sciences,* 151(2).

Boron Nitride Protective Coatings for Polymer Materials

S. S. GERASIMOVICH AND V. V. SLEPTSOV

Moscow State Aviation Technology University
27, Petrovka str
103767, Moscow, Russia

1. INTRODUCTION

The attention of many investigators has over the last few years been attracted to the generation of film materials simultaneously possessing excellent mechanical, optical and thermal properties as well as radiation stability. As is the case with diamond and diamond-like films, such requirements might possibly be met with boron nitride based films.

Boron nitride displays various allotropic modifications, depending on the specific thermodynamic conditions of synthesis. The cubic modification of boron nitride (c-BN) has a high microhardness and dielectric characteristics, excellent thermal stability and a high transparency. This is normally synthesized under conditions of high temperature and pressure [1]. Low-temperature ion-assisted deposition methods generating high pressure and temperature conditions on the molecular level, may also generate cubic boron nitride films [2,3].

The primary interest lies in the formation of BN films with a high percentage content of the c-BN phase by ion-assisted methods thus securing a high growth rate.

---

*Author to whom correspondence should be addressed.

85

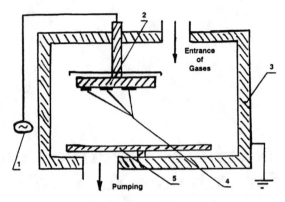

Figure 1. The schematic of the RF-diode system for deposition of BN films. 1—RF generator; 2, 5—electrodes; 3—working chamber; 4–substrates.

2. EXPERIMENTAL ASPECTS

BN films were obtained by RF-decomposition from borazine ($B_3N_3H_6$) and nitrogen (N_2) mixtures of different concentrations in an RF-diode system. A schematic of the experimental set-up is shown in Figure 1. One electrode of the diode system is grounded, the other connected to an RF-generator (13.56 MHz). The substrate, 130 mm diameter was fixed to the upper electrode. The specific power distributed to the RF-electrode ranged from 0.07 to 1.51 V/cm. The temperature of the substrates was no higher than 80°C. The distance between electrodes could be changed from 30 to 100 mm. The working (gas) pressure was $(5-7)*10^{-2}$ mm Hg.

3. EXPERIMENTAL RESULTS

The dependence of the film growth rate on RF-power for initial gas composition was investigated first. Depending on the specific conditions, it could be altered from 19.2 nm/s to 32.5 nm/s (5.5 μm/h–8.3 μm/h, respectively). These growth rates are higher than those obtained from directed ion beam deposition for the same experimental conditions [3]. A comparison of some of the parameters of the RF-diode and ion stimulated deposition methods are presented in Table 1.

The data in the table indicates that the RF-diode deposition method yields advantages in terms of growth rate and thus coating thickness with a simultaneous decrease in power and average energy of the particles forming the films.

Table 1. Characteristics of ion-assisted methods on BN deposition.

Ion-Assisted Method	Working Pressure (mm Hg)	RF Power (V/cm)	Average Energy (keV)	Growth Rate (nm/s)	Thickness (μm)
Directed ion-beam deposition (ion source IBAD)	$(1-8)\cdot 10^{-4}$	100–800	0.3–2.0	0.1–0.2	0.1–0.3
Radio-frequency chemical vapor deposition	$(5-7)\cdot 10^{-2}$	10–200	0.6–0.9	0.1–0.5	0.1–0.5

Optical characteristics of the BN films were measured in the 300–900 nm wavelength range (see Figure 2). The film thickness was 2 μm. The optical band gap value of the films was calculated from spectral transparency as 3 eV (curve 1) and 2.7 eV (curve 2). The films thus had a high transparency in the visible range (T = 90–95%). The transparency (T) of BN films depended on the specific growth conditions. It decreases with an increase of RF discharge power which causes alterations of the structure of the films, namely the content of the c-BN phase.

The structure of the films was examined by infrared spectrometry and

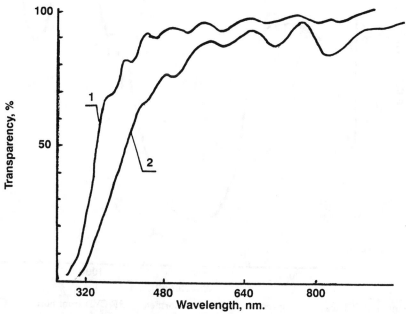

Figure 2. Optical transparency of 2 μm thick boron nitride films.

X-ray diffraction analysis. It was established that the film consisted of a mixture of the hexagonal (h-BN) and cubic (c-BN) modifications of boron nitride. The IR-spectrum of the films is shown in Figure 3. The presence of the cubic phase (c-BN) in the films was indicated by an absorption peak at 1100 sm^{-1}. The cubic phase content was calculated on the basis of the relation of the intensity of the peaks at 1100 cm^{-1} and at 1380 cm^{-1} [4] according to,

$$C = \frac{A_{1100 \, cm^{-1}}}{A_{1100 \, cm^{-1}} + A_{1380 \, cm^{-1}}}$$

The cubic phase content in films obtained by decomposition of borazine in the RF-diode system amounted to 25%.

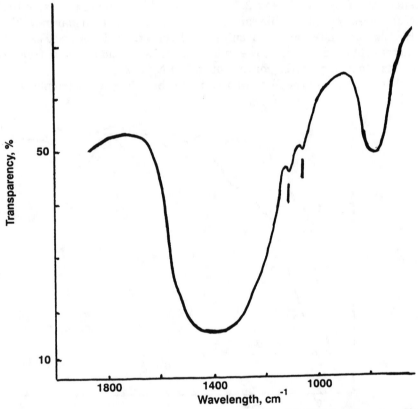

Figure 3. IR-absorption spectrum of BN films formed by RF CVD from borazine, film thickness was 1.5 μm

The same value was established by the X-ray phase analysis. The size of the c-BN regions were of the order 40–50 Å (4–5 nm). The c-BN content in films increased to 40% with an increase of nitrogen concentration in the gas mixture.

4. FILM EVALUATION

Abrasion wear tests of the BN films were carried out for polymer substrates (polycarbonate, polymethylmethacrylate) and glass substrates, with the Tabor test. Film thickness was 2 μm. The tests demonstrated that the samples withstood grinding by wheel abrasion at P = 1 kg loads and a number of turns, N = 100. The hardness of the BN film coated polymer surfaces was found comparable to that of a corundum surface, but without change of flexibility and elasticity of the polymer.

5. CONCLUSION

A method of radio frequency chemical vapor deposition (RF CVD) in diode systems allows the generation of boron nitride films, containing the cubic phase (c-BN) at a high growth rate (5.5–8.3 μm/h).

The method of production of the BN films allows an opportunity to alter the optical and mechanical properties of films due to a change in their structure.

Coatings on the basis of BN films are a reliable protection for polymer optical elements (lenses, mirrors, filters, prism, etc.) at reasonable cost.

REFERENCES

1. Samsonov, G. V. et al. 1960. "Boron, Its Compounds and Alloys," *Russian Academy of Sciences,* Eds., Kiev, p. 206.
2. Kuhr, M., S. Reike and W. Kulish. 1994. "Deposition of Cubic Boron Nitride with Inductively Coupled Plasma (ICP)," paper presented at Diamond 1994, to be published.
3. Grishko, L. B., V. M. Elinson, G. R. Ivanovski, V. V. Sleptzov and L. D. Holeva. 1986. "Thin Film Deposition from Directioned Particles of Plasma Flows," *E. P. (Elektronnaya Plasma)* v. 4, 152:6–11.
4. Reinke, S., M. Kuhr, W. Kulish and R. Kassing. 1994. "Recent Results in Cubic Boron Nitride Deposition in Light of the Sputter Model," paper presented at Diamond 1994, to be published.

Photoelectrochemistry of Synthetic Diamond Electrodes

YU. V. PLESKOV
A. N. Frumkin Institute of Electrochemistry
Russian Academy of Sciences
Leninsky prospekt 31
Moscow, 117071 Russia

INTRODUCTION

Synthetic polycrystalline diamond, owing to its enhanced chemical stability, may be thought to be a promising corrosion-stable electrode material, or protective coating for electrodes, e.g., for photoelectrochemical cells and sensors. On the other hand, electrochemical techniques can be used for the characterization of diamond films. In the present work we summarize our recent studies in the photoelectrochemistry of CVD diamond thin films. In particular, photocurrent spectra were recorded on the diamond/electrolyte solution contact (note that a semiconductor/electrolyte contact in many respects resembles the Schottky contact semiconductor/metal [1]); threshold energies were determined for the electron transitions in diamond, which can be related to the photocurrent formation [2]. Photopotential transients, upon illumination with nanosecond laser flashes, were recorded for the films under open-circuit conditions. The transients were analyzed using a simple equivalent circuit of the electrodes, whose elements were calculated by the photopotential decay curves [3].

EXPERIMENTAL

The electrodes were polycrystalline CVD diamond films, deposited onto tungsten substrates from hydrogen–methane gas mixture under the hot-filament activation (the deposition techniques were described in details earlier [2]). The film thickness was 2 to 10 μm. The films were boron-doped in the course of their growth; the boron concentration in the samples ranged from $5 \cdot 10^{17}$ to 10^{21} cm^{-3}. For sake of comparison, a few samples were also examined, whose conductance was caused, in all probability, by some unidentified impurity and/or crystal defect. In what follows, we shall call these samples "undoped." When mounting the samples as electrodes in the cell, we isolated the substrate surface with high-purity paraffin wax. All measurements were done in 1 M KCl or 0.5 M H_2SO_4 solution at room temperature.

Photocurrent spectra were recorded with a set-up designed and developed in the Palermo University (Palermo, Italy). The nearly monochromatic illumination of electrode surface (at a spectral resolution of 5 nm) was produced using a lamp–monochromator system. Photocurrent response under chopped illumination (at a frequency of 5 Hz) was detected by a lock-in amplifier (P.A.R., model 5206) as a function of wavelength and applied potential.

In recording photopotential transients, we used a set-up designed and developed in the Chemical Research Institute, Saratov University (Saratov, Russia). It is based on a nitrogen laser LGI-21 as a light source (the wavelength 337.7 nm). The laser produced flashes (width at a half-magnitude: 10 ns) repeated at a frequency of 10 Hz. We used a stroboscopic registration system which made it possible, by averaging the signal over large number of flashes, to increase signal-to-noise ratio. The signal from electrochemical cell passed through a matching preamplifier, then was amplified by an amplification unit and fed to the input of a stroboscopic memory oscilloscope S8-13.

RESULTS AND DISCUSSION

The electrodes exhibited photosensitivity in the vis.- and UV-parts of spectrum; the sign of open-circuit photopotential corresponded to a p-type semiconductor. Cathodic photocurrents were recorded for undoped samples in a broad spectral range (λ < 800 nm). Generally, for boron-doped films the photocurrent density was lower, than for undoped ones.

The photosensitivity of diamond films is caused, in all probability, by the optical electron transitions within the forbidden band of diamond; they result in formation of excess current carriers which participate in the charge transfer at the semiconductor (diamond)/ solution interface. To have a better insight into the nature of these transitions, we recorded photocurrent spectra of the diamond electrodes. Typical cathodic photocurrent spectrum for a boron-doped electrode (No. 306) is given in Figure 1. We estimated threshold energies for the electron transitions by plotting the spectra in $(i_{ph} \, h\nu)^n$–$h\nu$ coordinates, with $n = 1/2$ for indirect transitions (see insert to Figure 1). Two thresholds were often observed, one in the high-energy, the other in the low-energy part of the spectrum.

The particular values of threshold energies for undoped films were: 3.9–4.0 eV (high-energy threshold) and 1.7–1.9 eV (low-energy threshold). For boron-doped films, the high-energy and the low-energy threshold was 2.7–3.0 eV and 1.15–1.8 eV, respectively.

A tentative explanation of the results obtained can be given by using an energy diagram of the diamond/electrolyte contact (Figure 2), where band edges in diamond are shown, along with energies of some impurity levels in diamond and electrochemical potential levels for

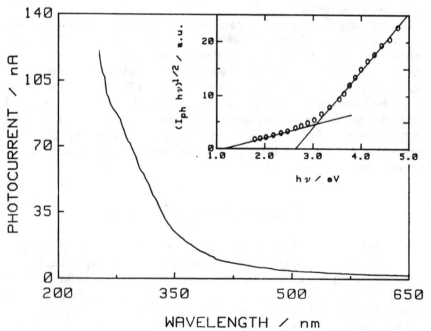

Figure 1. Typical cathodic photocurrent spectrum for a boron-doped electrode.

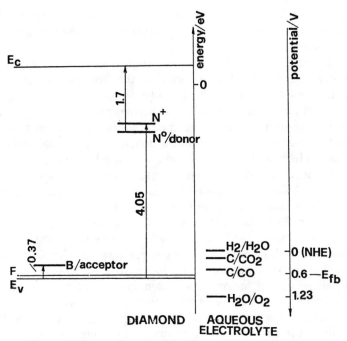

Figure 2. Energy diagram of the diamond/electrolyte contact.

some redox-systems in the electrolyte. One may speculate that the cathodic photocurrent of undoped samples in the high-energy part of spectrum (with the 3.9–4.0 eV threshold) is mainly contributed to by optical transitions of valence band electrons to nitrogen centers (note that nitrogen takes the form of N^+ in partly compensated diamond crystals). We further assume that the nitrogen levels, at their sufficiently high concentration, form a narrow band inside the forbidden band of diamond, through which the electrons reach the diamond/solution interface, where they pass to the solution (via the electrochemical reaction of hydrogen evolution). However, the very existence of the low-energy threshold, as well as of the lower threshold energies in the boron-doped samples suggests that mechanisms other than the above-discussed one are also involved.

The photopotential transient (V-t) observed upon the electrode illumination with laser pulses consists of three stages: (1) a peak that practically follows the shape of the laser pulse, its magnitude being 5 to 10 mV (which probably is caused by very fast recombination of photogenerated electron–hole pairs); (2) a relatively fast decay which lasts several microseconds; and (3) a slower decay which lasts, depending on the load resistance value, up to several tens of milliseconds. The positive sign of photopotential corresponds to a p-type semiconductor.

In discussing the stages of transient photo-processes in a diamond electrode, we shall use an equivalent circuit of the electrode shown in Figure 3. In this circuit, the process of photogeneration of nonequilibrium charge carriers is represented by a photocurrent generator i_{ph} with infinitely large impedance. The capacitance of space charge region is denoted as C_{sc}. It gets charged up upon illumination, due to the separation of photogenerated electron–hole pairs, and starts discharging right after the light pulse end. A process corresponding to stage (2), with a time constant of ca. 1 μs, is likely to be caused by redistribution of the photogenerated charge between the space charge region and moderately fast surface states, whose differential capacitance is C_{ss}. Unlike stage (1), this process requires a finite time $\tau_{ss} = R_{ss}(C_{sc}^{-1} + C_{ss}^{-1})^{-1}$, where R_{ss} is the resistance representing the slowness of surface states charging. The photopotential decay during stage (2) appears to be approximately one-half of the peak photopotential during stage (1), which means that the capacitances C_{ss} and C_{sc} are of the same order of magnitude.

Finally, both capacitances discharge up through an external load R_1 which equals the sum of inner resistance of the electrochemical cell R_s and input resistance of the oscilloscope (which was shunted, in the course of measurements, with an additional variable resistor). In the case when the time constant of stage (3) $\tau = R_1(C_{sc} + C_{ss})$ exceeds, by the order of magnitude, the relaxation time τ_{ss} of the surface states, the charge in these states has a chance to come into equilibrium with the space charge during the above-discussed stage (2).

We plotted the third part of the overall photopotential decay curve in semilogarithmic coordinates (Figure 3); by the slope of straight lines we determined the decay time constant τ for different values of the load resistance R_1 in the external circuit. Further extrapolation of the τ-R_1 lines thus obtained to $\tau \to 0$ gives the diamond bulk resistance R_s; from the slope of the line we calculated the full capacitance $C = C_{ss} + C_{sc}$. Then, assuming that $C_{ss} \cong C_{sc}$, we estimated the surface-states-charging resistance R_{ss} from the experimentally measured time τ_{ss} required for the charge redistribution, during the stage (2), to proceed: $\tau_{ss} = R_{ss} C$, where $C = (C_{sc}^{-1} + C_{ss}^{-1})^{-1}$. In particular, for the (undoped) sample No. 280 τ_{ss} = 2.2 μs and $C = 0.48 \mu$F cm^{-2}, hence, R_{ss} = 18 Ohm cm^2.

For larger R_1, the "parallel" resistance R_p, which relates to the charge transfer across the electrode/electrolyte interface, can be by no means ignored. Indeed, here the capacitance C discharges over both resistances R_1 and R_p.

We have measured values of R_s, R_p, and C_p also by the impedance spectroscopy techniques. In the table below we compare values of R_s, R_p, and full capacitance C for boron-doped films, determined by photopotential transient (PPT) and impedance spectroscopy (IS) techniques.

No. of Sample:	No. 262		No. 280		No. 306	
	PPT	IS	PPT	IS	PPT	IS
R_s/Ohm cm^2	3.6	2.7	4.5	5.0	—	10.8
C/μF cm^{-2}	0.70	0.67	0.48	0.60	0.36	0.56
R_p/kOhm cm^2	—	—	21	—	98	299

Comparing the data measured by the two different techniques, we conclude that they are in a reasonable agreement. A small discrepancy between them is mainly due to a still existing, even if very small, frequency dependence of the capacitance measured (which is not allowed for in the equivalent circuit, Figure 3).

We conclude that photoelectrochemical methods can give additional information, concerning properties of semiconductor diamond thin films, similar to that produced with optical and electrophysical techniques.

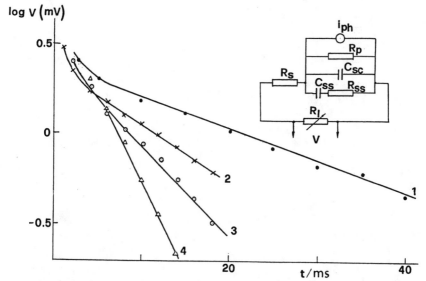

Figure 3. Photopotential decay curves. R_e (kOhm): 1—1000; 2—208; 3—90.9; 4—10.4. Insert equivalent circuit of the electrode.

ACKNOWLEDGEMENTS

Contribution to this study from my colleagues, A. V. Churikov, F. DiQuarto, S. Piazza, A. Ya. Sakharova, C. Sunseri, I. G. Teremetskaya and V. P. Varnin, is sincerely acknowledged. This work was supported in part by the Russian Foundation of Basic Research, project 96-03-34133.

REFERENCES

1. Pleskov, Yu. V. and Yu. Ya. Gurevich. 1986. *Semiconductor Photoelectrochemistry*, New York: Consultants Bureau.

2. Sakharova, A. Ya., Yu. V. Pleskov, F. DiQuarto, S. Piazza, C. Sunseri, I. G. Teremetskaya and V. P. Varnin. 1995. *J. Electrochem. Soc.*, 142:2704.

3. Pleskov, Yu. V., V. P. Varnin, I. G. Teremetskaya and A. V. Churikov. 1997. Ibid. 144: no. 1.

Photocharacteristics of a Metal-Carbon-Silicon Structure

T. V. Semikina and V. A. Semenovich
The National Technical University of Ukraine
(Kiev Polytechnical Institute)
Microelectronic Department
Kiev, pr. Pobeda 37
252056, Ukraine

A. N. Smyirieva
Institute of Superhard Materials
National Academy of Sciences of Ukraine

ABSTRACT: Capacitance-voltage (C-V) photocharacteristics of a metal (Ni)-carbon-semiconductor (Si) structure were examined in the 100 kHz to 2.5 MHz range under conditions of illumination and in the dark. The diamond-like carbon films were prepared by magnetron sputtering at different nitrogen concentrations in the gas mixture. The samples with inversion range were obtained. The density of the trapped charge as extracted from the measurements and was found to be of the order of 10^{10} cm^{-2} eV^{-1}.

KEY WORDS: diamond-like carbon films, heterojunction, inversion range, trap charge.

Diamond-like carbon films have received much attention recently as a potential material for electronic applications. Applications being evaluated include use as thin insulating films for heat dissipation in integrated circuits, on-chip waveguides, and as a material for luminescence and photodetection in the visible-ultraviolet [1].

97

In this work we have studied the possibility of preparing diamond-like carbon films towards new types of photodetectors sensitive in the visible and ultraviolet spectral ranges.

Diamond-like carbon (DLC) films were made by r.f. magnetron sputtering on silicon substrates of n- and p-type conductivity. The thickness of the films was 0.6–1.0 micron. The nitrogen concentration in the reactant gas mixture was changed from 25 to 45% during film deposition. The pressure in the chamber would be changed from 1 to 45 torr.

A metal (Ni with underlayer of Mo)-DLC-silicon n- and p-type (MCS) structure has been investigated. Ohmic metal contacts were made by electron-beam sputtering using a mask.

Capacitance was measured using the standard methodology. Frequency from 100 kHz to 2.5 MHz was applied to the device at d.c. biases ranging from 4–10 V. Capacitance-voltage (C-V) characteristics were measured in the dark and under illumination in the range $(3–30) \cdot 10^3$ lk.

The quality and parameters of metal-insulator-semiconductor (MIS) structures are determined by the insulator (or wide bandgap semiconductor)-semiconductor interface properties, including the built-in charge and energy spin density surface states. The samples, which were made at nitrogen concentration of 25% in the gas mixture, have shown an inversion range at –1.5 V bias for DLC/n-Si [Figure 1(a)] and at a 0.5 V bias for DLC/p-Si structure [Figure 1(b)] under illumination, the frequency was 450 kHz. At illuminations above $35 \cdot 10^3$ lk the capacitance increases until it reaches the value at accumulation. As is known, the rise of capacitance as a result of approaching the inversion range in the semiconductor range depends on whether the inversion electron or hole concentration has time to follow the a.c. signal. With this we measured the capacitance. As long as we carried out our measurements at high frequencies, the interface-trapped charge cannot follow the a.c. signal and hence the effect of such charge on the MCS capacitance becomes negligible. Hence, we cannot observe the inversion regime under dark field conditions. The C-V characteristics at different frequencies are shown in Figure 2. We observed that these characteristics have a constant capacitance region beyond $| 6 |$ V. The frequency dependence on C-V characteristics is negligible. Similar effects were obtained by K. K. Chan et al. [1]. They suggested a trap lifetime of the order of 10^3 s for MCS structures.

The capacitance-frequency characteristics at fixed bias for samples with different nitrogen concentration are shown in the Figure 3. It is interesting to note that the curve of the sample which gives the inversion, is similar to the dependence reported by K. K. Chan et al.

We investigated the influence of nitrogen concentration on the C-V characteristics. Figure 4 shows this dependence at 1.15 MHz. The rise of

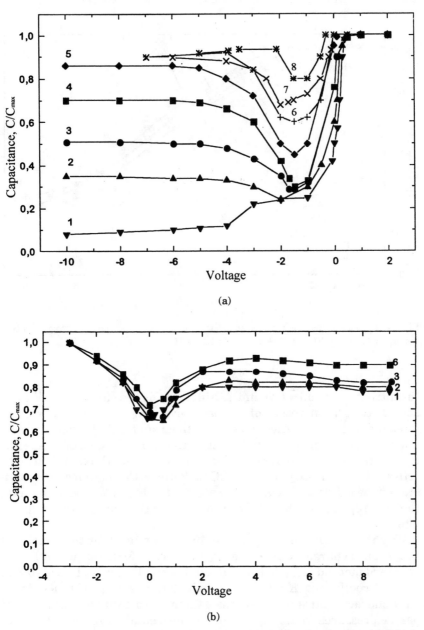

(a)

(b)

Figure 1. Capacitance-voltage characteristics in dark field and under illumination, $f =$ 450 kHz: (a) n-type Si; (b) p-type Si; 1—dark; 2—illumination is $3 \cdot 10^3$; 3—$6 \cdot 10^3$ lk; 4—$4 \cdot 10^4$ lk; 5—$1.7 \cdot 10^4$ lk; 6—$2.1 \cdot 10^4$ lk; 7—$3.1 \cdot 10^4$ lk.

99

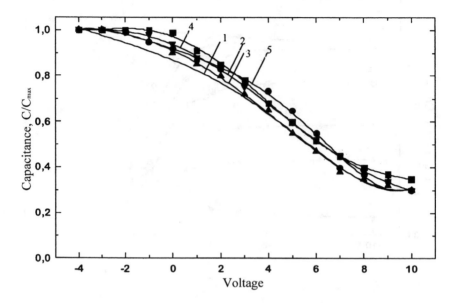

Figure 2. Capacitance-voltage characteristics (sample with p-type Si, nitrogen concentration is 25%). 1: 130 kHz; 2: 450 kHz; 3: 1.15 MHz; 4: 2 MHz; 5: 2.35 MHz.

the nitrogen concentration in DLC-films yields the change in C-V characteristics. The increases of nitrogen give rise to the classic voltage stretch-out characteristics and some of them approach the value at accumulation. Nitrogen is known to create deep donor levels and at specific concentrations also cluster combinations. This gives the changing properties at the interface junction, DLC/Si. Perhaps the motion of free carriers is impeded either because they are not freely discharged, or because they are replaced at the DLC/Si interface, so that charge will accumulate.

We obtained an average value of $4 \cdot 10^{10}$ sm^{-2}eV^{-1} for the density of the trapped charge from C-V characteristics at MHz frequency by the standard method [2,3]. The obtained flat band voltage shift is $|0.3$ to $2.0|$ V. A coefficient ($K = C_{max}/C_{min}$) which shows the possibility of varistor characteristics with decreases from 10 to 1 under conditions of partial pressures from 25 to 45 Torr in the frequency range 100 kHz to 2.5 MHz.

The investigation of the MCS structures shows a high photosensitivity $K_{ph} = 4.8 \cdot 10^{-6}$ lk^{-1}. The C/C_{max} ratio decreases from 0.2 to 0.9 in the illumination range (3–30) $\cdot 10^3$ lk. As a result of the measurements we can

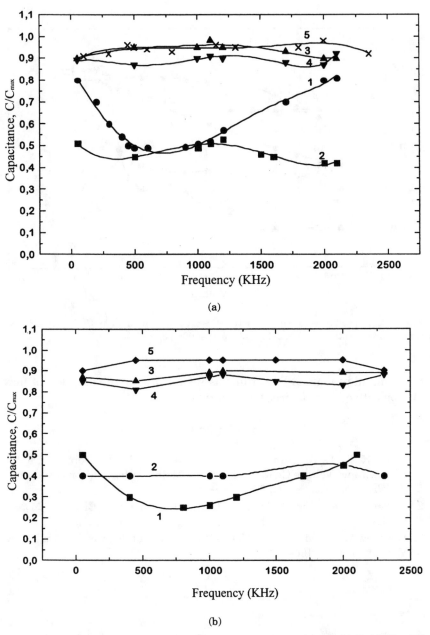

Figure 3. Capacitance-frequency characteristics of samples with n-type Si; (a) $U = 0$ V; (b) $U = -3$ V. Nitrogen concentration 1: 25%; 2: 30%; 3: 45%; 4: 00%; 5: 45%.

101

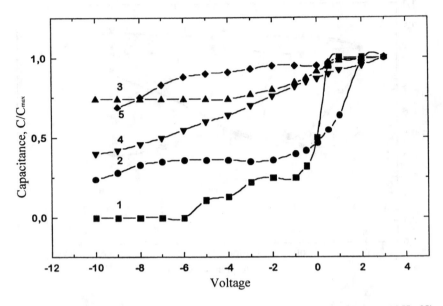

Figure 4. Capacitance-voltage characteristics of samples with n-type Si; f = 450 kHz. Nitrogen concentration 1: 25%; 2: 30%; 3: 45%; 4: 00%; 5: 45%.

recommend preparing DLC-films as insulators for the creation of new photoelectric MCS-devices with an inversion level.

REFERENCES

1. Chan, K. K., G. A. J. Amaratunga, S. P. Wong and V. S. Veerasamy. 1993. "Capacitance-Voltage Characteristics of a Metal-Carbon-Silicon Structure," *J. Solid-State Electronics,* 36(3):345–349.
2. Maller, R. S. (ed.) 1989. "Electronic Devices for Integration Circuits," M. Mir, pp. 470–512.
3. Vorob'ev, Yu. V. (ed.). 1988. "Methods of Studies of Semiconductors," Kiev, Vischa Skola, pp. 180–183.

Ion-Beam Induced Heterostructures in Diamond

A. A. GIPPIUS

P. N. Lebedev Physical Institute of the Academy of Sciences of Russia
Leninsky prospect 53, Moscow 117924, Russia

ABSTRACT: The work of various research teams on defect engineering in ion implanted diamond is reviewed with special attention given to ion beam induced modification of properties within the regions of various dimensionality.

1. INTRODUCTION

Ion implantation is a violent process inflicting large amounts of radiation damage upon the target by energetic ions. In the case of ion doping (which is by far the most commonly employed version of ion-beam modification) we generally speak in terms of defect control when radiation damage is considered as a negative factor to be minimized by a proper choice of an implantation-annealing route to grant that the properties of the material are dominated by the implants and not by radiation defects. If we consider ion-beam modification in a broader sense, then it is pertinent to speak of "defect engineering" which employs radiation damage to create new properties of a material.

In this paper the work of various research teams on defect engineering in ion implanted diamond is reviewed with special attention given to ion beam induced modification of properties within the regions of various dimensionality. Section 2 refers to ion-implantation doping where macroscopic properties of doped layers are crucially determined by microscopic "zero-dimensional" surrounding of implanted ions. Section 3 deals with the properties and possible applications of ion tracks which are highly damaged one-dimensional regions (with diameter of several nanometers and length up to hundred microns) formed along the paths of high energy (>1 MeV/a.m.u.) ions. Section 4 refers to two- or three-dimensional defect engineering, that is the formation of graphite layers due to the phase transitions in strongly damaged diamond.

2. ION DOPING LIMITATIONS: ZERO-DIMENSIONAL STRUCTURES

The range of ion energies used for ion implantation lies between about 30 keV and several megavolts. In this case most of radiation damage is created via mechanism of "nuclear stopping" when the moving ions undergo elastic collisions with the atoms of the target material. One of the key features of ion implantation doping is the very large amount of radiation damage, the number of vacancies and interstitials in the implanted layer exceeding the number of implanted impurity atoms by two orders of magnitude for ion energies around 100 keV. We deliberately use the word "amount" instead of "density" since we believe that it is the inhomogeneity of damage and, in particular, high local density of defects within disordered regions produced by ions, that strongly affect the processes of impurity-defect interaction within implanted layers.

In the course of development of doping techniques the choice of dopants for diamonds was made based upon the analogy with Si and Ge. In this way a number of group III, V and I elements have been tried both experimentally and theoretically [1] with most of the data referring to boron acceptor and some results concerning Li and Na donors. Any efficient scheme of ion doping implies that an implanted atom should be activated, i.e., placed at the correct lattice position (for instance, substitutional for boron and interstitial for lithium) and the effects of lattice disorder introduced by implantation should be minimized. Since the conditions necessary to activate substitutional or interstitial impurities must be different, we shall consider first the case of boron.

It should be noted that quantitative data on the efficiency of activation of implanted boron (i.e., the ratio of the amount of acceptor centres to that of implanted atoms) are scarce. The value of sheet resistance determined in conditions where its temperature dependence at least contains a component with the activation energy corresponding to boron ionization energy, is often taken as a measure of efficiency of activation. To increase this efficiency various conditions of implantation and annealing have been tried, the most extensive experimental and theoretical studies performed by Prins and co-workers [2]. The "cold implantation–rapid annealing" sequence, suggested by Prins, produced so far the best results in boron implantation doping, but it is the physical approach used that needs some comment in relation to general problems of defect control. The assumptions made by Prins and co-workers in their theoretical treatment and experimental studies are the following: (a) "frozen" defects produced by cold implantation are evenly mixed ("a fair, macroscopic approximation" [2]); (b) "dopant interstitials" will diffuse and interact with vacancies in, approximately, the same way as self-interstitials during annealing; (c) spatial distribution of implanted ions is more symmetric and peaks deeper below the surface than the vacancy distribution.

In the search for an effective "implantation–annealing" sequence, it was argued that the difference between the profiles of implanted atoms and radiation damage limits the efficiency of activation of substitutional impurities, since dopant interstitials "compete at a disadvantage for vacancies with self-interstitials" which are "more intimately mixed with vacancies." It was concluded that it should be helpful to introduce additional damage (for example, by C+ co-implantation) to provide more vacancies needed to activate the implanted boron atoms [2].

Let us note first of all that simple comparison of distributions of vacancies and implanted atoms seems somewhat misleading, if we take into account that integrated profile area of vacancies is two orders of magnitude larger than that of impurities. Additional damage (produced by co-implantation) which overlaps the impurity profile, introduces equal amounts of vacancies and self-interstitials and should not improve the situation as far as competition for vacancies is concerned.

Even more important is the inhomogeneity of damage which was not taken into account in [2]. It is known that considerable fraction of defects created by ion implantation is concentrated within disordered regions produced by displacement cascades which consist (for comparatively light ions) of a small amorphous nucleus, a highly disordered but crystalline boundary layer and a crystalline matrix containing a more dilute solution of point defects [3]. Due to dynamics of the cascade formation the inner part of a disordered region is amorphous or vacancy-rich, while a large fraction of interstitials is created predominantly at the periphery of the cascade [4]. It should be noted that this structure of disordered regions implies that there must be enough vacancies in the closest proximity of an implanted atom to activate a substitutional impurity. Moreover, with vacancy-rich nucleus of a disordered region and interstitials at its periphery the competition for vacancies can hardly be to a disadvantage of an implanted atom. Locally the situation can hardly be described simply as "dopant-interstitial" even for comparatively light ions. As far as heavier ions are concerned, it was shown in our experiments that the defect structure surrounding an implanted Ni atom should be understood not in terms of point defects but rather in terms of gross lattice disorder, which survives the annealing temperatures close to the limit of stability of diamond at normal pressure (1650°C) [5]. Kalish et al. found that implanted In atoms are surrounded by strongly damaged region up to the annealing temperature of 1800°C [6]. These data probably explain the failure of attempts to effectively dope diamond by implantation of relatively heavy atoms and stress the importance of light dopants. In the case of light implants—B, Li, Na(?) the local damage is not that severe, so that the lattice can be restored by proper annealing.

The disordered regions produced by ions represent "zero-dimensional" modification of the properties of diamond which strongly limits the advantages of ion doping. It is this defect distribution which is frozen in cold implantation experiments. The relative success of the "cold implantation–rapid annealing" regime (which gives the temperature dependence of sheet resistance with single activation energy close to the ionization energy of boron and with reasonable, after additional annealing at 1500°C, value of resistivity, comparable with the values found in natural type IIb diamonds) is largely due to processes which occur within or close to disordered regions, where considerable fraction of radiation damage is frozen and where annihilation of its components can be most effective, if fast transition from low to high temperature is realized.

Interstitial impurities, such as Li, require conditions for activation different from those for substitutional boron. In this case the presence of great amount of vacancies, particularly their high density within disordered regions, is a negative factor. That is why the efficiency of Li activation under usual implantation conditions must be inherently low. To be activated, Li atoms should be somehow separated from the region of high density of vacancies. This is confirmed by the experimental data presented by Lebedev Institute and Technion groups [7,8]. In both studies the n-type conductivity due to Li donors could be found only in the tail of the implant distribution (in some cases up to about 8 times deeper than the damage range [7]). In view of these data it might be promising to perform implantation into aligned samples to reduce displacement of host atoms, particularly at low temperature, which may allow the impurities to remain interstitial [1].

3. ION TRACKS: ONE-DIMENSIONAL STRUCTURES

When an ion energy is large compared to 25 keV/a.m.u. the ion has only little interaction with individual atomic nuclei of the target and moves on an essentially straight path through the material. For fast heavy ions of energies around 1–10 MeV/a.m.u. the electronic excitation dominates by two to three orders of magnitude, so the radiation damage in this range is mainly due to electronic stopping. The high density of electronic energy losses (~ 1 keV/Å) creates along the ion path a region of high electronic excitation, which can be quickly transferred to atomic motion resulting in the formation of trail of damage called latent tracks [9]. The damage mechanism is still unclear. Among various possibilities, that is the thermal spike [10], the ionic spike [9] (with atomic motion induced by the electrostatic repulsion of close neighbor ionized atoms) or more refined models, including inner-shell electron excitation with subsequent multiple ionization via Auger process [11], the thermal spike model looks preferable [12], in particular since it allows for the observation of latent track formation in crystalline and amorphous metals and semiconductor materials.

Various experimental techniques have been used to study the structure of latent tracks. Based upon the results of Rutherford back scattering (RBS) and Mossbauer spectroscopy, high resolution electron microscopy (HREM) and optical absorption in garnet $Y_3Fe_5O_{12}$ and hexaferrites $Ba(Sr)Fe_{12}O_{19}$ [13] the latent tracks represent three-shell cylindrical structures consisting of: (1) an amorphous core, (2) a zone disturbed by the burst-out of atoms which generate subsequent atomic displacements without amorphyzing the material, (3) an external strained region. The dimensions of amorphous core (sometimes with elliptical cross sections, oriented with respect to the crystal lattice) are from several to ten nm. Let us note that this three-shell structure is just one-dimensional version of (zero-dimensional) disordered regions produced by ions with energies typical for ion doping regime when nuclear stopping mechanism dominates. In yttrium garnet $Y_3Fe_5O_{12}$ it was found that in the interval of electron stopping $(0.5 \div 5)$ keV/Å several types of damage morphology can be observed ranging from spherical ($R \leq 6$ Å) discontinuous defects through extended cylindrical but still discontinuous to continuous cylindrical ($R \approx 30$ Å) with a homogeneous damage density [13,14].

The presence of this type of defects in solids is bound to strongly modify their properties and can constitute an important component of defect engineering. In diamond the extended defects or tracks must be of particular importance as it was suggested by Zaitsev [15]. The properties of the material within these defects and in their immediate proximity are different from those of the undamaged matrix. Lower density of the material within the tracks along with the strain existing in their external regions can both stimulate the diffusion of impurities and serve as "channels" for doping deep layers of diamond by impurities introduced into the surface layer by, say, usual implantation technique. Moreover, the tracks themselves, after proper doping and annealing may represent rod-like nanometer-sized semiconducting areas which might be of interest for diamond-based electronic devices with submicron scale elements [15]. Experimental data concerning the properties of extended defects or tracks introduced by high energy implantation of diamond are scarce. Internal strains in diamond implanted with 82 MeV carbon ions were detected spectroscopically using Raman and luminescence techniques. In diamond implanted with 124 MeV xenon ions it was found that helium atoms, introduced in the surface layer by ion implantation, can migrate into the crystal up to the distance determined by Xe ions range [15].

4. DIAMOND-GRAPHITE TRANSFORMATION: TWO- AND THREE-DIMENSIONAL STRUCTURES

Due to the metastability of diamond (under ambient conditions the stable phase of carbon is graphite) it tends to transform to graphite if the lattice damage density exceeds some critical value. In this case a sufficient number of diamond sp^3 bonds are broken which is a prerequisite for transformation (enhanced by heating) into a graphite phase with sp^2 bonds. This puts a severe limitation on ion implantation doping since it excludes remelting and resolidification or solid state epitaxial regrowth of ion-damaged layers by means of usual (at normal pressure) technological regimes employed in semiconductor technology (for example, rapid thermal annealing). On the other hand, ion beam induced graphitization opens new prospects of defect engineering. Using proper combinations of implants (neutral, electrically and/or optically active), ion energies, doses and annealing conditions it is possible to produce properly patterned diamond-base heterostructures comprising insulating, semiconducting, luminescent and conducting (graphitized) layers.

Formation of buried graphite layers have been observed in diamond subjected to ion bombardment both after laser annealing [16,17] and thermal annealing [18]. The existence of graphite layer was established by optical microscopy and channeling contrast microscopy [16,17] and by electrical and optical measurements [18]. In [18] the graphite layer was found to be formed (after annealing at 1400°C) in diamond samples bombarded with doses exceeding some critical value ($2.9 \cdot 10^{16} cm^{-2}$ for 350 keV He^+ ions). At doses which are slightly above critical a thin (~70 nm) graphite layer is formed at the depth close to the maximum of radiation damage distribution calculated using TRIM program for energy loss simulation. The analysis of optical measurements taking into account the absorption in the buried damaged layer and interference in the part of the crystal between the surface and the buried layer provided the data on the depth and the thickness of the layers as well as their conductivity at optical frequencies. The latter was found to be close to that of graphite. The values of the depth and thickness of graphitized layers for the fluence $4.1 \cdot 10^{16}$ cm^{-2} of 350 keV He$^+$ ions were respectively 685 and 150 nm [18].

The possibility to graphitize selected areas of diamond subjected to radiation damage provides an additional degree of freedom in the development of diamond based electronic devices. One can think of built-in graphite resistors, capacitors, connections between elements of circuits, etc. Certainly, extensive material research is needed before these suggestions can be realised. Of particular importance are the properties of the cap layer, that is the part of the crystal which is between the surface and the buried damaged layer. The residual damage in the cap must be a limiting factor of at least some of the possible applications. Another problem refers to the properties of the heterojunction between diamond and the graphite layer, for instance the sharpness of the boundary "conductor (graphite)–insulator" and the lateral inhomogeneity of the boundary layer.

5. CONCLUSION

Defects related problems of microscopic nature set definite limits to ion implantation as a component of diamond technology. On the other hand, the specific properties of diamond, particularly its metastability, opens new prospects of defect engineering and development of diamond based devices with elements of various dimensionality.

ACKNOWLEDGEMENTS

This work was supported by the Russian Foundation for Basic Research (project No. 95-02-04278-a).

REFERENCES

1. Kajihara, S. A., A. Antonelli and J. Bernholc. 1993. *Physica* B185:144.
2. Prins, J. F. 1992. Prins, *Mater. Sci. Rep.* 7:271.
3. Kimerling, L. C. and J. M. Poate. 1975. *Inst. Phys. Conf. Ser.* 23:126.
4. Nelson, R. S. 1973. *Instr. Phys. Conf. Ser.* 16:140.
5. Gippius, A. A. 1992. *Mater. Sci. Forum* 83–87:1219.
6. Kalish, R., M. Deicher, E. Recknagel and T. Wichert. 1979. *J. Appl. Phys.* 50:6870.
7. Vavilov, V. S., M. A. Gukasyan, M. I. Guseva and E. A. Konorova. 1972. *Soviet. Phys. Semicond.* 6:741.

8. Prawer, S., C. Uzan-Saguy, G. Braunstein and R. Kalish. 1993. *Appl. Phys. Lett.* 63:2502.

9. Fleischer, R. L., P. B. Price and R. M. Walker. 1975. *Nuclear Tracks in Solids* (University of California, Berkeley).

10. Bonfiglioli, G., A. Ferro and A. Mojoni. 1961. *J. Appl. Phys.* 32:2499.

11. Tombrello, T. A. 1984. *Nucl. Instr. Meth.* B2:555.

12. Meftah, A., F. Brisard, J. M. Costantini, E. Dooryhee, M. Hage-Ali, M. Herveien, J. P. Stoquert, F. Studer and M. Toulemonde. 1994. *Phys. Rev. B* 49, No. 18, 12457.

13. Toulemonde, M. and F. Studer. 1988. *Philos. Mag.* A58:799.

14. Meftah, A., N. Merrien, N. Nguen, F. Studer, H. Pascard and M. Toulemonde. 1991. *Nucl. Inst. and Meth.* B59:605.

15. Zaitsev, A. M. 1992. *Mater. Sci. and Engin.* B11:179.

16. Prawer, S., D. N. Jamieson and R. Kalish. 1992. *Phys. Rev. Let.* 69, No. 20, 2991.

17. Jamieson, D. N., S. Prawer, S. P. Dooley and R. Kalish. 1993. *Nucl. Instr. Meth.* B77:457.

18. Khmelnitsky, R. A., V. A. Dravin, A. A. Gippius, this volume.

NUCLEATION AND GROWTH

Production and Properties of Low-Temperature Tetrahedral Carbon Films

O. I. KONKOV, E. I. TERUKOV AND I. N. TRAPEZNIKOVA

A. F. Ioffe Physical-Technical Institute
Russian Academy of Sciences
194021 St. Petersburg
Russia

(Received March 28, 1997)
(Accepted April 25, 1997)

ABSTRACT: Films of high-tetrahedral amorphous carbon have been produced by glow discharge decomposition of methane in Ar/H_2 at temperatures of about 300 K. Structural, optical and electron transport properties of as-deposited films have been investigated. Optically transparent amorphous carbon films with refractive indices, 1.5–2.2 and density 2.75–2.8 g/cm^3 were obtained. The concentration of (optically) active hydrogen atoms was about 21–28 at.%. The sp^3/sp^2 ratio ranged between 0.8–0.85. The optical bandgap was 3–3.6 eV, electron mobility <2 $cm^2/V \cdot s$, and the resistivity 0.01–10^{16} Ohm·cm at 300 K. Photoluminescence spectra had a wide intensity band with a maximum at 487–493 nm and a half-width of 0.6–0.8 eV.

KEY WORDS: DLC, high-tetrahedral carbon films, photoluminescence, structure, electron properties.

INTRODUCTION

Hydrogenated amorphous carbon materials occupy an important place in the field of amorphous semiconductors. The material results from a variety of bonding hybridization schedules of carbon atoms, relative concentrations of the different bonds (sp^3, sp^2, sp^1) and hydrogen concentrations in the amorphous networks. Such materials possess high hardness and density, high electrical resistivity and chemical inertness, and optical transparency, and are called diamond-like carbon (DLC). They are usually produced by a variety of ion-beam methods with different carbon sources [1]. Some authors classify carbon material with a high sp^3/sp^2 ratio and high density, but with smaller DLC hardness as high-tetrahedral amorphous carbon (ht-a-C) [2]. Such a material is an analogue of tetrahedral amorphous silicon and germanium and is widely used in different microelectronic application as wear-resistant coatings and microelectronic device films.

In the case of plasma deposition, the most widespread technique for producing amorphous semiconductors, the resultant material has a smaller sp^3/sp^2 ratio and a lower hardness [3]. Such properties do not permit consideration of this material as ht-a-C.

The aim of this work was the production of high-tetrahedral amorphous hydrogenated carbon films by a low-temperature plasma method and the study of their properties.

EXPERIMENTAL

Our films were produced by decomposition of a 12% methane and argon/hydrogen mixture in a glow discharge high-frequency plasma. We have used a high generator frequency, 40 MHz, and a special chamber known as "system with quasi-closed working volume." Such a modification of the deposition system was made to increase ionization of the molecules of the working gas, enhance the degree of excitation of the resulting radicals and ions and, as a consequence, to grow films with a high density and high sp^3/sp^2 ratio. In the system, the region of glow discharge was mechanically limited by voltage-bearing and grounded electrodes and a cylinder of an insulating material (optical quartz glass) placed between them. The working gas was pumped only through a special system of apertures in both electrodes. The electrodes and cylinder set was placed inside the vacuum chamber assuring a pressure of $5 \cdot 10^{-3}$ torr during deposition. The pressure within the chamber was $5 \cdot 10^{-2}$ torr. This value corresponded to a minimum of the Paschen curve. Optical fused quartz and single-crystal p-silicon with a resistivity 20 Ohm·cm were used as substrates. The deposition temperature was below 150°C. The film thickness was 0.2–0.5 micrometer.

The structural properties of the films (i.e., the type of hydrogen and carbon bonding and the amount of the optically active CH_n bonds) were estimated using infrared (IR) spectroscopy. The film's optical transparency and optical absorption edge were investigated in the range of 200–1200 nm. The refractive indices were measured by an ellipsometry technique. Stokes photoluminescence (PL) spectra were studied with an Ar-laser at an excitation energy of $hv_{ex} = 2.54$ eV. The PL decay kinetics were studied using a pulsed nitrogen-laser with $hv_{ex} = 3.69$ eV and $\tau = 8$ ns. The anti-Stokes wing of the PL was excited by an He-Ne laser with $hv_{ex} = 1.96$ eV. All measurements were carried out at $T = 295$ K. Electron transport studies on a-C:H were carried out using a standard time-of-flight technique [4] on Pd/semiconductor/n-layer/c-Si (0.05 Ohm·cm) structures. The film densities were estimated by weighing. Films obtained by diluting the methane with hydrogen exhibited values that were close to the corresponding values for diamond-like carbon (DLC).

RESULTS AND DISCUSSION

Transparent hydrogenated carbon films with refractive indices in the range 1.5–2.2 and densities in the range 2.75–2.80 g/cm³ were deposited. These density values corresponded to values for DLC [3].

The IR transmission spectra of the films had a wide absorption band in the range 2800–3000 cm⁻¹ corresponding to the CH$_n$ stretching modes. This band can be deconvoluted into several Gaussian bands [5], with maxima at: 2870 cm⁻¹ (sp³-CH₃), 2925 cm⁻¹ (sp³-CH₂), 2960 cm⁻¹ (sp³-CH₃), 3050 cm⁻¹ (sp²-CH). The relative hybridization of carbon atoms was analyzed and the sp³/sp² ratio was estimated. The values were of the order 0.9–0.85 and the amount of the optically active bound hydrogen was 21–28 at.% (Figure 1).

It is known that IR-spectroscopy only gives information about optically active C-H$_n$ complexes. So the part of carbon atoms (as a rule of the interstitial type) can change the ratio of sp³/sp² bonds in DLC films. In order to prevent inaccuracies in characterizing the material, we also took into account the shape of the Auger spectrum of carbon, which, as is known, differs depending on whether sp³ (diamond) or sp² (graphite) is the dominant phase [6].

The optical absorption edge was also studied. The observed shape of the optical absorption edge showed the presence of the band tails in the density of states [7,8] typical for hydrogenated amorphous carbon. The width of the optical bandgap (E_g^{opt}), was estimated by expressing the dependence of the optical absorption coefficient, α, on photon energy,

$$(hv)^{1/2} \sim (hv - E_g^{opt}) \tag{1}$$

and then extrapolating it to $\alpha = 0$. The values ranged between 3.0 and 3.6 eV.

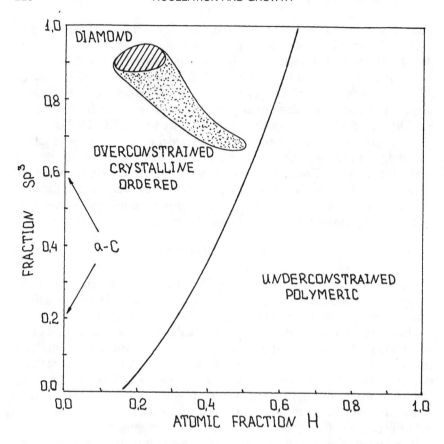

Figure 1. The influence of the sp^3 hybridized C-atoms on the hydrogen content (Angus diagram): the results obtained in this work for a-C:H films (dots) and ht-a-C (cross-hatched).

In typical PL spectra of low-temperature a-C:H an intense wide structureless band with a maximum in the spectral range (487–492 nm) and with a half-width (0.6–0.8 eV) is observed (Figure 2). The PL intensity (I) was considerably higher for films deposited in the presence of argon than for films obtained by adding hydrogen. This result agrees with the results of Reference [7] and fits into the framework of a model postulated there for amorphous carbon. It is thus understood that when argon is present during the film's growth, the deposited material contains a large number of light-emitting nanogranules of the graphite-like phase, which ensures a high PL intensity. The PL study in a wide temperature range (4.2–300 K) indicates that the PL band shape, its half-width and the position of the maximum, do not vary in a-C:H with temperature. Overlapping of the high-energy part of the PL spectrum with the Urbach tail [8] was observed.

Studies of the PL decay kinetics in a-C:H films excited by a nitrogen laser showed that it essentially follows an exponential law with an average decay time of 10 ns and is virtually independent of T_s and illumination time. The fact that the decay is exponential confirms the existence of an exciton-like recombination mechanism [7] involving nanogranules of the sp^2 phase.

Normalized PL spectra of the a-C:H films (Stokes wing) with excitation by 2.54-eV photons as a function of T_s, are shown in Figure 3. The energy position of the spectra and their half-width vary in the energy ranges E_m = 2.15–2.35 eV and Δ = 0.45–0.62 eV, respectively. No additional bands were detected in this spectral range. Emission from both as-prepared films and from a film subjected to prolonged illumination was essentially identical as regards the spectral shape, since it is determined by the density of states in the corresponding energy interval.

Anti-Stokes PL emission spectra of a-C:H films grown at different T_s temperatures were measured under excitation by $h\nu_{ex}$ = 1.9-eV photons. The spectra are shown in Figure 4. The anti-Stokes wing basically follows an exponential dependence of the form,

$$I(h\nu) \sim \exp(h\nu/E_o)\exp(-h\nu/kT) \qquad (2)$$

where E_0 is some energy parameter which pertains to the density gradi-

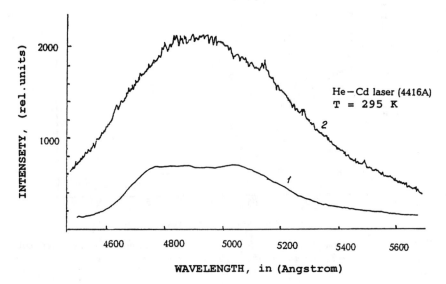

Figure 2. PL spectra of ht-a-C films deposited under (1) hydrogen dilution and (2) argon.

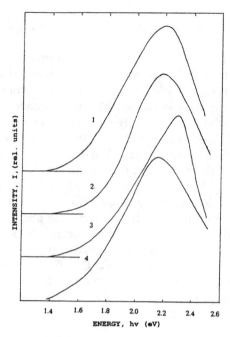

Figure 3. Stokes photoluminescence spectra of a-C:H films grown at different temperatures T_s(°C): 430 (1), 250 (2), 175 (3), 25 (4) ($T = 295$ K, $h\nu_{ex} = 254$ eV, $E = 10$ W/cm^2).

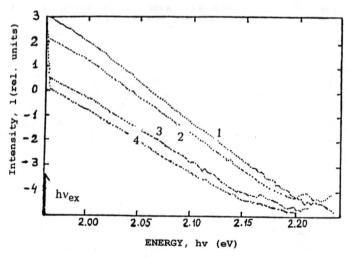

Figure 4. Anti-Stokes photoluminescence spectra of a-C:H films grown at different substrate temperatures T_s (°C): 430 (1), 250 (2), 175 (3), 25 (4) ($T = 295$ K, $h\nu_{ex} = 196$ eV, $E = 5$ mW/cm^2).

118

ent of the localized states and which characterizes the distribution of photoexcited carriers [7]. The figure shows that the intensity of anti-Stokes emission is higher for films deposited at high T_s than that for films obtained with $T_s < 100°C$ (curves 3,4). A similar behavior is observed for the characteristic energy in Equation (2).

The results confirm that the dominant channel for radiative recombination is the channel involving nanogranules of the sp^2 phase. Their concentration is higher in a-C:H films deposited at high temperatures (T_s) than in films prepared at $T_s < 100°C$. PL quantum efficiency is higher in the films prepared at high T_s.

The electron transport parameters of the deposited carbon films were investigated by a standard time-of-flight technique [9]. The shape of the transient current pulse was typical for the dispersed transport of a-C:H films [4]. We advance, therefore, the most popular interpretation of the drift mobility measurement data in amorphous semiconductors, namely the multiple trapping on localized states distributed exponentially near the conduction band edge. However, our results fail to agree completely with all the assumptions of this model. First, we did not observe equality of the dispersion parameters α_b and α_a before the transit time [at the initial part of the $I(t)$ curve] and after this time [at the terminal part of the $I(t)$ tail], respectively. This occurs because α_b does not exceed 1, but α_a, in our experiment, varied up to 5. Secondly, the curves of the temperature dependence of the dispersion parameters $\alpha = \alpha(T)$ do not cross the point $T = 0$ and are not straight lines. In the framework of the trapping model this behavior of the dispersion parameters is explained by the variation in the trapping cross section of localized states moving into the mobility gap [10]. Finally, a strong dependence of the dispersion parameters on the exterior field is observed in our measurements. For a-C:H deposited at 25°C the increase in the field from $1.5 \cdot 10^3$ to $3 \cdot 10^4$ V/cm leads to a decrease in α_a from 4 to 1.22. A feasible explanation of the experimental data may be a certain modification of the multiple trapping model, if, e.g., it were assumed that the tail of the density of states in the conduction band obeys the Gaussian law,

$$N(e) = Nc \exp[-(e / kT_g)^2] \tag{3}$$

where T_g is the characteristic temperature of the distribution.

This qualitative agreement inspired us to make quantitative estimates also. The characteristic temperature for Gaussian distribution for a-C:H films is about 1000 K. On the basis of these results we can obtain information about the distribution of localized states in the tail [4].

The electronic drift mobility was obtained from the time dependence of the transient current as $2 \text{ cm}^2 \cdot \text{V}^{-1} \cdot \text{s}^{-1}$.

CONCLUSION

The results of the study of a-C:H properties are shown on the diagram offered by Angus [11] for characterization of the carbon materials (Figure 1). We conclude that there is a real possibility for the deposition of high-tetrahedral amorphous carbon films by low-temperature methane decomposition plasma methods.

ACKNOWLEDGEMENTS

This work was supported in part by Arizona University and the Russian Foundation for Basic Research (grant RFBR N 96-02-16851).

REFERENCES

1. McKenzie, D. R. 1991. *Phys. Rev. Lett.,* 67:773–781.
2. McKenzie, D. R., D. Muller, B. A Pailthorpe, Z. H. Wang, E. Kravtchinskaia and D. Segal. 1991. *Diam. Rel. Mater.,* 1:51–61.
3. Tamor, M. A. 1991. *Appl. Phys. Lett.,* 58:592–599.
4. Konkov, O. I., I. N. Trapeznikova and E. I. Terukov. 1994. *Semiconductors,* 28:1406–1409.
5. Dischler, B., A. Bubenzer and P. Koidl. 1983. *Sol. State Commun.,* 48: 105–11.
6. Mikhailov, S. N., I. N. Trapeznikova and E. I. Terukov. 1991. *Proceedings of the Russian Seminar on Amorphous Semiconductors and Their Applications,* Leningrad, 72 (in Russian).
7. Chernyshov, S. V., E. I. Terukov, V. A. Vassilyev and A. S. Volkov. 1991. *J. Non-Cryst. Sol.,* 134:218–221.
8. Babaev, A. A., M. Sh. Abdulvagabov, I. N. Trapeznikova and E. I. Terukov. 1991. *Neorganicheskie Materialy,* 27:2205–2207 (in Russian).
9. Spear, W. E. 1983. *J. Non-Cryst. Solids,* 59–60:1–24.
10. Cohen, J. D. and D. V. Lang. 1982. *Phys. Rev. B,* 25:5321–5329.
11. Angus, J. C. 1991. *Diamond and Related Materials,* 1:61–63.

Boron Doped Diamond Films

R. KH. ZALAVUTDINOV,* V. KH. ALIMOV, A. E. ALEXENKO,
A. E. GORODETSKY AND A. P. ZAKHAROV
Institute of Physical Chemistry
Russian Academy of Sciences
Leninsky prospect 31
117915 Moscow, Russia

ABSTRACT: Boron doped diamond films obtained by a CVD method on single-crystal Si and AlN ceramics have been investigated by means of Electron Probe Microanalysis (EPMA), Secondary Ion Mass Spectrometry (SIMS), X-Ray Diffraction (XRD), Transmission Electron Microscopy (TEM) and Scanning Electron Microscopy (SEM). SIMS analyses of the films grown at different conditions have shown a B/C atomic ratio range from 4.3×10^{-4} to 3.5×10^{-3}. The films are comprised of octahedra (2–5 μm) with a block structure of 0.1–0.3 μm in size. Resistance of the films varied in the range 10^3 to 10^6 Ohm. Luminescence under electron beam excitation is light-green, which differs from that of undoped diamond films which luminesce in the blue.

KEY WORDS: diamond films, boron, oxygen, composition, structure, resistance.

1. INTRODUCTION

Doped diamond films, obtained by different methods, are considered to be the most promising materials, for the fabrication of powerful and compact electronic and optoelectronic devices capable of operating under extreme conditions [1]. Doping of diamond films with boron during their growth from the vapor phase seems to be the only way to prepare semiconducting diamond material of satisfactory quality. In this investigation we have studied the chemical composition, structure, morphology, and resistance of boron doped CVD diamond films.

---

*Author to whom correspondence should be addressed.

121

2. EXPERIMENTAL

The boron doped diamond films were grown from the vapor phase by DC arc discharge excited between needle-shaped refractory cathode and a substrate holder in a hydrogen-methane mixture at a pressure of about 10^4 Pa. Polished single-crystal Si ($12 \times 12 \times 1$ mm) and AlN ceramics ($11 \times 9 \times 1$ mm) were used as substrates. The growth process was initiated by introduction of seed diamond particles not more than 1 μm in diameter, dispersed all over the substrate surface. Fused boron oxide B_2O_3 or neocarborane $C_2B_{10}H_{12}$, placed within the growth chamber, were used as boron sources. Thickness of the films was estimated by weighing and resistance was measured by means of a two point probe made of W-Re alloy with 2 mm distance between the probes.

Chemical composition, structure and morphology of the films were analyzed by Electron Probe Microanalysis (EPMA), Secondary Ion Mass Spectrometry (SIMS), X-Ray Diffraction (XRD), Transmission Electron Microscopy (TEM) and Scanning Electron Microscopy (SEM). In EPMA, SEM and TEM a primary electron beam with energy in the range 10 to 50 keV and with diameter of 1–50 μm was used.

Boron bulk and oxygen surface contents of diamond film were measured by SIMS using surface sputtering. The SIMS facility was installed in a special UHV system with a typical base pressure $<10^{-7}$ Pa; background pressure was mainly due to hydrogen and carbon monoxide with a small amount of water. To attain a uniform erosion rate during SIMS measurements, the 4 keV Ar ion beam was focused to about 100 μm and was scanned over an area of 0.5×0.4 mm^2 at 45° to the surface. The current of the Ar-ions was 0.5 μA. Rim effects from the sputtered crater were avoided by reducing the area of the measurement to 0.3×0.2 mm^2 (in the center of sputtered spot). The sputtering interval was 15 minutes, measurement time was about 1 minute.

3. RESULTS AND DISCUSSION

The growth rates of the films varied between 2–5 μm/h at the substrate temperature of 1173–1273 K. Films were 2–6 μm thick and consisted of (100) and (111) crystals of the same order of magnitude in diameter. Resistance of the films ranged between 10^3–10^6 Ohm depending on doping level. Data on the samples are given in Table 1.

EPMA of boron doped diamond films has not revealed boron at a primary electron energy of 10 keV. This indicated that the boron concentration was lower than 0.5 at%. SIMS was applied to analyze the impurity content (Figure 1). The ratio of the combined intensities of $^{10}B^+ + {}^{11}B^+$

Table 1. Substrate, boron source, thickness, and resistance
of boron doped diamond films.

Diamond Film Number	Substrate Type	Boron Source	Thickness (μm)	Resistance (Ohm)
13-360	Si	B_2O_3	2.0	1×10^4
13-364	Si	B_2O_3	3.6	7×10^5
13-365	Si	B_2O_3	2.5	6×10^3
13-371	Si	B_2O_3	5.7	3×10^4
13-372	Si	B_2O_3	5.6	2×10^5
13-426	AlN	$C_2B_{10}H_{12}$	5.1	1×10^3

signals to the intensity of the $^{12}C^+$ signal was used for the measurement of boron concentration. To minimize the influence of adsorbed molecules (containing oxygen atoms) on the $^{10}B^+$, $^{11}B^+$ and $^{12}C^+$ yields, the $(^{10}B^+ + {}^{11}B^+)/^{12}C^+$ ratio was measured only after surface sputtering with Ar-ions for 15 minutes (Table 2). Boronized graphites with boron content in the range from 5×10^{-2} to $1.5 \times 10^{+1}$ at% were used as control samples for the determination of the relationship between $(^{10}B^+ + {}^{11}B^+)/^{12}C^+$ intensity ratio and B/C atomic ratio. This relationship was measured after Ar-ion sputtering for 30 minutes and was found to be linear (Figure 2). The

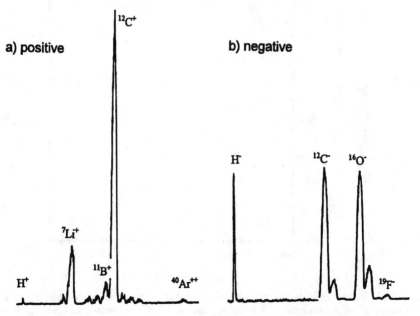

Figure 1. Positive (a) and negative (b) SIMS spectra for diamond film 13-365.

Table 2. Boron and oxygen contents in the diamond films.

| | B/C Atomic Ratio (%) | | | O/C Atomic Ratio (%) | |
Diamond Film Number	After Sputtering for 15 min.	After Sputtering for 30 min.	On the Surface	After Sputtering for 15 min.	After Sputtering for 30 min.
13-360	0.063	0.052	32	8	5
13-364	0.170	0.150	25	7	6
13-365	0.290	0.350	28	6	4
13-371	0.050	0.048	27	7	5
13-372	0.045	0.043	26	6	3
13-426	0.089	0.086	13	4	3

$^{16}O^-/^{12}C^-$ intensity ratio was used for the determination of adsorbed oxygen content on the surface. We assumed that maximum oxygen amount after prolonged exposure of a sample to air was 1.22×10^{15} O-atoms/cm^2 [2]. This fact allowed us to relate $^{16}O^-/^{12}C^-$ intensity ratio of O/C atomic ratio.

XRD analysis of sample 13-426 has shown that the diamond film had a lattice parameter of 0.3567 nm and a block structure (0.1–0.3 μm).

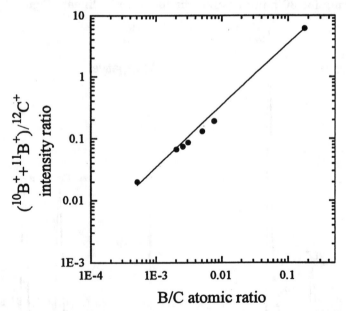

Figure 2. Ratio of the intensity of boron isotope signals to the intensity of carbon signals as a function of B/C atomic ratio for boronized graphites with boron content in the range from 5×10^{-4} to 1.5×10^{-1}.

TEM patterns obtained from tips of the diamond crystals consisted of spot reflections corresponding to (111), (220) and (311) planes. Some decrease in lattice parameter value with increase of boron content in the films has been observed.

Luminescence under electron beam excitation was light-green and differed from undoped diamond films which luminesce in the blue.

Composition and structure analyses of the boron doped diamond films have shown that the films do not contain precipitations of boron, boron oxides or carbides. We suggest that boron atoms are located in substitutional positions. Oxygen, registered by SIMS, is part of chemisorbed molecules resulting from their interaction with air.

4. CONCLUSIONS

Study of diamond films boron doped by different techniques has shown that B/C atomic ratios changed between 4.3×10^{-4} to 3.5×10^{-3} and that the films are comprised of octahedrons (2–$5\,\mu$m) which have a block structure with size dimensions in the range 0.1–$0.3\,\mu$m. Resistance of the films ranged between 10^3–10^6 Ohm. Luminescence under electron beam is light-green and differs from the blue luminescence of undoped diamond films. Our data give us grounds to conclude that boron atoms locate in substitutional positions in the diamond structure.

5. ACKNOWLEDGEMENTS

This work was supported by the U.S. Department of Energy, under Contract AN-8800 with Sandia National Laboratories.

REFERENCES

1. Alexenko, A. E. and B. V. Spitsyn. 1992. *Diamond and Related Materials,* 1:705.
2. Matsumoto, S. and N. Setaka. 1979. "Thermal Desorption Spectra of Hydrogenated and Water Treated Diamond Powders," *Carbon,* 17:485–489.

CVD Diamond Growth on Porous Si

ALEXANDER N. OBRAZTSOV,* IGOR YU. PAVLOVSKY
AND VICTOR YU. TIMOSHENKO
*Physics Department of Moscow State University
119899 Moscow, Russia*

ABSTRACT: The ultrafine diamond powders (UFD) were incorporated into Si substrates by using electrochemical etching in UFD contained HF:C₂H₅OH: H₂O solution. Produced by this way thin layers have shown typical for porous silicon optical and photoluminescence properties. On the treated Si wafers the significant increasing of diamond nucleation density for CVD diamond grown comparing with substrates prepared by common scratching method and anodized in electrolyte without UFD was obtained.

The results of the optical, IR, Raman, X-ray, SEM investigations of the samples at the different stages of substrate preparation and diamond films growth processes are presented.

The technology could be useful for both the diamond nucleation increasing and the porous Si layers protection from the environment influence by the grown CVD diamond films coating.

KEY WORDS: ultrafine diamond, porous Si, CVD diamond nucleation.

1. INTRODUCTION

The nucleation of diamond crystallites is the crucial step in the chemical vapor deposition (CVD) synthesis of diamond thin film, because the nucleation behavior determines the initial rate and coverage of deposition, and contributes to the structure and properties of the final film [1]. Many factors have been shown to contribute to nucleation. The mechanical pretreatment (polishing with a lap, abrading

---

*Author to whom correspondence should be addressed.

with diamond paste or ultrasonic irradiation with the diamond powder suspension) can produce many scratches, and thus increase the density of diamond nucleation by several orders of magnitude [1,2]. One of the important factors in the increase of diamond nucleation is the substrate topography. The preferential nucleation has been shown to occur at the convex features like edges and apexes [2–5].

Mechanical treatment is rather difficult to quantify and reproduce. The voltage biasing of the substrate surface has been shown to provide the effective increase of the number of the diamond nucleation sites [6]. The chemical pretreatment of the silicon substrate through the hydrogen (H_2) etching at high temperature ($>900°C$), the treatment in HF-HNO_3 solution [3], or the electrochemical etching creating the porous Si layer [7], were supposed to produce many defects promoting the diamond nucleation. But this opinion contradicts the data about the extremely small density of the point defects in the porous layer due to dangling bond saturation by hydrogen [8–10].

To clarify the problem of CVD diamond deposition on the porous Si we have performed some comparative experiments on diamond deposition onto the virgin and the porous Si; and onto the Si substrate prepared by combination of porous layer formation and seeding with the ultrafine diamond particles.

2. EXPERIMENTAL DETAILS

The mirror polished wafers of (100), 10 Ω cm, boron doped Si were used in our experiments. Porous Si layers were formed by standard anodization in HF:C_2H_5OH:H_2O = 1:2:1 electrolyte at the current density of 30 mA cm^{-2} for 5 min. A part of the porous Si films was fabricated using the same electrolyte with addition of the ultrafine diamond (UFD) powder. The UFD powder has been produced by the explosive method and contained the diamond grains with dimension of 4 nm [11].

An arc discharge plasma CVD reactor was used for the experiments. The diamond deposition conditions were as following: the gas ratio H_2:CH_4 = 100:3; the total pressure 100 Torr. The substrate temperature (850–$900°C$ or 500–$550°C$) was measured with a thermocouple attached to the substrate holder and with a pyrometer. The discharge current was 5 A, and the discharge voltage (between cathode and anode electrodes) was of 700 V. The substrate dimensions were 20 mm × 20 mm.

The quality of the diamond films was estimated by Raman spectroscopy, Scanning Electron Microscopy (SEM), and X-ray diffractometry.

We studied the Photoluminescence (PL) with the argon laser 488 nm line excitation and infra-red (IR) (by Perkin Elmer IR spectrometer) properties of Si porous layer before and after diamond deposition.

3. RESULTS AND DISCUSSION

The first set of CVD diamond films deposited at the standard conditions and the temperature of 850–900°C has clearly shown the increase of the diamond nucleation density after Si anodization. The CVD films grown onto the anodized Si after 1-hour growth were continuous and consisted of the well-faceted diamond crystallites in contrast to the films grown onto the virgin Si wafer where only separate diamond crystallites appeared.

It should be noted that substrate temperature was controlled by using the thermocouple attached to the substrate back side. At the same time we detected the significant difference in the color of the substrate surfaces for the virgin and anodized Si. It could occur due to the increase of the presurface (porous) layer temperature due to the difference in the thermal conductivity of the porous and bulk Si layers. Such thermal conductivity difference was observed in our previous investigation and found to be significant [12]. In such case the real temperature of the substrate surface might be much higher than thermocouple indication.

For the previous investigations of CVD diamond films grown onto porous Si [7] and onto the nanometer-scale Si wires [3] this effect has not been taken into account. In [3], the authors reported that, in spite of predictions, the nucleation occurred preferably not on the tips nor the tip edges of the nanowires for the particular processing conditions. Also no nucleation has been observed on the sides of the wires. Nucleation occurred at sites across the base of the substrate, at the same plane from which the nanowires emanated. The nucleation density for substrates was low, less than 10^4 cm^{-2}, similar to that for the pristine silicon. Apparently in all these cases surface temperature was much higher than it was necessary for the effective diamond growth.

To clarify this effect, the second set of CVD films was deposited either onto the untreated or anodized Si at the same conditions (except the temperature, which was decreased down to 500–550°C). No traces of diamond have been observed. However, we had no way to estimate the real temperature onto the substrate surface. Probably, the temperature of porous layer was too high in the second set.

The noticeable difference in the CVD diamond growth was detected for substrates fabricated by electrochemical etching in solution contained UFD powder. The diamond nucleation density was only slightly higher for the films produced at 850–900°C and at the temperatures 500–550°C we have obtained the continuous optically smooth films. Raman study of the samples have shown that the films had partly amorphous, partly crystalline structure comprising both the diamond-like and the non-diamond carbon (see Figure 1) and could be referred to as the "nanocrystalline" diamond films. Figure 2(a) and 2(b) show the SEM images of planar and cross-sectional views of the diamond films grown onto the porous Si layer with the embedded UFD powder.

Porous Si is known as the material with high efficiency of photoluminescence in the visible region. It was surprising that after the diamond film deposition the investigated samples have demonstrated the PL band being only slightly decreased in the intensity. It could be so in case of relatively high optical transparency of the film to visible light. The same effect was observed for IR region through the compari-

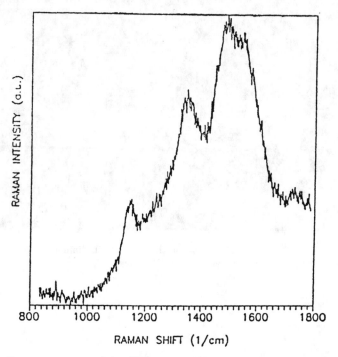

RAMAN SHIFT (1/cm)

Figure 1. Raman spectrum of the CVD diamond film on a silicon substrate with the porous layer with the embedded UFD powder. "Diamond" peaks near 1140 and 1330 cm⁻¹ could be seen.

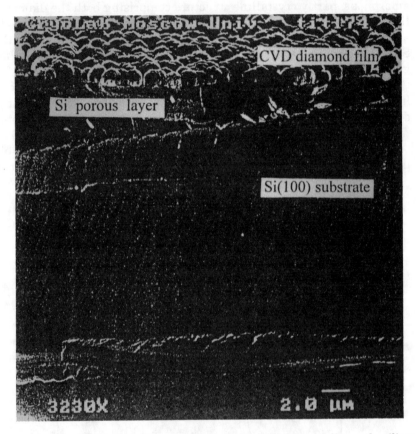

Figure 2a. The SEM image of the cross-section of the CVD diamond film on the silicon-substrate with porous layer filled with the UFD powder.

130

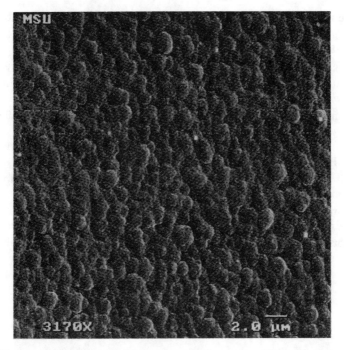

Figure 2b. The SEM image of the CVD diamond film on the silicon substrate with the porous layer filled with the UFD powder.

son of IR spectra of the samples before and after the diamond film deposition.

Thus, the possibility of the diamond film growth on the porous silicon at low temperature has been established. In spite of the poor quality of the fabricated films it was shown that the proposed method of the substrate pretreatment could provide the effective deposition of thin diamond films. At the moment, the investigation in this direction is in progress. We are seeking for the experimental conditions providing the better quality of the deposited diamond films.

REFERENCES

1. Popovichi, G. and M. A. Prelas. 1992. *Phys. Status Solidi A,* 132:233.
2. Sun, Zh., Zh. Zeng and N. Xu. 1994. *Diamond Films and Technology,* 3:227.
3. Dennig, P. A., H. I. Liu, D. A. Stevenson and R. F. W. Pease. 1995. *Appl. Phys.,* 67:909.

4. Dennig, P. A. and D. A. Stevenson. 1991. *Appl. Phys. Lett.,* 59:1562.

5. Dennig, P. A., H. Shiomi, D. A. Stevenson and N. M. Johnson. 1992. *Thin Solid Films,* 212:63.

6. Yugo, S., T. Kani, T. Kimura and T. Muto. 1991. *Appl. Phys. Lett.,* 58:1036.

7. Spitzl, R., V. Raiko, R. Heiderhoff, H. Gnaser and J. Engemann. 1995. *Diamond and Related Materials,* 4:563.

8. Konstantinova, E. A., P. K. Kashkarov and V. Yu. Timoshenko. 1995. *Phys. Low-Dim. Struct,* 12:127.

9. Konstantinova, E. A., Th. Dittrich, V. Yu. Timoshenko and P. K. Kaskarov. In Press. *Thin Solid Films.*

10. Dittrich, Th., K. Kliefoth, I. Sieber, J. Rappich, S. Rauscher and V. Yu. Timoshenko. In Press. *Thin Solid Films.*

11. Obraztsov, A. N., M. A. Timofeyev, M. B. Guseva, V. G. Babaev, Z. Kh. Valiullova and V. M. Banina. 1995. *Diamond and Related Materials,* 4:968.

12. Obraztsov, A. N., H. Okushi, H. Watanabe and V. Yu. Timoshenko. In Press. *J. Appl. Phys.*

A Comparative Characterization of Low-Pressure Discharges for Diamond and Diamond-Like Film Deposition and Other Applications

A. A. SEROV
RRC "Kurchatov Institute"
Kurchatov sq., 123182
Moscow, Russia

1. INTRODUCTION

At present, **low-pressure** discharges (LPD) are used successfully for various technological applications. Among them deposition and dry etching of thin films, coatings and structures. Up-to-date needs in such specific fields as microelectronics make increased demands on technology [1]. This requires elaboration of methods of new material film synthesis, such as diamond, diamond-like carbon, SiC, β-C_3N_4. Demands on coating qualities are: uniformity over the widest area, smoothness, lack of influence of underlying structures by the treatment process. This, leads to increased demands on discharge plasma properties, such as decreasing the energy at the surface to reduce substrate temperature, T_S. Diamond CVD has normally been realized at substrate temperatures $T_S = 800$–$1000°C$ by means of a thermal plasma as high-pressure discharges (HPD). A decrease of T_S to about $100°C$ [2] can be realized through low pressure discharges (LPD) since the intrinsic plasma property is highly nonequilibrium. Another property of LPD in comparison to HPD, is an increased area of plasma spatial uniformity. At the same

133

time, typical for LPD plasmas, high values of plasma electron temperature $T_e \approx 5$–10 eV cause superthermal plasma density, and far more LPD chemical effectiveness as compared to HPD plasma. It enables LPD to produce dense fluxes of chemically active radicals in spite of low working pressures. At low-pressures only reactions of dissociation take place in a bulk while reactions of synthesis occur on the surface being supplied by a plasma with chemically active radicals in various states of excitation, which are able to deliver additional energy. Surfaces under treatment are also subjected to the action of charged particles as well as electromagnetic irradiation. The parameters of the fluxes of the above mentioned species can be changed by varying the discharge parameters, and are specific for each type of discharge.

The impact of the various energetic species to the surface under treatment is specific; being determined by the energetics of the surface process. Such processes as anisotropic dry etching or diamond-like film deposition which need a flux of energetic ions ($E_i \approx 10$ eV–1.5 keV) incident to the surface. The most specific is the growth of diamond films. It is known that diamond grows in HPD at high rates although with a columnar, rough structure. Lowering the discharge pressure causes the diamond film structure to become more fine but the growth rate is decreased. Attempts at the growth of diamond in LPD, borders on the primary formation of the graphite phase. There are some growth models that take into consideration the stabilizing role of H and O, and graphitization which take place during transformation of the sp^2 bond into sp^3. Realization of diamond growth in LPD depends on both growth mechanisms (which are not presently known in detail) and LPD-plasma conditions which must fit the mechanisms. Some promising experimental results on diamond growth in plasma systems at $T_S < 800°C$ have recently been reported [3].

2. PROPERTIES OF LOW-PRESURE DISCHARGES

The chemical effectiveness of an LPD plasma is determined by the degree of dissociation, a_d, of the initial gas flux which passes across a plasma layer ($a_d = 1 - n/n_0$; n_0, n are the densities of initial gas molecules before and after the layer passage, respectively). If one considers a plasma layer of limited size, L, having a constant value of $<\sigma_d V_{Te}>$ (σ_d is the dissociation cross section, v_{Te} is the electron thermal velocity $v_{Te} = \sqrt{8kTe/\pi m}$, Te is a plasma electron temperature, m is mass of electron) then a_d is determined by

$$\alpha_d = 1 - \exp\left(-\langle \sigma_d \cdot v_{Te} \rangle \zeta / v_{T0}\right) \qquad (1)$$

where

$$\zeta = \int_L n_p(z)\,dz = \overline{n}_p \cdot L$$

is a plasma parameter, $n_p(z)$ is the plasma density spatial distribution across the layer, $\overline{n}_e$ is averaged over L and V_{T0} is the gas molecules' thermal velocity. The value of $\langle \sigma_d v_{Te} \rangle$ is dependent on Te which is latent parameter of Equation (1).

Different types of LPD are to be compared to each other from the point-of-view of their chemical effectiveness, i.e., the possibility of production of chemical radicals. Furthermore, one has to compare the parameters of Equation (1): T_e, n_e, ζ, L, as well as mechanisms of energy deposition in the plasma. We have limited ourselves to the discussion of the major LPD properties, together with some other important ones for surface treatment, such as area of treatment and flux of energetic particles toward the surface.

Nonequilibrium of LPD results from the increase of free-path length which enables electrons to increase their energy in fields acting within the plasma. Collisions with heavy particles take away only a small fraction of the electron energy $\Delta\varepsilon \approx 2em/M$ (where M is the mass of the heavy particle, and m is the mass of the electron) and mainly randomize electron velocities. After many collisions, time-space averaging gives a mean energy: T_e. Inelastic losses make T_e decrease its value which is characteristic, especially for dense plasmas (ECR, Beam-Plasma Discharges). For low-density plasmas, where elastic collisions dominate (glow, RF discharges), increased values of T_e are characteristic. The electron distribution function for this case has increased compared to the maxwellian high-energetic "tail" which cause plasma generation, mainly by direct ionizing collisions. In dense plasmas, cross sections for ionization and dissociation are increased by the influence of excitation of molecules. The ratio of ionization rates for the cases of ionization from excited state (stepwise ionization) and from the ground state by direct electron impact at maxwellian distribution gives $R_I = \dot{n}_{st}/\dot{n}_{di}$, from consideration of a two-level system,

$$R_i = (k_{st}/k_{di})\,n_e; \qquad k_{st}/k_{di} \approx (I/T_e)^{7/2} \tag{2}$$

where k_{st}, k_{di} are the rate coefficients of the respective ionization processes, and I is the ionization potential from the ground state.

The estimation of T_e for several LPD's can be done with the Boltzmann equation, which leads to an equation for the electron distribution function,

$$\frac{1}{v^2}\frac{\partial}{\partial v}\left(\frac{e^2 E_{eff}^2}{3m^2 v}v^2\frac{\partial \bar{f}_0}{\partial v}\right) + St\left(\bar{f}_0\right) = 0 \qquad (3)$$

where E_{eff} is the effective electric field strength, v_t is the transport collision frequency, $St(f_0)$ is a collision term. If elastic collisions predominate over inelastic ones, one gets an expression for the distribution function:

$$\bar{f}_0 = const \cdot \exp\left(-\int_0^v \frac{mvdv}{T_0 + M(eE_{eff}/mv_t)^2/3}\right) \qquad (4)$$

and as in the case of a DC electric field (DC discharge) $\Delta T = T_e - T_0$ is expressed as,

$$\Delta T = (eE/mv_t)^2 M/3 \qquad (5)$$

for an AC electric field (RF, Microwave, ECR, Beam-Plasma Discharge) one gets,

$$E = E_0 \cos \omega t, \quad E_{eff}^2 = E_0 v_t /2(v_t^2 + \omega^2) \qquad (6)$$

$$\Delta T = e^2 E_0^2 M/(6m^2(\omega^2 + v_t)) $$

From Equations (5) and (6) it can be seen that at low values of ($\omega \ll v_t$) ΔT decreases as v_t (pressure) increases, while at $\omega \gg v_t$, ΔT depends only on ω. This is one of the reasons for the difference between DC (and also RF discharges) on the one hand and microwave discharges on the other hand. Another difference is in the values of E acting in the plasma; for Glow and RF discharges plasmas $E_0 \approx 1$–10 V/cm, the main voltage drop being concentrated mainly near the electrodes; for the microwave case E_0 in the plasma reaches up to 1 kV/cm. The above estimates, although indicating true behavior of ΔT, are excessively high values, especially for dense plasmas where inelastic collision cross sections are increased by vibration excitation. The last one is characterized by the vibration temperature T_v ($T_0 < T_v < T_e$), and causes increased values of dissociation cross sections σ_d, Equation (1), as well. Quenching of the vibration states takes place due to such processes as molecular vibration exchange (V-V), vibration-temperature relaxation (V-T), and dissociation. The (V-T) process is responsible for a gas temperature (T_0) increase for which the rate is: $dT_0/dt \sim n_0 k_0 \exp(-BT_0^{-1/3})$, with a possible $k_0 \approx 10^{-10}$ cm^3/s, and $B \approx 10^2$ is determined by the Massey parameter [4]. The T_0 increase may begin at low pressures, $P \approx 10^{-2}$–10^{-1} Torr and reach $T_0 \approx 1$–3 eV at $\approx$10 Torr. In this case the plasma starts substrate-heating.

At present, there are many LPD's of different types: DC glow discharge, Penning cells, Magnetrons, RF, Microwave and Beam-Plasma Discharge, and their modifications. Some of them will be characterized by their properties, below.

3. RADIO-FREQUENCY DISCHARGES (RFD)

RFD are capacitively (CRFD) and inductively (IRFD) coupled ones [5,6]. Conventional RFD gives $n_p \approx 10^{11}$ cm^{-3}, $T_e \approx 5$–20 eV, $\xi \approx 10^{11}$–10^{12} cm^{-2}, EDF having an energetic tail. IRFD has a size uniformity of $\approx$10 cm, which for CRFD is up to 0.5 m. The operational pressure region is 10^{-3}–10^{-2} Torr. CRFD has the advantage of the existence of a self-bias V_S adjustable from 20–1500 V, though it varies together with n_p and T_e by changing the input RF power (excessive).

4. MICROWAVE DISCHARGES (MD)

MD due to high frequencies $(f > 1$ GHz) of electromagnetic waves introduced in the plasma from an external source the parameters are $T_e \approx$ 2–5 eV, $n_p^{max} \approx 1.25 \cdot 10^{-8} f^2$ cm^{-3} which are dependent mostly on f, Equation (6). Values of n_p may exceed n_p^{max} due to an enlarged size of the skin-layer $(\delta \approx 1.4 \ (\nu/(2\pi f))^{1/2})$ at pressures >10 Torr, where discharge becomes self-heating for the substrate [4]. Typical values of ζ are $10^{11} \cdot 10^{12}$ cm^{-3}, $V_S = 5$-20 eV (ambypolar). The pressure range is 10^{-4}–10^3 Torr. The size of plasma uniformity is <10 cm, depending on pressure and frequency.

5. ELECTRON CYCLOTRON RESONANCE DISCHARGE (ECRD)

ECRD refers to microwaves acting in a magnetic field, effectiveness being defined by the electric field $E_{eff} = E_0/(1 - (\Delta\omega/\nu_t)^2)^{1/2}$, where $\Delta\omega = \omega_{He} - \omega$ (ω_{He} is the cyclotron frequency, $\omega = 2\pi f$ is the microwave frequency) [7]. At $\Delta\omega = 0$, power absorption is determined by a "static" electric field: $P = n_e e^2 E^2/(2 \, m\nu_t^2)$ [5], and ΔT can reach high values, especially at low pressures $<10^{-4}$ Torr. Electron-neutral collisions decrease ΔT. At $>10^{-2}$ Torr, they broaden $\Delta\omega$, preventing resonance. Nevertheless, ECRD exists in a wide pressure region 10^{-5}–10^2 Torr at input powers of 10–10^4 W, producing dense plasmas with $n_p \approx 10^{12}$–10^{13} cm^{-3}, $T_e \approx 5$–100 eV (depending on pressure and magnetic confinement) and $\zeta \approx 10^{13}$ cm^{-2}. ECRD can produce a flux of energetic ions to a substrate and be self-heating. Drawbacks of conventional ECRD, besides $\Delta\omega$ tuning, are: low area of uniformity limited to ≈ 50 cm^2, and difficulties in independent variation of the plasma parameters.

6. STATIONARY BEAM-PLASMA DISCHARGE (BPD)

BPD in a longitudinal magnetic field is the new discharge type for plasmas in chemical applications. The intrinsic energy deposition mechanism is the "reverse Landau damping" one, which causes effective excitation of Langmuir plasma waves at their resonant interaction with beam electrons propagating through the BPD plasma [8]. Due to the beam electrons energy loss, plasma waves increase their amplitude $E \sim \exp(\gamma t)$, where t is time, growth rate $\gamma = \pi\omega_{pe}V/(2n_p)(\delta f/\delta V)$, ω_{pe} is the plasma frequency, V the beam electron velocity, and f is the beam electron distribution function. Energy deposition is being saturated at $\delta f/\delta V = 0$, plasma waves absorb 60% of beam energy transforming it into T_e according to the above mentioned mechanisms Equations (3), (4), (6) and producing a dense plasma. Due to the above mechanism power is inserted into the plasma from the inside. There is no limit to n_p for BPD, as is the case for microwave discharges ($\omega_{cutoff} = \omega_{pe}$). Stationary BPD operates in a wide pressure region 10^{-5}–10 Torr. Power deposition can be easily varied in the range $W = J_b \cdot U_b = 0.1$–10 kW (J_b is beam current, U_b is acceleration voltage), generating a plasma of density $n_p \approx 10^{13}$–10^{14} cm^{-3}, $\zeta \approx 10^{13}$–10^{14} cm^{-3}; $T_e = 2$–5 eV which is close to optimal for effective realization of plasma chemical reactions. If necessary, BPD can operate in a pulse-periodical regime to adjust T_e in the region 5–100 eV. The plasma layer can be formed a plain "sheet-like" configuration, 30–50 cm wide and a few meters long, allowing treatment of large area surfaces. Surfaces under treatment have very small self-bias ($V < 0.5$ eV), and are not subjected to ion bombardment and plasma heating.

7. WAYS OF IMPROVING DISCHARGE PROPERTIES

Fitting discharge properties into complex processes is a different matter. Comparison of the described types (and their possible modifications) show them to be different from each other by their intrinsic parameters suitable to solve separate technological problems, on the one hand, but reveal their drawbacks on the other. The traditional way is to modify each discharge type by variation of its intrinsic parameters and configurations. Another way is combining different types of discharge which are compatible and complementary in their properties. An example of such a successful combination is the Beam Plasma-RF discharge (BPRFD) which combines BPD and CRFD [9]. This combined discharge has a high plasma chemical effectiveness that enables one to provide surfaces with mixed fluxes of chemically active radicals (q_R), and ions (q_i) of variable ratio q_R/q_i and ion energy. Ion energy as well as n_p and T_e can be varied almost independently of each other. Flexibility of discharge is important for the realization of complex processes including: formation of the precursor-mixtures, gas-transport, realization of selected reaction channels both in plasma and on the surface, excitation of surface layers and enhancement of surface molecule mobility. The importance of such a discharge flexibility is indicated, by recent reports of the successful (to some extent) low-temperature diamond growth in LPD [2,3,10]. It was noted that the lower the pressure the more narrow the region of the discharge parameters enabling diamond film growth.

This report was a result of work performed with the financial support of the Russian Fund for Fundamental Research (Grant 96-02-19062a).

REFERENCES

1. 1995. *Phys. Plasmas*, 2(6):2164.

2. 1995. *Diamond Films and Technology*, 5(2):1.

3. 1995. *Diamond Films and Technology*, 5(2)29.

4. Rusanov, V. D. and A. A. Friedman. 1984. *Physics of Chemically Active Plasma* [in Russian], Moscow: Nauka.

5. 1994. *Phyz. Plasmy* [in Russian], 20(3):322.

6. 1993. *J. Vac. Sci. Technol.*, A11(1):152.

7. Wilhelm, R. 1991. *Microwave Discharges: Fundamentals and Applications*, New York: Plenum Press, p. 161.

8. Ivanov, A. A. 1995. ICPIG XXII, *AIP Conference Proceedings 363*, New York: AIP Press, p. 41.

9. Ivanov, A. A., L. N. Knjazev and A. A. Serov. 1995. ICPIG XXII, Contrib. pap. 4, p. 51.

10. 1995. *Diamond and Related Materials*, 4:149.

The Role of Oxygen Atoms in sp2-sp3 Conversion on the Growing Surface of Carbon Film

A. P. DEMENTJEV AND M. N. PETUKHOV*

IRTM, RRC Kurchatov Institute
Kurchatov sq.
Moscow 123182
Russian Federation

ABSTRACT: A chemical interaction of oxygen atoms with a surface of sp2-hybridized carbon solids (graphite, soot and C_{60}) has been studied by XPS and X-ray-excited AES methods. The transition of carbon atoms from the sp2 state to the sp3-hybridized state has been observed. The measured concentration of the sp3-hybridized carbon atoms on the surface is in excess of the C—O bonds concentration. On the basis of these data, the following model of a carbon film growth is proposed: decay of double carbon bonds by oxygen atoms and formation of the sp3-hybridized C—C and C—O bonds. It is suggested that the key role of oxygen atoms in the diamond films growth is sp2 → sp3 conversion C—C bonds on a surface under deposition by flame and CVD method.

KEY WORDS: sp2, sp3 bonding, oxidation, electron spectroscopy, modeling

1. INTRODUCTION

The growth of diamond particles and films by chemical vapor deposition (CVD) in $H_2/O_2/C_nH_m$ systems is an area of active research. Bachmann et al. [1] on the basis of a large body of experimental data have plotted C—H—O phase diagram providing a common scheme for all major CVD methods. There are many opinions about roles of H and O atoms in diamond growth. Opinions vary and the evi-

---

*Author to whom correspondence should be addressed.

140

dence is not clear. Spitzyn et al. [2] were the first to suggest that hydrogen atoms remove only graphite rather than diamond-like carbon from the growing film. The same role is assumed for oxygen atoms [1]. The etching of graphite phase is the most commonly encountered standpoint of DLC films growing mechanism. Beyond that point a chemical kinetic model was developed by Franklach and Hai Wang [3] to explain the kinetic competition between diamond and non-diamond phases. According to this model [3], the key role of H and O atoms is the suppression of aromatic species formation in the gas phase. Beckman et al. [4] supposed: the main role of oxygen is the transformation of active carbon into the more stable carbon monoxide. Olson et al. [5] showed that a diamond film could be grown using pure hydrogen and carbon fluxes, and hence H and C atoms are all that are required for diamond phase synthesis to occur. The evaporation of only carbon atoms does not lead to diamond formation [5,6]. It becomes evident that technological advancement requires a detailed understanding of O and H atoms roles in the diamond phase nucleation on a surface. Obviously, this information may be obtained by surface-sensitive methods.

Here we report results of the XPS and X-ray excited AES (XAES) studies of interactions of graphite, soot and fullerene (C_{60}) with oxygen atoms. The XPS method was used for identification of the C—O interaction. The N(E) C KVV Auger lineshapes for graphite (sp2 hybridization) and diamond (sp3 hybridization) have different characteristic features and are useful to identify sp2 and sp3 states of carbon atoms in solids [7]. A chemical reaction on the surface of sp2 carbon solids has been supposed on the basis of obtained data. This chemical reaction has been assumed as the model surface reaction occurring in CVD method.

2. EXPERIMENTAL

The XAES and XPS data were obtained using MK II VG Scientific spectrometer. Photoelectron and Auger processes were excited using an Al Kα source with a photon energy of 1486.6 eV, and the vacuum in the analysis chamber was maintained at 5×10^{-10} Torr. Spectra were collected in the analyzer constant energy mode, with a pass energy Ep = 20 eV for XPS and Ep = 50 eV for XAES analysis. The C1s XPS spectra were acquired with a 0.1 eV step size, and 0.3 eV step size was used for the C KVV Auger spectra. The Auger spectra were obtained with an overall collecting time tc = 20 ÷ 30 min and a signal-to-noise ratio of 500.

A calibration of N(E) C KVV Auger spectra was made according to C1s energy positions Eb (C1s) = 284.4 eV for diamond and Eb(C1s) = 284.5

eV for graphite [7]; $Eb(C1s)$ = 284.6 eV was measured for soot and $Eb(C1s)$ = 284.8 eV was obtained for C_{60} sample. Background subtraction for the raw Auger spectra was made on the basis of Shirley's method by using a standard VG Scientific program.

The XPS and XAES methods have different information depth. In order to equalize the information depth of both methods, the XPS spectra were taken off at angle of electron collection of 20°.

The standard Auger spectra of the pure substrates were obtained before investigation. The graphite spectrum was obtained from a plane of highly oriented pyrolytic graphite. A surface of an artificial diamond specimen $3 \times 3 \times 1$ mm^3 was prepared by polishing with a diamond paste and ultrasonically washing in acetone immediately before insertion into the spectrometer [7]. A film of soot of 1000 A in thickness was deposited on an aluminium foil from a flame of candle. A fullerene sample was made by pressing of a C_{60} powder into an indium metal substrate and inserted into a vacuum chamber.

The interaction of the graphite and soot samples with oxygen atoms was performed inside a Penning cell. A mixture of Kr and 1% O_2 gases was fed into the cell under 10^{-4} Torr. The voltage of discharge was 4 kV. Only the neutral oxygen atoms can achieve the sample surface under these conditions. An ultraviolet radiation during oxygen exposition was also presented in the cell. The samples were exposed by oxygen atoms inside the cell from 30 s to 5 min. The fullerene sample was oxidized under vacuum evaporation of the C_{60} powder in oxygen at 10^{-4} Torr. Then the samples were transferred into an analysis chamber of spectrometer for pursuance of measurements. The surfaces of soot, graphite and fullerene are inert to an ambient air oxygen. Detectable amount of oxygen on the surface of samples was measured only after heating at T > 400 K.

3. RESULTS

Figure 1 shows the C1s core-level emission from the graphite after interaction with the oxygen atoms. Two main features are evident at 284.5 and 286.5 eV binding energy. The features have full widths at half maximum (FWHM) of 1.5 and 1.7 eV. The lower binding energy feature originates from C atoms with no O attached [8] (denoted C—C in Figure 1), while the second feature arises from C atoms single-bonded to O (denoted C—O) [8]. Integrated emission intensities obtained by curve fitting in Figure 1(a) yield a ratio of C—C to C—O of 2,8 indicating a relative concentration of C—C and C—O bonds in the near-surface re-

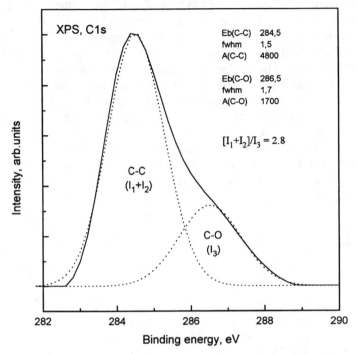

Figure 1. The C1s emission from an oxidized graphite surface and peak analysis of spectral features, Eb—binding energy of a peak; fwhm—full width at half maximum; A—integrated intensity of emission. The A(C—C)/(C—O) ratio gives the $[I_1 + I_2]/I_3$ ratio (see text).

gion of the sample. Obviously C atoms bonded to O atoms are sp3-hybridized.

The C1s levels of graphite (sp2 hybridization) and diamond (sp3 hybridization) have equal binding energy and the XPS method cannot separate the sp2 and sp3 states of C—C bonds [7]. The C KVV Auger line shapes for graphite and diamond have different characteristic features and are very sensitive to the concentration of sp2 and sp3 states of carbon atoms [7,9].

The N(E) C KVV Auger spectrum of the oxidized graphite surface after correction for charging effect and background subtraction is shown in Figure 2. The N(E) C KVV spectrum of the oxidized graphite surface may be considered as a synthesis of the diamond spectrum and the initial graphite Auger spectrum [9]. A ratio of areas of the synthesizing spectra gives the estimation of sp2 and sp3 states concentration. The ratio of sp2 to sp3 states presented in Figure 2 spectrum is 0.4.

The ratio of C—C to C—O bonds from the XPS measurements should be equal to the ratio of sp2 to sp3 states from XAES experiment in the case that sp3 states originate only from C—O bonds. In our case the comparison of XPS and XAES data shows that the total concentration of sp3 of carbon atoms on the surface of samples is far in exccess of sp3 states due to C—O bonding. By this is meant that a part of sp3 states originates from C—C bonds. So the intensity of the C1s peak due to C—C interaction consists of an intensity due to sp2-hybridized C—C bonds and an intensity due to sp3-hybridized C—C bonds.

Let I_1 be the number of sp2-hybridized C—C bonds; I_2—number of sp3-hybridized C—C bonds; and I_3—number of C—O bonds (obviously sp3-hybridized). We can determine the ratio of C—C to C—O bonds from XPS analysis:

$$X = [I_1 + I_2] / I_3 \qquad (1)$$

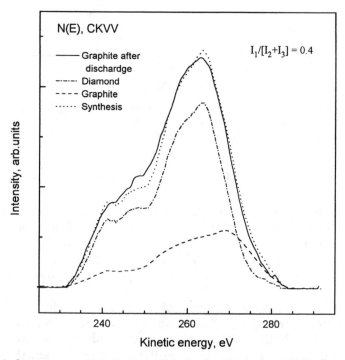

Figure 2. Least-squares-fit synthesis of the background corrected C KVV signal from an oxidized graphite surface using the diamond and the initial graphite spectra as components. The ratio of the graphite spectrum area to the diamond spectrum area gives the $I_1/[I_2 + I_3]$ ratio.

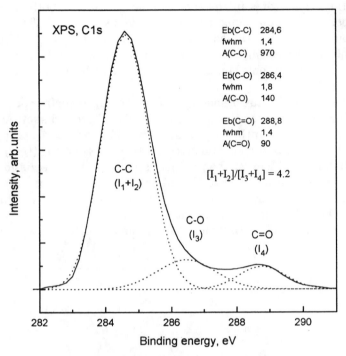

Figure 3. The C1s emission from an oxidized soot film and peak analysis of spectral features, Eb—binding energy of a peak; fwhm—full width at half maximum; A—integrated intensity of emission. $[I_1 + I_2]/[I_3 + I_4]$—the ratio of C—C bonds to C—O bonds (see text).

The ratio of sp2 to sp3 states from XAES experiment becomes

$$A = I_1 / [I_2 + I_3] \qquad (2)$$

Combination of Equations (1) and (2) gives the ratio of the sp3 states due to C—C bonds to the sp3 states due to C—O bonds:

$$I_2/I_3 = [X - A]/[A + 1] \qquad (3)$$

For the oxidized graphite surface I_2/I_3 equals to 2.

Figure 3 shows the C1s core-level emission measured after exposing the soot film to O atoms from the Penning cell. The spectrum feature Eb = 284.6 eV originates from C—C bonds, and the second feature having Eb = 286.4 eV occurs due to single-bonded C—O interaction. The third feature at 288.8 eV in Figure 3 demonstrates the arising of double C=O bonds [8]. The production of C=O bonds are caused by the presence of sp-hybridized C atoms in the soot sample. Obviously, both C—O and C=O bonded carbon atoms are sp3-hybridized. A ratio of integrated emission intensities obtained by curve fitting in Figure 3 $[I_1 + I_2]/[I_3 + I_4]$ equals 4.2 indicating a relative concentration of C—C bonded carbon atoms and carbon atoms bonded to O atoms (here I_4—number of double C=O bonds).

The $N(E)$ C KVV Auger spectrum of the oxidized surface of the soot film after correction for charging effect and background subtraction is shown in Figure 4. Synthesis of this spectrum by the initial spectrum of soot and the diamond spectrum gives a ratio of sp2 and sp3 $(I_1/[I_2 + I_3 + I_4])$ of 0.3. Comparison of the photoemission and Auger emission data shows that the total concentration of sp3-hybridized carbon atoms on the surface of soot sample is far beyond sp3 states originating from C—O and C=O bonds. The evaluated ratio $I_2/[I_3 + I_4]$ is 3.

The XPS and XAES results for the C_{60} film deposited in vacuum with

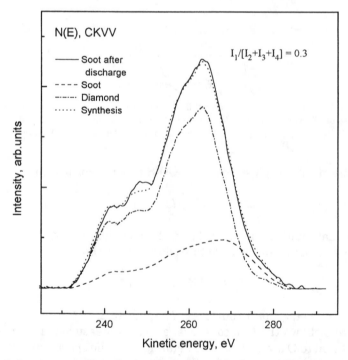

Figure 4. Least-squares-fit synthesis of the background subtracted C KVV spectrum from an oxidized soot film using the diamond and the initial soot spectra as components. The ratio of the graphite spectrum area to the diamond spectrum area gives the $I_1/[I_2 + I_3 + I_4]$ ratio.

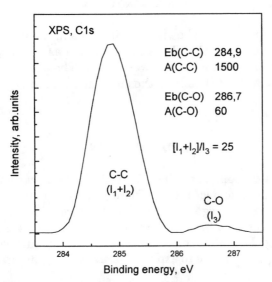

Figure 5. C1s XPS spectrum obtained after evaporation of C_{60} powder in oxygen. The A(C—C)/A(C—O) ratio gives the $[I_1 + I_2]/I_3$ ratio (see text).

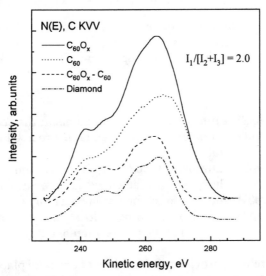

Figure 6. N(E) C KVV Auger spectrum of oxidized C_{60} after background correction. The $C_{60}O_X$—C_{60} line is the difference between the oxidized C_{60} spectrum and the normalized initial spectrum of C_{60}. The Auger spectrum of diamond is shown for comparison with $C_{60}O_X$—C_{60} lineshape. The ratio of the C_{60} spectrum area to the $C_{60}O_X$—C_{60} spectrum area gives the $I_1/[I_2/I_3]$ ratio.

low oxygen impurity are shown in Figures 5 and 6 and eliminates the
similar behavior of C—C and C—O interaction as for graphite, but in
this case an ultraviolet radiation was absent. The obtained ratio of the
sp3 states due to C—C bonds to the sp3 states due to C—O bonds I_2/I_3
equals 8.

4. DISCUSSION

Our results demonstrate that the sp3 states of carbon atoms are
formed in the near-surface region after interaction of the sp2-hybridized
carbon solids with oxygen atoms. The relative concentration of the
formed sp3 states is far beyond the concentration of the sp3 states due to
C—O bonding. The following scheme of the carbon surface–oxygen interaction may be proposed to explain these results (graphite as an example):

1. Decay of double carbon bonds by oxygen atoms
2. Migration of oxygen on the surface or desorption of oxygen
3. Subsequent reaction of oxygen atoms with neighbor carbon atoms

These processes may be summarized in the following equation:

$$nC(sp2) + mO \rightarrow (n - 2m - k)C(sp2) + m[2C(sp3)]O + kC(sp3) \quad (4)$$

where

$nC(sp2)$—number of the carbon atoms in the reactive zone
mO—number of the oxygen atoms reacted with the surface
$(n - 2m-k)C(sp2)$—number of sp2-hybridized carbon atoms which
have not interacted with oxygen atoms
$m[2C(sp3)]$—number of carbon atoms forming C—O bonds
$kC(sp3)$—number of sp3-hybridized carbon atoms after migration and/or desorption of oxygen

So, among etching, previous chemical reactions take place on the surface of carbon film and chemical equation for graphite phase etching
(Equation (2) from Reference [1]):

$$nC(graphite) + mE * \leftrightarrow C_n E_m \text{ (gas)}$$

where E*—O atoms must be supplemented by Equation (4).

5. CONCLUSIONS

Cooperative XPS and XAES analysis yields unique information about surface chemical reactions for carbon solids with oxygen; identifies the sp2 → sp3 transition under oxidation of graphite, soot and C_{60} and separates sp3-hybridized states of carbon atoms due to C—C bonds from sp3 states of C—O bonds. One of the key roles of oxygen atoms in diamond films growth is sp2 → sp3 conversion of C—C bonds on a surface under deposition by flame and CVD methods.

REFERENCES

1. Bachman, P. K., D. Leers and D. U. Weichert. 1991. "Diamond Thin Films: Preparation, Characterization and Selected Applications," *Ber. Bunsenges. Phys. Chem.*, 95:1390–1400.

2. Spitzyn, B. V., L. L. Bouilov and B. V. Darjaguin. 1981. "Vapor Growth of Diamond and Other Surfaces," *J. Crys. Growth*, 52:219.

3. Frenklach, M. and H. Wang. 1991. "Detailed Surface and Gas-phase Chemical Kinetics of Diamond Deposition," *Phys. Rev.*, B43:1520.

4. Beckman, R., S. Reinke, M. Kuhr, W. Kulisch and R. Kassing. 1993. "Investigation of Gas Phase Mechanisms during Deposition of Diamond Films," *Surf. Coatings Techn.*, 60:506.

5. Olson, D. S., M. A. Kelly, S. Kapoor and S. B. Hangstrom. 1993. "Sequential Deposition of Diamond from Sputtered Carbon and Atomic Hydrogen," *J. Appl. Phys.*, 74:5167.

6. Foord, J. S., L. K. Ping and R. B. Jackman. 1995. "Molecule Beam Epitaxy for the Growth of Thin Film Diamond," *The 6th European Conference on Diamond, Diamond-like and Related Materials*, (abstract 3.3), Barcelona, Spain.

7. Dementjev, A. P. and M. N. Petukov. 1996. "Comparison of X-ray-excited Auger Lineshapes of Graphite, Polytheylene and Diamond," *Surf. Interface Anal.*, 24:517.

8. Beamson, G. and D. Briggs. 1992. "High Resolution XPS of Organic Polymers," *The Scienta ESCA 300 Database*, ed. by John Wiley Sons Ltd., England, pp. 54–99.

9. Petukhov, M. N. and A. P. Dementjev. "The Development of C KVV X-ray-excited Auger Spectroscopy for Estimation of the sp2/sp3 Ratio in DLC Films."

Colloidal Diamond Thin Films

S. V. KUKHTETSKII AND L. P. MIKHAYLENKO
Institute of Chemistry and Chemical Metallurgical Processes SD RAS
ul. K. Marksa 42
Krasnoyarsk, 660049
Russia

Many new materials (mainly as oxides) with unique properties have been synthesized using the sol-gel technique: films, coatings, glasses and porous materials [1]. The extension of this technology to diamond systems could promote the synthesis of a new class of diamond materials and composites. The application of this has been restrained by the fact that so far there were no simple and productive methods for the preparation of a diamond sol with the necessary properties: stability, rather small sizes of particles and relative narrow particle size distributions. Diamond hydrosols satisfying to a large extent the listed conditions and suitable for use in sol-gel processes have been obtained and investigated by the authors [2].

Explosive synthesis ultradisperse diamond (UDD) powders [3,4] are the raw materials for the colloidal diamond preparation. These powders result from detonation products of explosives with a negative oxygen balance. Research has shown that UDD powders consist of aggregates of primary cubic diamond crystallites. The average crystallite size measured with the low-angle X-ray scattering method depends on synthesis conditions but varies from 2 to 5 nanometers [5]. The size of (strong) aggregates in various commercial batches of UDD powders reaches, however, up to a hundred microns. These peculiarities of nano-sized diamond powders strongly limit their application in the sol-gel technology.

An aqueous suspension of UDD powders can be readily made. It results in a muddy opaque liquid, which is stable from several minutes to several hours only. The suspensions usually have a rather complex fractional composition. It has been shown [6] that primary diamond crystallites, of 4–8 nanometers, form the strong clusters of 20–60 nanometers. These clusters, in turn, form large aggregates of 1000 nanometers and more in size. Suspension stability and cluster strength to a great extent depend on chemical structure of the particle surface, which, in turn, is determined by the technology of the synthesis and the purification of diamonds in manufacture.

We have investigated various industrial batches of UDD powders and found that some technologies of the synthesis and the purification [7] allowed one to obtain very thin stable diamond hydrosols [2].

The hydrosols were prepared through ultrasonic dispersion of weighted amounts of UDD powder in distilled water. The suspensions were then treated in a centrifuge for a long period of time. The obtained colloid was a transparent and opalescent solution. The color of the solution varied from pale yellow to brown depending on a particle concentration. Up to 10% of the powder may be transferred into the hydrosol with this method. Such solutions remain stable for two years and more, provided that there are no essential external influences. According to the X-ray spectrochemical and diffraction data of the solid phase the hydrosol consists of diamond particles.

The investigation of UDD hydrosol by scanning tunneling microscopy (STM) has shown, that it consists of the single particles of 5–6 nanometers for the most part. The technique of the thin colloid research by STM and the microscope design have been reported [2,8]. The principal feature of this technique is the diamond particle images that are obtained in a thin film of the colloid directly in the course drying. It is possible to avoid aggregation of particles during observation when samples are sufficiently diluted.

The experimental setup is shown in Figure 1. A newly cleaved surface of highly oriented pyrolytic graphite (HOPG) was used as a substrate. Scanning of the dry substrate was carried out in order to find an area free of coarse defects [Figure 1(a)]. Then a drop of the greatly diluted diamond colloid was placed onto the substrate [Figure 1(b)]. During evaporation of the drop the substrate surface was continuously scanned to eliminate the aggregation of particles under the STM tip. Distinct images of the colloidal particles on the substrate were observed starting at a certain moment in time [Figure 1(c)]. Depending on the rate of drying, the images of colloidal particles could be observed for 10–15 minutes. For this period it was possible to obtain several pictures with good reproducibility.

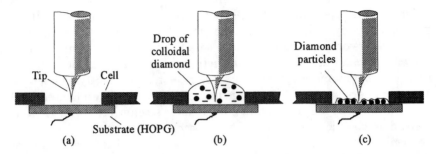

Figure 1. Experimental setup of colloidal diamond investigation by the scanning tunneling microscope.

An image of single diamond particles is shown in Figure 2. The size of these particles is about 5–6 nm. This value corresponds to estimates of the colloidal particle size based on sedimentation and diffusion data. Approximately the same value of the primary diamond crystallite size was obtained by X-ray low-angle scattering method [3–5].

Hydrosol is a unique method to conserve colloidal diamond for future applications such as synthesis of glasses, composites and films. The authors have deposited thin films with regular 2-D closely packed structure on the HOPG substrate by slow drying of the colloid in the presence of an electric field. These films have been investigated by the STM method, at a tunnel voltage of 5 mV, and a current of 2 nA. The image of this film is shown in Figure 3. There is a defect at the right-hand corner. Resulting from the touching of the surface by the STM tip. It can be seen that particles are arranged very regularly and form a 2-D structure with a hexagonal symmetry. The film is a single layer of particles.

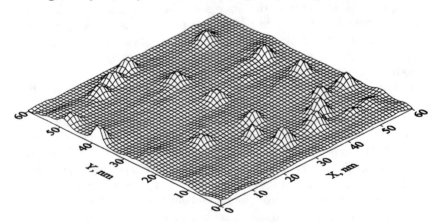

Figure 2. STM image of single diamond nanoparticles.

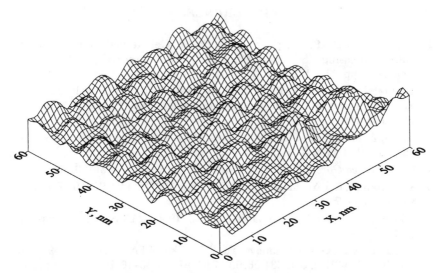

Figure 3. The colloidal diamond thin film structure.

From the regularity of the film structure one may conclude that the particles have a rather narrow size distribution, i.e., they are practically monodispersed. Distance between the particle centres is of the order of 8–9 nanometers. They are a little larger than the primary diamond crystallites determined by an X-ray low-angle method and the single diamond particle sizes noted above. This is probably due to the presence of water layers around diamond particles. Dried at room temperature, colloidal diamonds (macroscopic samples) contain up to 50% of linked water, which can be removed at temperatures above 150°C only. The films were obtained at room temperature, therefore they may contain approximately the same quantity of water.

The mechanism of the regular structure formation for this film is probably similar to the known mechanism of periodic colloidal structure formation for colloids with rather larger particles [9]. This mechanism is caused by far-acting forces between charged colloidal particles.

It could be shown that certain methods of UDD powder processing allow the extraction of free single diamond nanoparticles. It is possible to save them in form of an ultrathin colloidal solution for a long time (several months and even years). The generation of diamond films with the regular colloidal structure has been realized.

This work was supported by the Russian Research Program, "Fullerenes and Atomic Clusters."

REFERENCES

1. Mackenzie, J. D. 1986. "Application of the Sol-Gel Method: Some Aspects of Initial Processing," *Science of Ceramic Chemical Processing.* NY, Toronto, Singapore, pp. 113–122.

2. Kukhtetskii, S. V. and L. P. Mikhailenko. 1996. "Study of Ultradisperse Diamond Hydrosols by Scanning Tunneling Microscopy," *Colloidal Journal,* 58(1):137–139.

3. Staver, A. M. and N. V. Gubareva, et al. 1984. *Phisica Goreniya i Vzryva* (in Russian), 20(5):100–104.

4. 1987. *Phisica Goreniya i Vzryva* (editorial) (in Russian), 23(5):3.

5. Anisichkin, V. F., I. Yu. Malkov and V. M. Titov. 1988. *DAN SSSR* (in Russian), 303(3):625–627.

6. Sakovich, G. V. and V. D. Gubarevich, et al. 1990. *DAN SSSR* (in Russian), 310(2):402–404.

7. Chiganov, A. S., G. A. Chiganova, Yu. V. Tushko and A. M. Staver. 1993. Patent RF N2004491 cl. C 01 331/06, 02.07.91. BI. No. 45–46 (in Russian), p. 85.

8. Kukhtetskii, S. V. and L. P. Mikhaylenko. 1994. *Herald of Russian Acad. Tech. Sci.,* Part A, 1(7):150.

9. Efremov, I. F. 1971. "Chimiya," *Periodicheskiye Colloidniye Structure,* L (in Russian).

Nucleation and Growth of Anisotropic Clusters

V. I. SIKLITSKY

A. F. Ioffe Physico-Technical Institute
Polytechnicheskaya 26, St. Petersburg, 194021, Russia

ABSTRACT: The influence of the probability that spherical particles are interconnected during the nucleation and growth of fractal cluster is considered in analyzing cluster formation in Diamond Like Carbon (DLC).

INTRODUCTION

Close attention has been paid recently to the processes of self-assembling of different kinds of clusters under various external conditions [1]. Usually, simulation of large single clusters has been performed using the Diffusion Limited Aggregation (DLA) model [1]. Statistical trends of cluster growth have not been taken into account, so it has been impossible to compare the theoretical results with a vast body of experimental data. In this paper formation of clusters in early stage (nucleation) of their origin is considered in terms of the DLA model of cluster growth in analyzing experimental results on cluster formation in DLC.

COMPUTER SIMULATION

Clusters were formed by non-lattice three dimensional Diffusion Limited Aggregation (DLA) model [1]. We assumed that the fractal cluster is formed by spheres connected together with different sticking probabilities. The probability of interconnection is a function of the surface area of the particle, determined in the Oxyz coordinate system. All directions of the coordinate system are equivalent. The relation for

155

probabilities $P_z < P_y < P_x = 1$ was taken into account. For every grown cluster a calculation of size was performed. Namely, the normalized sum of squared coordinates for each axes was determined. Thereafter, the obtained sizes were normalized to their maximum value.

Histograms of the dependence of normalized cluster sizes on the number of particles that have the same form, i.e., the same normalized size are shown in Figures 1–3. Such distribution functions determine strictly three groups of clusters: needles, scales, and flakes. The number of particles belonging to each of the groups is determined by different probabilities P_x, P_y, P_z. It is seen that with increasing number of elements in the fractal particle a transition occurs between these groups of clusters, namely, needles→ scales→flakes. Thus, if N is large enough, particles have ellipsoidal form.

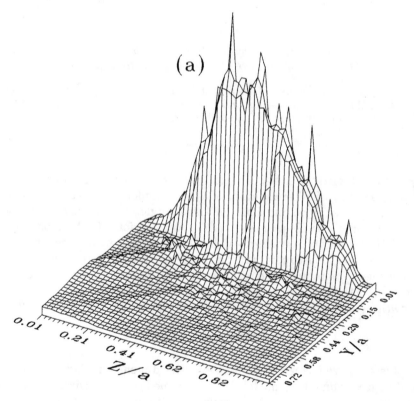

Figure 1. 3D Distribution function of anisotropic cluster forms obtained by nonlattice three dimensional DLA model. Probabilities of interconnection of spheres in the fractal are $P_x = 1.0, P_y = 0.5, P_z = 0.5$. Number of spheres in the fractal are: (a) $N = 10$; (b) $N = 14$; (c) $N = 17$; (d) $N = 50$.

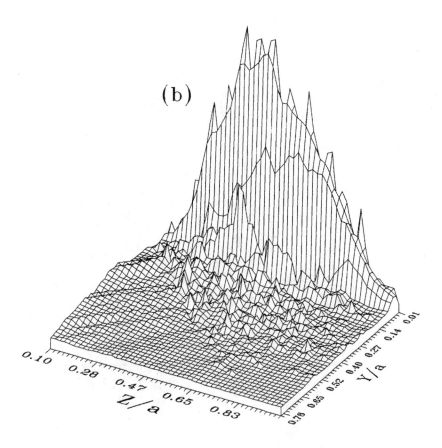

(b)

Z/a
Y/a

Figure 1 (continued). 3D Distribution function of anisotropic cluster forms obtained by nonlattice three dimensional DLA model. Probabilities of interconnection of spheres in the fractal are $P_x = 1.0$, $P_y = 0.5$, $P_z = 0.5$. Number of spheres in the fractal are: (a) $N = 10$; (b) $N = 14$; (c) $N = 17$; (d) $N = 50$.

157

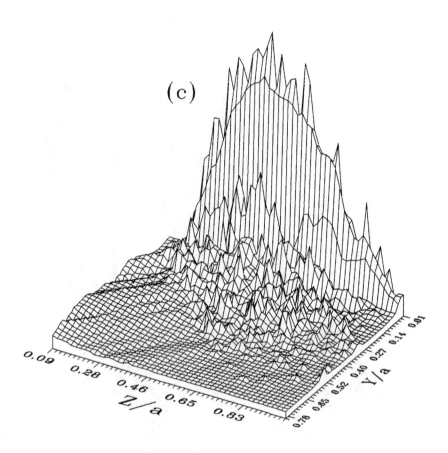

(c)

Figure 1 (continued). 3D Distribution function of anisotropic cluster forms obtained by nonlattice three dimensional DLA model. Probabilities of interconnection of spheres in the fractal are $P_x = 1.0$, $P_y = 0.5$, $P_z = 0.5$. Number of spheres in the fractal are: (a) $N = 10$; (b) $N = 14$; (c) $N = 17$; (d) $N = 50$.

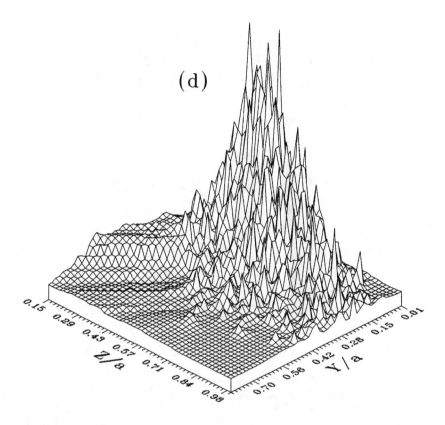

(d)

Figure 1 (continued). 3D Distribution function of anisotropic cluster forms obtained by nonlattice three dimensional DLA model. Probabilities of interconnection of spheres in the fractal are $P_x = 1.0$, $P_y = 0.5$, $P_z = 0.5$. Number of spheres in the fractal are: (a) $N = 10$; (b) $N = 14$; (c) $N = 17$; (d) $N = 50$.

159

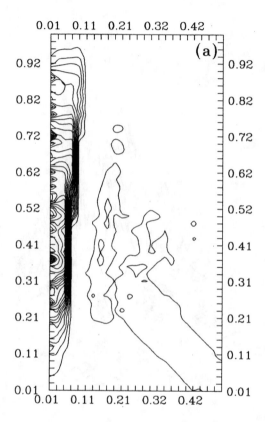

Figure 2. Contour lines obtained for 3D distribution function of anisotropic cluster forms (parameters choiced for simulation are the same ones to Figure 1). Number of spheres in the fractal are: (a) $N = 10$; (b) $N = 14$; (c) $N = 17$; (d) $N = 50$.

160

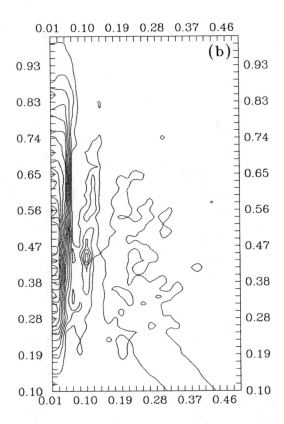

Figure 2 (continued). Contour lines obtained for 3D distribution function of anisotropic cluster forms (parameters choiced for simulation are the same ones to Figure 1). Number of spheres in the fractal are: (a) $N = 10$; (b) $N = 14$; (c) $N = 17$; (d) $N = 50$.

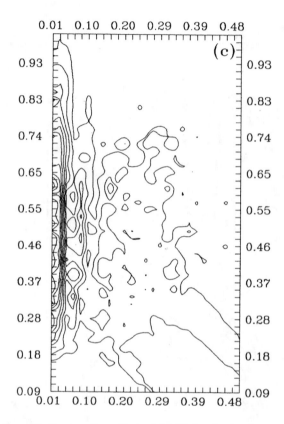

Figure 2 (continued). Contour lines obtained for 3D distribution function of anisotropic cluster forms (parameters choiced for simulation are the same ones to Figure 1). Number of spheres in the fractal are: (a) $N = 10$; (b) $N = 14$; (c) $N = 17$; (d) $N = 50$.

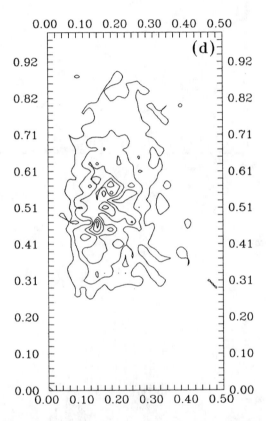

Figure 2 (continued). Contour lines obtained for 3D distribution function of anisotropic cluster forms (parameters choiced for simulation are the same ones to Figure 1). Number of spheres in the fractal are: (a) $N = 10$; (b) $N = 14$; (c) $N = 17$; (d) $N = 50$.

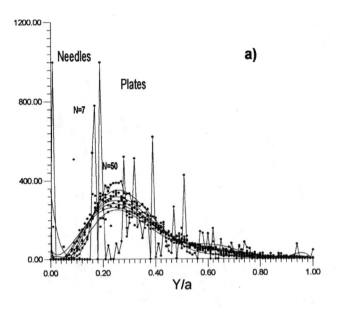

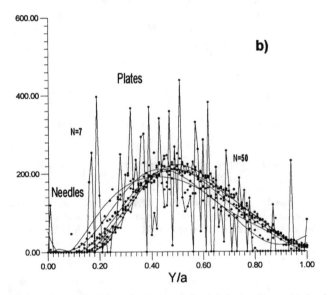

Figure 3. 2D distribution function of anisotropic fractal cluster forms obtained by nonlattice two dimensional DLA model. Probability of interconnection of spheres in the fractal are: (a) $P_x = 1.0$, $P_y = 0.2$, $P_z = 0.01$; (b) $P_x = 1.0$, $P_y = 0.5$, $P_z = 0.01$; (c) $P_x = 1.0$, $P_y = 0.8$, $P_z = 0.01$.

164

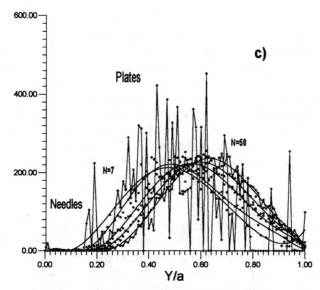

Figure 3 (continued). 2D distribution function of anisotropic fractal cluster forms obtained by nonlattice two dimensional DLA model. Probability of interconnection of spheres in the fractal are: (a) $P_x = 1.0, P_y = 0.2, P_z = 0.01$; (b) $P_x = 1.0, P_y = 0.5, P_z = 0.01$; (c) $P_x = 1.0, P_y = 0.8, P_z = 0.01$.

DISCUSSION

The nucleation and further growth of various forms of clusters in DLC films taking place as a result of selfassembling which occurs in the course of their preparation was observed in Reference [2]. In particular, anisotropic clusters including spherical (form factor $f = 1/3$) and nonspherical ($f = 0.5$) particles were optically registered in DLC films [2]. It was found that in the course of cluster formation process at least two kinds of conductive cluster networks were registered by analyzing both optical data and electrical conductivity percolation thresholds. By means of small angular X-ray scattering it was possible to study the fractal dimensionality of the structural element, i.e., the scatterer of the conductive network, depending on the concentration of conductive phase, varied from 0 to 24 at%, and vacuum annealing [3,4]. The intensity of scattered radiation may change strongly the monotonous character of the scattering curve [3,4].

A dependence of the X-ray scattering intensity on scattering angle for different copper concentrations was considered in [3]. The peculiarities of the scattering curve may be associated with the nonspherical character of scatterers [5,6]. It should be noted that these peculiarities are observed at lower copper contents: the contribution of nonspherical particles decreases with increasing copper content. This fact may be ascribed to the possible spherical forms of the fractal scatterers formed at higher copper concentrations. The copper-doped DLC is a fairly complicated object because a size-distribution function of scatterers [7]. A fractal particle was chosen in References [3,4] as a model of scatterer. Changes in its fractal dimensionality were studied as a function of copper concentration in DLC. The scatterer form was not taken into account. It should be noted that a more correct expression for the scattering intensity $I(q)$ for a scatterer with definite form can be presented as follows:

$$I(q) = S(q) \cdot F(q) \tag{1}$$

where $S(q) = q^{-\alpha}$ is the structure factor and $F(q)$ is the form factor ($q = 4\pi\lambda^{-1}\sin(\theta/2)$ is wave vector, λ is wavelength, θ is scattering angle). This suffices to fit the experimental data presented in Reference [3]. $F(q)$ is a function of alignment and form of the scatterer. An example of a calculated distribution function is presented in Figures 1 and 3. The form factor of aligned particles of ellipsoidal form is presented in Reference [5] where a full theoretical description was given of electromagnetic wave scattering by a media containing aligned anisotropic scatterers with typical dimensions of the order of the wavelength of the initial electromagnetic wave. In particular, analytical expressions were presented for calculating Stokes parameters of radiation. It was also shown that the presence of similar particles in the scattering media suffices to generate various forms of polarization of scattering radiation, such as linear polarization and circular one.

CONCLUSION

A simple model of nucleation of the fractal particle performed within DLA shows a pronounced discrimination through the produced clusters as a result of self-assembling process and can be used to analyze experimental data on cluster formation in DLC.

ACKNOWLEDGEMENTS

Author would like to thank Prof. S. Mitura and Dr. Yastrebov for useful discussion, Prof. V. I. Ivanov-Omskii for his encouragement. This work was supported in part by the Arizona University grant and the Ministry of Science and Technological Policy of Russia within the framework of the National Interdisciplinary Program "Fullerenes and Atomic Clusters" (grant No. 94007).

REFERENCES

1. Pietronero, L. and E. Tosatti, eds. 1988. *Fractal in Physics*. Amsterdam: North-Holland.

2. Ivanov-Omskii, V. I., A. V. Tolmatchev and S. G. Yastrebov. 1996. "Optical Absorption of Amorphous Carbon Doped with Copper," *Philos. Mag. B.*, 73:715–722.

3. Ivanov-Omskii, V. I., V. I. Siklitsky and M. V. Baidakova. In Press. "Fractal Structure of Copper Clusters Embedded in DLC," *Diamond (and other wide band-gap) Composites,* NATO Advanced Research Workshop, St. Petersburg.

4. Faleev, N. N., V. I. Ivanov-Omskii, V. I. Siklitsky and S. G. Yastrebov. 1995. "Fractal Structure of DLC and Embedded Copper Clusters," Book of Abst., *MRS 1995 Fall Meeting,* Symposium M, "Disordered Materials and Interfaces: Fractals, Structure and Dynamics," Boston: U.S. Paper M9.22:406.

5. Dolginov, A. Z. and V. I. Siklitsky. 1992. "Stokes Parameters of Radiation Propagating through an Aligned Gaseous-Dust Medium," *Mon. Not. R. Astr. Soc.,* 254:369–382.

6. Guinier, A. and G. Fournet. 1955. *Small-Angle Scattering of X-Rays.* New York: John Wiley & Sons, Inc.

7. Ivanov-Omskii, V. I., A. V. Tolmatchev, S. G. Yastrebov, A. A. Suvorova and A. A. Sitnikova. 1996. "Optical and TEM Study of Copper-Born Clusters in DLC," *The 3rd Int. Symp. on Diamond Films,* St. Petersburg.

Carbyne Crystallization by Impulse Electron Beam into Carbon Thin Films Grown by Ion Beam Sputtering

A. P. SEMENOV AND N. N. SMIRNYAGINA

Buryat Institute of Natural Sciences of Siberian Division
Russian Academy of Sciences
670047, Sakhjanova St.
6, Ulan-Ude, Russia

Carbyne thin films have been obtained by ion assisted carbon vapor condensation [1], and ion beam sputtering of graphite [2]. In this study we present the results on two stage formation of carbyne thin films, with electron and ion beams.

The ion beam sputter deposition system used for the present preparation, is illustrated in Figure 1. The carbyne thin films growth was carried out in a home-made beam-forming arrangement [3] with standard vacuum equipment (Vacuum Universal Post -VUP-4) and ion plasma source [4]. The installation consisted of an ion source 1, the target 2, a holder 3 with substrates 4. Substrate heating was adjusted from 300 to 673 K. The accelerated ion beam 5, was inclined at 40–45° to the target. Carbon layers have been deposited on (111) Si substrates with deposition temperatures not exceeding 673 K. The graphite disk target >99.9999% pure and 55 mm in diameter, was sputtered by ion beams (Ar and H_2 mixtures). The ion beam current and energy were 5–10 mA and 4000 ev, respectively. The growth rate of the car-

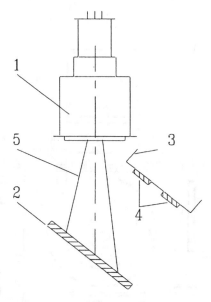

Figure 1. Scheme of the ion beam sputter deposition system.

byne thin films was typically 0.10–0.50 nm/s, with a gas pressure of 2.3×10^{-3} Pa.

Thin film processing by electron beam has been carried out in an electron-vacuum installation with a powerful industrial type electron gun EPA-60-0.42 [5]. The installation also contains control blocks, BUEL [6], a high-voltage rectifier V-TPE-2-30 K-2 UHL4, and a working chamber with vacuum aggregates AVP-250, AVP-160 and NVPR-40-066. The installation is extremely compact in design (Figure 2) encompassing several technological processes. The pressure in the working chamber did not exceed 2×10^{-3} Pa.

The deposited thin films were studied by X-ray diffraction, IR and Raman spectroscopy. The X-ray powder diffractometer DRON-II, using CuKα-radiation, was employed for phase analysis, and the determination of lattice parameters. The films were examined by IR absorption spectrometry (UR-20) in the 700–4000 cm⁻¹ range to detect the presence of carbon bonds. The Raman spectra were obtained with the spectrometer DPHS-24, using the 632.8 μm line of a helium-neon laser.

A short duration electron beam treatment was carried out because the thin films after sputtering contained *i*-C. The scanning electron beam powder was 50–200 W. The crystallization of different carbon phases: diamond, graphite, carbyne and others as well as silicon carbide, were

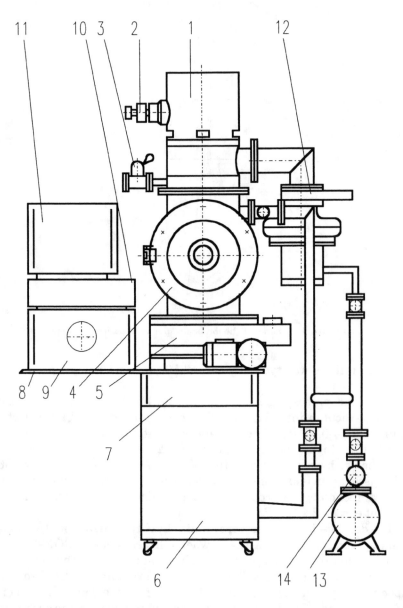

Figure 2. Schematic of the electron beam installation. 1. Cathode chamber; 2. high-voltage entry; 3. vacuum shutter; 4. vacuum chamber; 5. vacuum unit AVP-250; 6. equipment-racks; 7. control panels; 8. equipment cover; 9. controls; 10. operator console; 11. vacuum gauge VIT-2; 12. vacuum unit AVP-160; 13. pump NVPR-40-066; 14. safety valve.

170

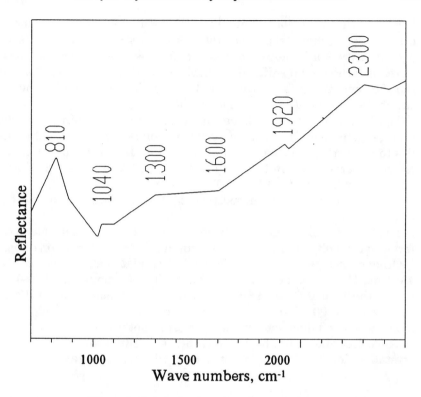

Figure 3. Infrared spectra for carbyne thin films.

observed after electron beam annealing. It was determined that in varying the ion sputtering parameters and the electron beam annealing regime, it is possible to control the phase content in polycrystalline carbon thin films.

Carbynes are regarded as linear carbon polytypes with the *sp*-configuration and carbon-carbon chains, either conjugated triply bonded or cumulated doubly bonded, parallel to the *c*-axis, with bonding achieved by intermolecular linkage [7,8]. Evidence for the existence of C-C chains in carbyne has been found with X-ray diffraction, various spectroscopic methods, such as IR and Raman spectroscopy. Carbyne crystalline structures contain impurity atoms (Si, Fe, Al, Mg) at the end of the chains. They play the part of catalyst in carbyne crystallization. The cumulene-type carbynes with an intermolecular linkage between the chains due the free bonds will be more stable than the polyyne-types carbynes. In IR spectrum the polyyne carbon-carbon chains is identified always indicated with a line at 2100–2300 cm^{-1} (very strong) due to the triple -C≡C-bond, as well as deformation fluctuations at 800 cm^{-1}. Cumulative carbon-carbon chains show peaks at 1950 (very strong), and 1070 (average intensity) cm^{-1} due to cumulene bonds =C=C= which are raised because of carbon bond shifts along triple bonds. The absorption band in the 1600 cm^{-1} range is connected with the fundamental bond of the zig-zag chain configuration of the four carbon atoms.

The IR reflectivity and Raman data for the carbon thin films after processing by electron beam are shown in Figures 3 and 4. The data testifies to the carbyne formation in the thin film. In this sample, one can clearly observe peaks in the IR spectrum at 810, 1040, 1920, 2300 cm^{-1} and activity in the area 1300–1600 cm^{-1}.

Figure 4 shows the Raman spectrum of the thin films. We observe two peaks near 1645 and 2143 cm^{-1}.

The IR and Raman data have been interpreted as primarily due to the formation of polyyne and cumulene structure in the carbyne thin films.

Other carbon phases were not found. In IR and Raman spectra peaks at 1330 and 1580 cm^{-1}, the asymmetric band with a maximum at 1500–1550 cm^{-1} or the band at 1350 cm^{-1}, as well as the high intensity bands at 1350 ± 10 cm^{-1} and 1600 ± 20 cm^{-1}, which are characteristic for diamond, graphite, i-carbon and diamond-like thin films, are not present.

The effect of electron beams in carbon thin film growth leads to the crystallization of an hexagonal structure, with lattice parameter $\alpha = 51$

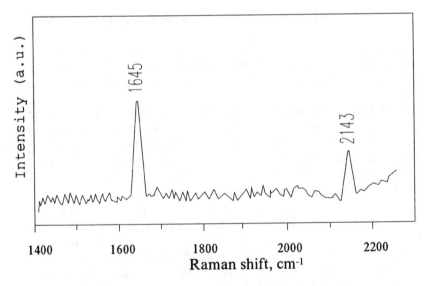

Figure 4. Raman spectra for carbyne thin films.

μm. The diffraction patterns show only hk0 reflections indicating that the carbon chains are oriented perpendicular to the substrate surface. It was found that the hexagonal unit cell parameter of thin film carbyne changes with the electron beam energy. It is known, that carbyne structures are formed by a crystallographic arrange of C–C-chains with either conjugated triple or cumulated double bonds and that the c-axis length is a function of the number of carbon atoms in the chains. The chain length of carbynes are a function of the temperature of formation. Higher temperatures leads to cracking the chains. The number of carbon atoms per unit cell is closely related to the synthesis temperature. The higher temperature the smaller the chain lengths.

In summary, it could be shown, that deposition of carbon thin films onto Si substrates by ionic beam sputtering (Ar and H_2 mixture) of a graphite target, followed by scanning electron beam treatment, makes possible the crystallization of films with a carbyne structure.

REFERENCES

1. Kudryavtsev, Yu. P., S. E. Evsyukov and M. B. Guseva, et al. 1993. *Izv. Acad. Nauk-USSR*, Ser. Khim., 3:450–463 (in Russian).

2. Kudryavtsev, Yu. P., S. E. Evsyukov and V. G. Babaev, et al. 1992. *Carbon*, 20:213.

3. Semenov, A. P. 1993. *Pribory i Teknika Eksperimenta*, 2:220–221 (in Russian).

4. Semenov, A. P. 1993. *Pribory i Teknika Eksperimenta*, 5:128–133 (in Russian).

5. Grigoryev, Yu. V., V. I. Karlov and A. S. Murashov, et al. 1989. *Pribory i Teknika Eksperimenta*, 2:228 (in Russian).

6. Grigoryev, Yu. V., Yu. G. Petrov and V. I. Pozdanov. 1990. *Pribory i Teknika Eksperimenta*, 2:236–237 (in Russian).

7. Heimann, R. B., J. Keliman and N. M. Salansky. 1983. *Nature*, 306(5938):165–167.

8. Heimann, R. B., J. Kleiman and N. M. Salansky. 1984. *Carbon*, 22(2):147–156.

MODELING AND PHASE EQUILIBRIA

Temperatures and Hydrogen Atom Concentrations in a Microwave Plasma Used for Diamond Deposition

A. GICQUEL,* Y. BRETON, S. FARHAT AND K. HASSOUNI
Laboratoire d'Ingénierie des Matériaux et des Hautes Pressions
CNRS-UPR 1311-Université Paris-Nord
Avenue J. B. Clément 93430, Villetaneuse, France

M. CHENEVIER
Laboratoire de Spectrométrie Physique
CNRS-URA 008, Université Joseph Fourier de Grenoble
B. P. 87, 38402, Saint Martin d'Hères, Cedex, France

ABSTRACT: A spectroscopic analysis of the plasma is presented. The variations of H-atom temperatures and of H-atom mole fraction as a function of the methane percentage introduced in the feed gas and of the power density are shown. A comparison between excited H-atom and ground state H-atom temperatures measured respectively by Two Photon Allowed Transition (TALIF) and Optical Emission Spectroscopy (OES) is made. A study on the validity of the actinometry method (OES) for the H atoms is carried out based on theory and TALIF measurements. Excited H-atom temperature and ratios of emission intensities (I_H/I_{Ar}) are shown to constitute interesting spectroscopic parameters for measuring gas temperature and ground state H-atom mole fraction by OES.

1. INTRODUCTION

Spatially resolved spectroscopic analysis of the plasma provides measurements of local characteristics of the plasma and of the

---

*Author to whom correspondence should be addressed.

plasma/surface interface (temperatures, concentrations) which are responsible for diamond growth. Their knowledge allows the validation of models and an increased understanding of the phenomena. The spectroscopic analysis also provides a mean for optimizing and monitoring the reactors. Since the control of industrial reactors by laser spectroscopy is unrealistic, the development of measurements by Optical Emission Spectroscopy (OES), associated to their calibration by laser diagnostics techniques is necessary.

In low pressure diamond deposition reactors operating with H_2 + CH_4 mixtures, the role of the H atoms and of the carbon containing species radicals (in particular CH_3 and C_2H_2) have been reported by several authors [1–3,6]. Their detection and the measurements of their concentrations and of the parameters controlling their production and destruction are of prime importance. In plasma reactors, we already showed that the H-atom concentration depends on the electron temperature, the gas temperature, the percentage of methane introduced in the feed gas and the surface reactions such as etching and catalytic recombination processes [4–6]. We report here measurements of H-atom temperatures and mole fractions, in the plasma volume, by means of Two Photon Allowed Transition (TALIF) and OES.

The measurements of ground state and excited state H-atom translational temperatures are seen to be both almost equal to the gas temperature measured by CARS [1] or calculated by a 1 D diffusive H_2 plasma flow model developed in our laboratory [5]. They strongly increase as a function of the power density (W cm^{-3}) while they remain rather constant as methane percentage is introduced in the feed gas at up to 5%.

We also report a study owing to establish the domain of validity of actinometry method for H atoms under conventional diamond deposition conditions [6]. Actinometry is a method implying an introduction of, usually, a rare gas as an impurity in the feed gas, which constitutes the actinometer [9]. The excitation mechanism from ground state to an excited state of both actinometer and the X species under study (here H atoms) must be a direct electron impact. In addition, the excitation cross-sections must have the same energy threshold and must be proportional in the range of electron energy corresponding to the discharge. Under these conditions, and knowing the cross-sections of quenching of the excited species (X and actinometer A) by ground state species (here mainly H_2 molecules and and H atoms), the X species mole fraction can be related to the emission intensity ratio (I_X/I_A) via the following relationship: $x_X = F x_A f(Q,p,f(T_e))I_X/I_A$, where x_X and x_A represent X and A species mole fractions, F an optical system calibration term, and $f(Q,p,f(T_e))$, a function depending on the quenching cross-sections term (Q), the pressure (p) and the ratio of the electron-species collision excitation rate constants of X and A species, which is electron temperature dependent ($f(Te)$) [6]. Our work consisted in (1) determining the conditions of validity of actinometry for H atoms over the range of diamond deposition conditions (electron temperature, H-atom density), and (2) finding the function $f(Q,p,f(Te))$. We have in particular estimated, from experiments, the quenching cross-section for the collisions between excited Ar-atom with ground state H_2 molecules [6]. The quenching cross-sections for the collision between excited Ar-atom and H-atom with ground state H atoms have been calculated [6] and the quenching cross-section for the collisions between excited H-atom and ground state H_2 molecules have been found in the literature [10,11]. On these bases, the variations of relative H-atom mole fraction as a function of the power density and of the percentage of methane were obtained. Absolute variations were obtained after calibration of the system. It has been realized by comparing the depletion of H atoms owing to methane dissociation observed as the percentage of methane is increased under low power density conditions, to that calculated by a 0 D H_2 + CH_4 plasma model.

2. TECHNICAL DESCRIPTION

2.1 Reactor

The reactor consists of a silica bell jar low pressure chamber [4], fed with a mixture of 0–5% CH_4 diluted in hydrogen which is activated by either a 1200 W or a 6 kW, 2.45 GHz microwave generator. The averaged input power density is varied from 4.5 to 37 W cm^{-3}, by simultaneous variations of power and pressure respectively from 800 to 14000 and from 400 to 2400 W. The substrate temperature is adjusted, independently to the plasma parameters, from 700°C to 900°C.

2.2 Optical Emission Spectroscopy (OES)

A Jobin Yvon THR 1000 mounted with an 1800 groves per mm grating, blazed at 450 nm and equipped with a photomultiplier (Hamamatsu R 3896) and a red region intensified OMA III (EGG-Princeton Instrument-1460) was used to measure emission intensities of H_α (λ = 656.5 nm), H_β (λ = 486.1 nm) and of the 4p → 4s argon transition (λ = 750.3 nm) (2p1 → 1s2) when argon was introduced at 1% in the plasma as an actinometer [6]. Equipped with a photomultiplier

(Hamamatsu R 3896) and a Pelletier effect cooling system, the optical system allowed us (with spectrometer slits of 1.5 μm) to reach a resolution of around 5 pm, making possible measurements of Doppler broadening on H atoms during the radiative decay from the $n = 3$ to $n = 2$ fluorescence line (H_α) and then determination of H_α-atom temperature ($T_{H\alpha}$) [4,6,8]. Stark and pressure broadenings have been considered for determining $T_{H\alpha}$. The light emitted from the plasma was collected by an optical collimator and transported via a one millimeter in diameter optical fiber to the entrance slit of the monochromator. This device enables a spatial resolution in the plasma emissive cylinder of approximately 2 mm in diameter. The optical system was mounted in a computer controlled moving table, allowing axial and radial measurements. Emission intensities profiles, averaged on the line-of-sight, were measured at 90° to the axis of the reactor. An Abel Inversion procedure could be applied on these profiles in order to obtain emission intensities radial distributions.

2.3 Two Photon L.I.F. Technique (TALIF)

Two photons allowed transition laser induced fluorescence (TALIF) [4,6,7] measurements were conducted in the reactor equipped with with two U.V. grade silica windows allowing transmission of 205 nm light. The whole reactor was accurately translated vertically and horizontally with respect to the laser beam in order to obtain axial and radial profiles. The system used to generate the light at about 205 nm is published in Reference [7]. Under the experimental conditions, the observed two-photon line profile is Doppler broadening dominated. Its full-width at half-maximum is directly related to the translational temperature of the atoms, and its area to their concentration.

3. RESULTS

The H-atom temperatures (T_H) obtained by TALIF were seen to be equal to the gas temperatures, measured by CARS ($T_{RH2(X)}$) or calculated with the help of a 1 D diffusive H_2 plasma flow model developed in our laboratory [5,6]. $T_{H\alpha}$ measured by OES is also very close to the gas temperature, and, owing to the uncertainties, it can be considered as in equilibrium with the molecules. Microwave diamond deposition reactors are characterized by very high temperatures. They are very sensitive to the averaged power density, and increase from 1550 K

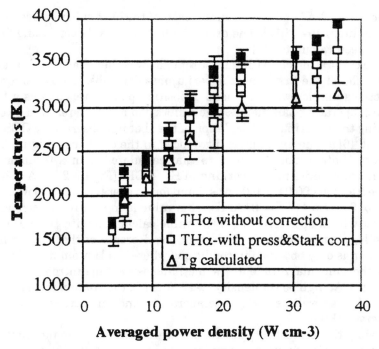

Figure 1. T_H (TALIF) and line-of-sight $T_{H\alpha}$ (OES) (both from Doppler broadening) measured in the bulk of the plasma, where there is no dependence of temperatures on the location.

to around 3200 K as this increases from 4.5 to 37 W cm^{-3} (Figure 1). As the percentage of methane increases up to 5%, T_H and $T_{H\alpha}$ remain constant at 2150 to 2200 K.

A theoretical analysis of the process of production of H atoms in the $n = 3$ excited state allowed us to conclude that actinometry can be used under conditions where the electron temperature is less than 20,000 K, and the H-atom mole fraction more than few percents [6]. However, correcting factors must be taken in consideration as the experimental conditions are changed (pressure and power). The quenching correcting factors have been determined through a comparison of measurements from TALIF and OES or through calculations [6]. We have estimated the quenching cross section of the excited 4p (2p1) argon state by molecular hydrogen at $42 \pm 30 \text{\AA}^2$, and found that H$_\alpha$-atom by molecular hydrogen in the literature (around 60 $\text{\AA}^2$) [10,11]. The quenching cross sections of excited 4p (2p1) argon state and of excited H(n = 3) state by atomic hydrogen have been estimated, at respectively 50 Å^2 and 18 Å^2, on the basis of the hard sphere model found in the literature [6,12]. The excitation rate constants correcting factor have been calculated [6].

The calibration of TALIF measurements was then made. It allowed us to reach, for the first time, the variations of the absolute H-atom mole fractions in microwave diamond deposition plasma reactors as a function of operating parameters. The calibration has been realized by comparing the experimental depletion of the H atoms observed by OES and TALIF [6] as the methane is introduced in the discharge [Figure 2(a)], at low H-atom mole fraction, to that calculated with a 0 dimension chemical kinetics model working in H_2 + CH_4 [Figure 2(b)]. A H-atom mole fraction of 0.04 to 0.05 was found for an averaged input power density of 9 W cm^{-3} (Tg = 2150 K). This value is in relatively good agreement (within 50% error) with the value calculated with the 1 D diffusive H_2 plasma flow (6%). Note that the decrease of the H-atom mole fraction is only observed in conditions where the H-atom mole fraction has the same magnitude as the methane percentage introduced in the feed gas. At high power density ($>$ 15 W cm^{-3}), the H-atom mole fraction is higher as well as the gas temperature and an increase instead is observed (Figure 2).

Using the correcting factors and calibration, we found that the H-atom mole fraction increases from 0.01 ± 0.005 to around 0.32 ± 0.15 as the power density increases from 6 W cm^{-3} to 30 W cm^{-3} (Figure 3). The experimental H-atom mole fraction is seen to be in relatively good agreement with that calculated with the 1 dimension diffusive H_2 plasma (where energy and mass balance equations are solved and the surface reactions considered), within the experimental error. The dissociation mechanism of molecular hydrogen is mainly attributed, at low power density ($<$6 W cm^{-3}), to the efficiency of the electron-molecule collisions (Te high, Tg moderated), while, at high power density ($>$20 W cm^{-3}), to the thermal processes efficiency (Te is lower, Tg is higher).

4. CONCLUSION

H-atom temperature and mole fraction, measured by TALIF, have been compared respectively to excited H$_\alpha$-atom temperature and emission intensities ratio (I_H/I_{Ar}) measured by OES. Both T_H and $T_{H\alpha}$ were seen to be equal to the gas temperature (Tg). They strongly increase as the power density is increased and tend towards a plateau after 25 W cm^{-3}. They remain constant as the percentage of methane is increased

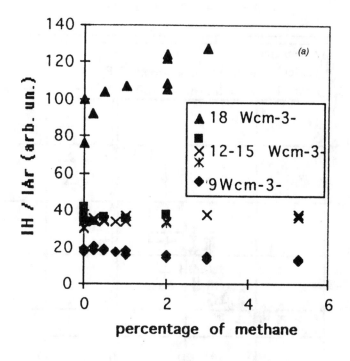

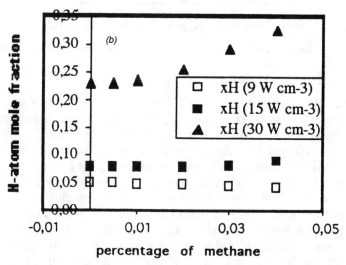

Figure 2. (a) Emission intensity ratio $I_{H\alpha}/I_{Ar}$ measured by actinometry and (b) calculated H-atom mole fraction as a function of the percentage of methane introduced in the discharge for two plasma conditions: 9 W/cm³: (600 W, 25 mbar) and 15 W cm⁻³:1000W, 52 hPa).

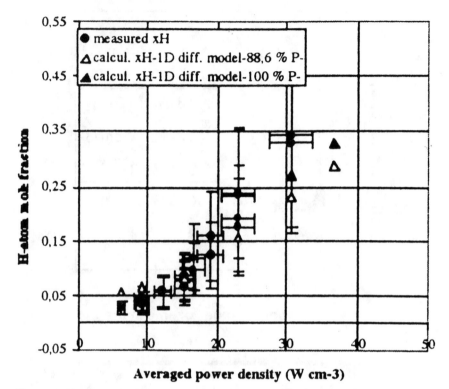

Figure 3. H-atom mole fraction as a function of the power density. Comparison with H-atom mole fraction calculated with a 1 D diffusive model.

at up to 5%, at least at relatively low power density (9 W cm⁻³). Absolute measurements of H-atom mole fraction have been obtained owing to a deep study involving TALIF, OES and calculations which allowed us to (1) validate the actinometry method under diamond deposition conditions and (2) calibrate our system. The H-atom mole fraction strongly increases with the averaged power density. Its variation as a function of the percentage of methane depends on its value relatively to the methane added, and on the gas temperature. For the lowest power densities tested (low Tg, low x_H), the H-atom mole fraction decreases as a function of the methane percentage, while it increases for power density higher than 15 W cm⁻³ (high Tg, high x_H).

ACKNOWLEDGEMENTS

This work was partially financed by DRET. The authors thank J. C.

Cubertafon and J. P. Booth for their contributions in TALIF measurements. A. Tserepi is thanked for her contribution in some OES experiments.

REFERENCES

1. Goodwin, D. G. 1993. *J. Appl. Phys.*, 74(11):6895–6906.
2. Connell, L. L., J. W. Fleming, H. N. Chu, D. J. Vestryck, E. Jansen and J. E. Butler. 1995. *J. Appl. Phys.*, 78(6):1–13.
3. Kovach, C. S. and J. C. Angus. 1996. "Generalized Reactor Model for Chemical Vapor Deposition of Diamond," submitted to *Diamond and Related Materials.*
4. Gicquel, A., K. Hassouni, S. Farhat, Y. Breton, C. D. Scott, M. Lefebvre and M. Pealat. 1994. *Diamond and Related Materials,* 3(4–6):581–586; A. Gicquel, M. Chenevier, K. Hassouni, Y. Breton, J. C. Cubertafon. 1996. *Diamond and Related Materials,* to be published; A. Gicquel, M. Chenevier, Y. Breton, M. Petiau, J. P. Booth and K. Hassouni. 1996. *J. Phys. III France,* 6:1167–1180.
5. Hassouni, K., S. Farhat, C. Scott and A. Gicquel. 1996. *J. Phys. III France,* 6:1229–1243.
6. Gicquel, A., K. Hassouni, M. Chenevier, C. Scott, S. Farhat and M. Lefebvre. 1996. "Modelling and Diagnostics in Plasmas," in *Handbook for Industrial Diamonds and Diamond Films,* Section 7.2, M. Prelas, G. Popovicci, K. Bigelow, eds., to be published.
7. Chenevier, M., J. C. Cubertafon, A. Campargue and J. P. Booth. 1994. *Diam. Rel. Mat.,* 3:587–592.
8. Ropcke, J. and A. Ohl. 1994. *Contributed Plasma Physics,* 34(4):575–586.
9. Coburn, J. W. and M. Chen. 1980. *J. Appl. Phys.,* 51(6):3134–3136.
10. Prepperneau, B. L., K. Pearce, A. Tserepi, E. Wurzberg and T. A. Miller. 1995. *Chemical Physics.*
11. Bittner, J., K. Kohse-Höinghaus and U. Meier. 1988. *Th. Just. Chem. Phys. Letters,* 143(6).
12. Kushner, M. J. 1988. *J. Appl. Phys.,* 63:2532.

A Three-Temperature Zero-Dimensional Model of a Hydrogen Microwave Plasma

CARL D. SCOTT,* SAMIR FARHAT, KHALED HASSOUNI
AND ALIX GICQUEL
Laboratoire d'Ingénierie des Matériaux et des Hautes Pressions
Université Paris Nord
93430 Villetaneuse, France

MARIO CAPITELLI
Dipartimento di Chimica
University of Bari
70126 Bari, Italy

ABSTRACT: A nonequilibrium plasma is numerically simulated in zero-dimension using a three-temperature model for a microwave powered plasma used for diamond film deposition. Two energy equations (electron and vibration) and seven species equations are solved with an approximate accounting for diffusion losses to a substrate surface. Simulations are run at conditions in which the size of the plasma is held constant while varying the pressure and microwave power. Results are given as function of power density and show that the difference in vibrational and translational temperature varies approximately as the inverse of the pressure squared. The electron temperature and electron mole fraction decrease, while the H-atom mole fraction increases with increasing power density. The prediction of H-atom mole fraction is in reasonable agreement with measurements. Also studied is the effect of a non-Maxwellian electron energy distribution function (EEDF), that shows that the H-atom mole fraction and the electron temperature is higher, whereas the electron mole fraction is lower, compared with a Maxwellian EEDF assumption. However, the difference in the H-atom predictions is within experimental ac-

---

*Permanent Address: 1014 Lynn Circle, Friendswood, TX 77546.

curacy, therefore one cannot use these data to validate the nonequilibrium EEDF.

KEY WORDS: hydrogen, plasma, nonequilibrium, modeling, CVD.

INTRODUCTION

The objective of simulating a microwave plasma is to predict operational characteristics with a view to improving performance and diamond film growth rate and quality. A phased approach has been taken, first to simulate the dominant species in a diamond film reactor—hydrogen. A schematic of hydrogen microwave plasma reactor is given in Figure 1. By first considering hydrogen we can verify the calculational techniques and then add hydrocarbons to the model in order to simulate specific diamond growth conditions. Since the electrons have a significant influence on the power absorbed by the plasma and possibly on the plasma chemistry we must include them in the simulation. At pressures considered in diamond microwave reactors the electrons are in strong nonequilibrium in the plasma. Likewise, depending on the conditions, vibrational energy may be decoupled from the translational modes. Some of the reaction rates depend on the distribution of vibrational energy. In principle, one could set up a system of Boltzmann equations to solve, but this would be very expensive in computational resources. Therefore, we have considered a model in which

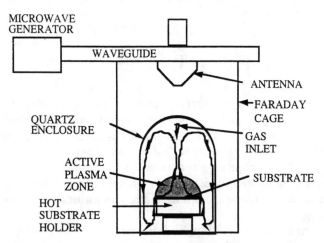

Figure 1. Diamond deposition apparatus showing microwave generated plasma ball adjacent to substrate.

each mode may be at different temperatures, but each energy mode is in Boltzmann distribution within itself. Energy flows from the microwave field to the electrons and then to the vibrational and translational modes. To assess the effect of a non-Boltzmann EEDF on the model results, a set solutions of the electron Boltzmann equation was computed [1] outside the present code. An effective electron temperature was found from the mean electron kinetic energy, $T_e = 2/3k \langle E\ f(E) \rangle$. The resulting EEDFs' influence are considered by recalculating rate and energy transfer coefficients, then with them, solving the equations that govern the model. The model has been described in detail in Scott et al. [2]. That paper describes how rates were determined from cross sections and describes the energy exchange terms in detail.

ANALYSIS

The model is represented by the following set of equations:

Species Mole Fractions

$$\frac{dy_i}{dt} = (\omega_i - \xi_i)\frac{W_i}{\varrho} \tag{1}$$

where ω_i = gas phase molar production rate of species i. The volume-equivalent loss rate is

$$\xi_i = \frac{S_w}{V_p}\sum_r \mu_{ir}c_i^w k_r \tag{2}$$

where S_w is the substrate surface area, V_p is the plasma volume, c_i^w and W_i are the wall molar concentration and molecular weight, respectively, of species i. k_r is the surface reaction rate coefficient calculated from the recombination probability γ and μ_{ir} are the stoichiometric coefficients for reactions r and species i. The wall molar concentration is determined from an estimate of the boundary layer thickness δ, the diffusive species flux, and the volume molar concentration c_i.

Measurements of the H-atom mole fraction and the gas temperature as functions of distance from the substrate were used to estimate δ. A constant value was assumed for all cases.

Vibrational Energy

$$\frac{de_v}{dt} = Q_{Tv} + Q_{ev}^{1v} + Q_{ev}$$ (3)

Electron Energy

$$\frac{de_e}{dt} = \alpha P_{MW} - (Q_{Te} + Q_{ev} + Q_{ev}^{1e} + Q_{ed}^{3} + Q_{ei})$$ (4)

where the Q terms on the right hand side of Equations (3) and (4) are the energy exchange terms between electrons (e), translation (T), and vibration (v). The latter two terms of Equation (4) refer to dissociation and ionization. The superscripts 1 and 3 refer to excitation of singlet and triplet states, respectively, of H_2. The terms on the right hand side of Equations (1)–(4) are shown explicitly in Reference [2]. The microwave power is P_{MW} and the fraction of power absorbed by the plasma is α. Equations (1)–(4) are solved for steady state using a time dependent relaxation scheme.

RESULTS

To determine trends in the behavior of the hydrogenic diamond film reactor with pressure and power, especially the power density, Equations (1)–(4) were solved for a set of cases corresponding to measurements of the gas temperature and atomic hydrogen density in the plasma. The wall temperature was assumed to be 1000 K. The recombination probability for ions was assumed to be unity, whereas, for H-atoms $\gamma = 0.1$ [3]. Cases were run assuming that 100% of the power is absorbed and that 88% of the power is absorbed. The latter efficiency was estimated from heat transfer considerations. The energy lost by the plasma must show up as heat transfer. Based on temperature and hydrogen atom gradient measurements, the heat transfer to the substrate was determined. It was assumed that this energy loss is provided by absorption of microwave energy by the plasma. The gas temperature was measured [4] using Doppler broadening of hydrogen lines. The H-atom mole fraction was measured using two photon allowed laser induced fluorescence (TALIF) and actinometry. Measurements were made at a distance from the substrate of about 2 cm. In previous measurements [2] and in calculations [5] it was estimated that the boundary layer thickness δ is approximately 1 to 2 cm. In these calculations we have used 1.3 cm because it seems to best represent the thickness as defined in Reference [2] and also in Reference [6]. The power and pressure were increased together so as to maintain a constant plasma radius of 2.5 cm. Solutions were obtained over a range of pressures of 1400 to 14,000 Pa for the Maxwellian EEDF assumption, and from 2500 to 11,000 Pa for the non-Maxwellian cases. Since the non-Maxwellian EEDF was obtained only for the 2500 Pa case the other cases are subject to somewhat more error due to possible differences in shape of the EEDF. However, it was found that the composition of the plasma had only a second order effect on the EEDF and the rates calculated from them. The measured gas temperatures used as input for the cases is shown in Figure 2. Solutions were obtained and the hydrogen atom mole fraction, known to be an important parameter in etching of graphite in diamond formation, is shown in Figure 3. We can see that there is fairly good agreement with experiments [4], especially in the overall trend with power density. We also see that the assumption of fraction of power absorbed and of the EEDF have a small effect, considering the experimental error. The abscissa P/V used in presenting the results is determined using the entire volume V of a sphere. It was experimentally observed that the temperature and atom fraction in the spherical ball of plasma discharge did not change appreciably when the substrate was moved into position, forming an *hemi*spherical ball. This may be due to a small difference in losses from the surface of the free plasma as compared with losses to the substrate surface.

The electron temperature is presented in Figure 4. It shows a trend toward lower temperature as the power density increases as expected, since there is more coupling of the electrons to the other energy modes at higher pressure. We can also see that the results of the non-Maxwellian EEDF are significantly higher than that of the Maxwellian EEDF. This probably results from the lower ionization fraction resulting from fewer high energy electrons that were seen in the non-Maxwellian EEDF. The electron mole fraction is given in Figure 5, where it is seen that the non-Maxwellian EEDF produces the fewest electrons. There is a significant reduction in ionization at the lower power density (88%) cases.

An interesting trend in the vibrational temperature is seen in Figure 6 where the difference $T_v - T$ is plotted. At $p = 2500$ Pa the difference is 100 K, which agrees with coherent anti-Stokes Raman spectroscopy, (CARS) measurements [2]. As expected, it can be seen

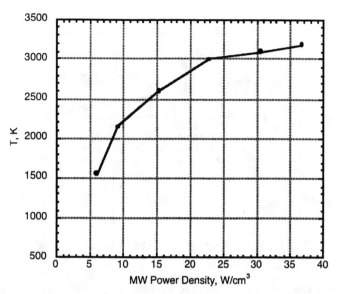

Figure 2. Measured gas temperature dependence on microwave power density in plasma for plasma of constant radius of 2.5 cm.

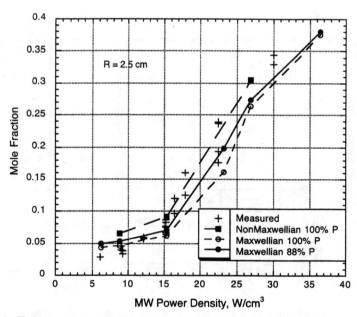

Figure 3. Hydrogen atom mole fraction measured and calculated dependence on power density for plasma of constant radius of 2.5 cm.

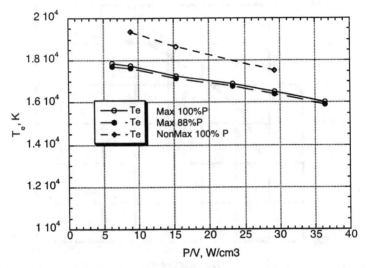

Figure 4. Calculated electron temperature dependence on microwave power density and EEDF for plasma of constant radius of 2.5 cm.

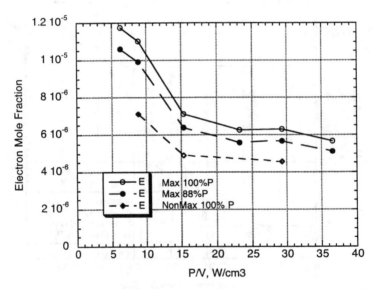

Figure 5. Calculated electron mole fraction dependence on microwave power density and EEDF for plasma of constant radius of 2.5 cm.

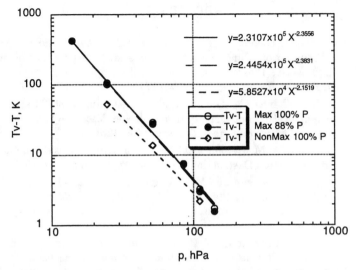

Figure 6. Difference in vibrational and gas temperatures as function of pressure for plasma of constant radius of 2.5 cm.

that the difference in temperatures is greatest at low pressure; but the difference decreases approximately as inverse of the pressure squared.

CONCLUSIONS

Zero-dimensional fluid species and energy equations for a three-temperature hydrogen plasma have been solved to find trends in the H_2 dissociation, ionization, electron temperature and vibrational temperature. The aim of the study is to advance this technique so that in future work hydrocarbons can be added and growth of diamond can be predicted and optimized. It was found that, given as parameters the measured gas temperature and absorbed microwave power, the vibrational and electron temperature equations coupled with the species equations, could be solved to obtain T_e, T_v, and mole fractions of all the species. The electron temperature follows a logical trend, decreasing with power density (increased pressure). The calculated hydrogen atom mole fraction agrees reasonably well with the measurements for both equilibrium and non equilibrium EEDFs, and within experimental error. Therefore, we are not able to establish whether the non-Maxwellian EEDF is more accurate for the cases studied. It was found that the ion fraction decreased with power density, whereas the atom fraction increased.

REFERENCES

1. Capitelli, M., G. Colonna, K. Hassouni and A. Gicquel. 1996. "Electron Energy Distribution Functions and Rate and Transport Coefficients of $H_2/H/CH_4$ Reactive Plasmas for Diamond Film Deposition," *Plasma Chemistry and Plasma Processing*, 16(2):153–171.

2. Scott, C. D., S. Farhat, A. Gicquel and K. Hassouni. 1996. "Determining Electron Temperature and Density in a Hydrogen Microwave Plasma," *Journal of Thermophysics and Heat Transfer*, 10(3):426–435.

3. Harris, S. J. and A. M. Weiner. 1993. "Reaction Kinetics on Diamond: Measurements of H-Atom Destruction Rates," *J. Appl. Physics*, 74:1022–1032.

4. Gicquel, A., M. Chenevier and M. Lefebvre. 1996. "Spatially Resolved Spectroscopic Analysis of the Plasma," in *Handbook for Industrial Diamonds and Diamond Films*, M. Prelas, G. Popovici and K. Bigelow, eds., Chapter 7-2, to be published.

5. Hassouni, K., C. D. Scott and S. Farhat. 1996. "Modeling of the Diffusional Transport of H_2 Plasma Obtained Under Diamond Deposition Discharge Conditions," in *Handbook for Industrial Diamonds and Diamond Films*, M. Prelas, G. Popovici and K. Bigelow, eds., Chapter 5, to be published.

6. Goodwin, D. G. 1993. "Scaling Laws for Diamond Chemical-Vapor Deposition. II. Atomic Hydrogen Transport," *J. Appl. Physics*, 74(11):6895–6906.

Modeling of Mass and Energy Transport in H₂ Plasma Obtained under Diamond Deposition Discharge Conditions

K. HASSOUNI,* S. FARHAT, C. D. SCOTT** AND A. GICQUEL

Laboratoire d'Ingenierie des Matériaux et des Hautes Pressions (LIMHP)
CNRS-Université Paris Nord
Av. J. B. Clément 93430
Villetaneuse, France

ABSTRACT: We present here a one-dimensional diffusion model which describes the species and energy transport in H₂ plasma obtained under moderate pressure discharge conditions. In this model, the H₂ plasma is considered as a three temperature thermochemically non-equilibrium flow with three different temperatures, T_g, T_v and T_e, and seven species undergoing 27 chemical reactions. The equations which govern the temperatures and the species densities were solved for several diamond deposition discharge conditions (power and pressure). The comparison with experimental measurements shows that the gas and H₂ vibration temperatures are well predicted by the model while a discrepancy was obtained for H-atom mole fraction. Results show that most of the microwave power absorbed by the electron is dissipated in the activation of H₂ molecules dissociation (58%) and energy transfer to heavy species translation-rotation and H₂-vibration modes (40%). They also show that the increase of the average microwave power density (MWP$_{av}$), in the range [9 W/cm³, 30 w/cm³] leads to the increase of the gas temperature and H-atom mole fraction and to the decrease of the electron temperature and ionization degree.

---

*Author to whom correspondence should be addressed.
**Permanent address: NASA Lyndon B. Johnson Space Center, EG3, Houston, TX 77058, USA.

1. INTRODUCTION

The knowledge of plasma flow characteristics is of prime interest for the control, the optimization and the scale-up of microwave plasma assisted diamond deposition reactors. Indeed, the growth rate and the quality of the deposited films are strongly dependent on the reactive species fluxes at the substrate surface. The estimation of these fluxes is also necessary for determining the growth and precursor species and understanding the diamond growth mechanism.

The species fluxes at the substrate surface are controlled by transport in the plasma (diffusion, convection and gas phase chemistry) and by surface reactions kinetics. Then, to estimate these fluxes one has to build a model taking both these phenomena into account.

Because of the numerous chemical species and reactions involved in H_2/CH_4 plasma, the modeling of species and energy transport in this plasma is quite complex. Furthermore, since H_2 represents the major part of the feed gas, the modeling of the deposition plasmas requires first a good description of pure H_2 plasmas obtained under diamond deposition condition.

2. THE TRANSPORT MODEL

2.1 Model Assumptions

Under diamond deposition discharge conditions, the heavy particles translation-rotation energy, H_2 vibration energy and electrons translation energy modes in an H_2 plasma are not in equilibrium and the major chemical species are H_2, H, H^+, H_2^+, H_3^+, H^- and e^- [1]. The transport of these chemical species and energy modes are coupled through the species chemical production rates which strongly depend on the different energy modes. As a consequence, the model presented here results in a description of the plasma by seven species transport equations coupled to three energy transport equations. It assumes that the distribution functions of the three energy modes are Maxwell-Boltzmann with three different temperatures (T_g, T_v and T_e).

2.2 Model Equations

The set of plasma mass and energy transport governing equations are summarized in the following subsections.

Species continuity equations

$$\frac{d}{dz}\left[\varrho \cdot D_s \frac{dx_s}{dz}\right] = W_s \qquad (s = 1,7) \tag{1}$$

ϱ is the plasma total mass density, x_s, D_s and W_s are the molar fraction, diffusion coefficient and chemical production rate of the species 's.' W_s was estimated from an H_2 plasma chemical kinetic model reported in Reference [1].

H_2 vibration energy equation

$$\frac{d}{dz}\left[\lambda_v \cdot \frac{dT_v}{dz}\right] - \frac{d}{dz}\left[\varrho \cdot D_{H2} \cdot E_{v-H2} \frac{dx_{H2}}{dz}\right] = -Q_{v-e} - Q_{v-t} \tag{2}$$

The two derivatives appearing in the left hand side (LHS) of Equation (2) are flux terms corresponding to the vibration energy conduction and to the diffusion of vibrationally excited H_2. The source terms Q_{v-e} and Q_{v-t} correspond to the rates of energy transferred from H_2 vibrational mode to the electron kinetic mode and to the heavy particles translation-rotation mode $\cdot$ λ_v is the vibration energy conduction coefficient and E_{v-H2} is the H_2 vibrational energy per unit mass.

Electron energy equation

$$\frac{d}{dz}\left[\lambda_e \cdot \frac{dT_e}{dz}\right] - \frac{d}{dz}\left[\varrho \cdot D_e \cdot h_e \frac{dx_e}{dz}\right] = \text{PMW} - Q_{e-v} - Q_{e-t} - Q_{e-chem} \tag{3}$$

The LHS of this equation accounts for the energy fluxes due to conduction transport and electron diffusion. The source terms, Q_{e-v} and Q_{e-t}, account for the energy transfer from the electron energy mode to the H_2 vibration energy mode and heavy particles translation-rotation mode. Q_{e-chem} is the rate of energy loss by chemical processes activation (dissociation of H_2 and ionization of H_2 and H). MWPD is the microwave power density absorbed by the electrons. The spatial distribution of MWPD is a model input parameter whose choice will be discussed in the next section. λ_e is the electron energy mode conduction coefficient and h_e is the electron enthalpy.

Total energy equation

$$\frac{d}{dz}\left[\lambda_e \cdot \frac{dT_e}{dz} + \lambda_v \cdot \frac{dT_v}{dz} + \lambda_{t-r} \cdot \frac{dT_g}{dz}\right] - \sum_S \frac{d}{dz}\left[\varrho \cdot D_s \cdot h_s \frac{dx_s}{dz}\right] = \text{PMW} - Q_{rad} \tag{4}$$

The fluxes term appearing in the LHS of Equation (4) correspond to the conduction of the three energy modes and to the diffusion of the enthalpic species. The source terms correspond to the microwave power absorbed by the electrons, MWPD, and to the energy loss by radiation, $Q_{rad} \cdot \lambda_{t-r}$ is the translation-rotation mode conduction coefficient and h_s the 's' species massic enthalpy.

The expressions of the transport coefficients and energy transfer source terms may be found in References [1] and [2].

3. SIMULATION PARAMETERS AND BOUNDARY CONDITIONS

The model described above was used to investigate the diamond deposition microwave plasma bell jar reactor described in Reference [3]. In this reactor the plasma volume is spherical in the absence of the substrate holder and almost hemispherical in the presence of this (Figure 1). The radial gradients of the species concentrations and the temperatures vanish at the plasma symmetry axis. As a consequence, the plasma transport on this axis may be described by a one dimensional model. The solution of the transport equations requires the boundary conditions for the species and temperatures at the inlet and at the substrate surface. It also requires the knowledge of the microwave power density axial distribution.

3.1 Boundary Conditions

In first approximation, at the inlet of the computation domain, out-

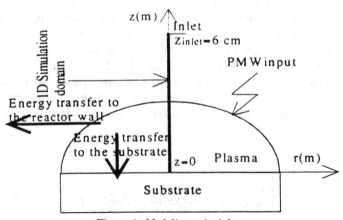

Figure 1. Modeling principle.

side of the discharge area, the boundary conditions are chosen by setting the gradients of all the variables to zero. The gas and H_2 vibration temperatures at the substrate surface are given by Coherent Antistokes Raman Spectroscopy (CARS) measurements near the surface [4]. The accommodation of the electron energy at the substrate is very weak [5], and a zero gradient of T_e is assumed at the surface. The species boundary conditions are derived from the surface reactions mechanism reported by Scott [1] with a recombination coefficient of 0.1 for atomic hydrogen [6].

3.2 Spatial Distribution of the Microwave Power Density

The axial distribution of the absorbed power is assumed to be the solution of a one dimensional wave equation where the wave is assumed to be injected at the top of the reactor and reflected back at the end of the cavity. The radial profile is assumed to be similar to the axial one and the plasma electrical permittivity constant over all the plasma volume. The absorbed power distribution is then given by Equation (5) which exhibits two parameters: P_0 and z_0. These parameters were estimated from optical emission spectroscopy and heat fluxes measurements [2].

$$\text{MWPD}(z,r) = P_0 \cdot \cos^2 [2\pi(z - z_0)/\lambda] \cdot \cos^2 [2\pi r/(\lambda^2/16 - z^2)^{1/2}] \quad (5)$$

4. RESULTS AND DISCUSSION

4.1 Comparison with Experiment

For the absorbed microwave power density axial profile presented in Figure 2 and corresponding to a total input power of 600 W and an average absorbed microwave power of 9 W/cm^3, the calculated electron temperature (Figure 2) is about 18000 K in the plasma bulk (0.5 cm < z < 2 cm) and increases in the vicinity of the substrate surface (0 < z < 0.3 cm). The calculated gas and H_2 vibration temperatures are nearly in equilibrium over all the plasma, the calculated axial profiles are in good agreement with those determined by CARS [4] (Tg_{exp} − Tg_{model} < 100 K). They show a sharp decrease of T_g from a value of 2000 K in the plasma bulk (z > 2 cm) to 1100 K at the substrate surface (Figure 3). The value of the thermal boundary layer thickness obtained from both the experiment and the model is about 2 cm. For the investigated discharge conditions, the calculated H-atom molar fraction in the plasma bulk is $6 \cdot 10^{-2}$. The calculated axial profile of the nor-

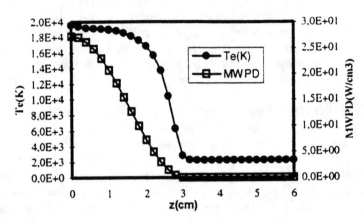

Figure 2. Axial profiles of the input microwave power density and the calculated electron temperature.

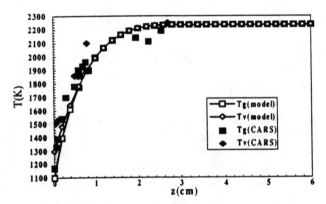

Figure 3. Calculated and measured axial profile of gas and vibration temperatures (MWP_{av} = 9 W/cm³).

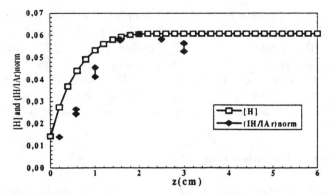

Figure 4. Calculated axial profile of H-atom mole fraction and measured axial profiles of the normalized IH/IAr (MWP_{av} = 9 W/cm³).

malized H-atom molar fraction: $[H]/[H]_{bulk}$, is given in Figure 4, where the experimental profile, obtained from actinometry measurements [4], of this ratio is also presented. The comparison between the measured and calculated profiles shows that the H-atom diffusion boundary layer thickness is well predicted by the model (δ = 2 cm). The calculated values of $[H]/[H]_{bulk}$ near the substrate (0.1 cm < z < 1 cm) are however much higher than those measured by actinometry (Figure 4).

The major ionic species are the electrons, H_3^+ and H^+ (Figure 5). The

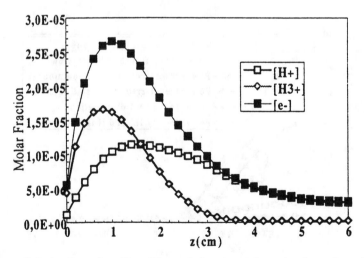

Figure 5. Calculated axial profiles of the mole fraction of the major charged species H^+, H_3^+ and e^-.

electron and H_3^+ molar fraction show similar evolution in the plasma, they rich maximum values of $1.8 \cdot 10^{-5}$ and $1.2 \cdot 10^{-5}$ at 10 mm and 8 mm from the substrate surface. The molar fraction of H^+ ion is nearly constant and is about $5 \cdot 10^{-6}$ in the plasma bulk (1 cm $< z <$ 3 cm) and decreases for $z <$ 1 cm. The H_2^+ and H^- ions, even very important for the kinetic modeling of this kind of plasma, has very low molar fractions which are characterized by maximum values of $5 \cdot 10^{-9}$ for H_2^+ and $2 \cdot 10^{-8}$ for H^-.

4.2 Energy Dissipation Channels

The total power densities gained and lost by the electron are approximately equal; the slight difference between these densities is due to the electron enthalpic flux gradient which also participates in the electron energy Equation (3). This weak difference shows that the electron energy may be well described by only considering source terms, this should lead to an important simplification of the mathematical problem, since the resulting electron energy equation is algebraic and not differential. Figure 6 shows that most of the energy gained by electron from the electromagnetic field is lost by activation of the H_2 dissociation which constitutes 58% of the total lost power at 1 cm from the substrate surface. The second channel for the electron energy dissipation consists of the transfer to the translational and vibrational modes of the heavy particles, e.g., heating of the gas (40%).

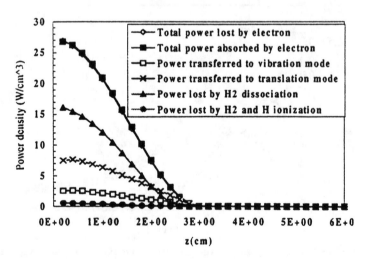

Figure 6. Power dissipation channels in the plasma (MWP$_{av}$ = 9 W/cm^3).

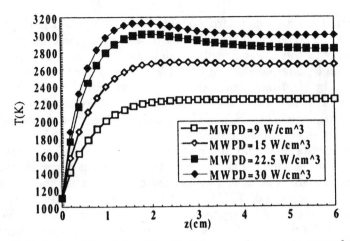

Figure 7. Evolution of T_g axial profile with the averaged microwave power density.

4.3 Effect of the Average Absorbed Microwave Power Density (MWP$_{av}$)

The increase of the average power density from 9 W/cm³ to 30 W/cm³ leads to a strong increase of the gas temperature which varies from 2300 K to 3100 K in the bulk of the plasma ($z > 3$ cm) (Figure 7). The variation of T_g with MWP$_{av}$ is non-linear and is attenuated at high power density ($\Delta T_g / \Delta$MWP$_{av}$ decreases with MWP$_{av}$).

The increase of T_g with MWP$_{av}$ results in a sharp increase of H-atom

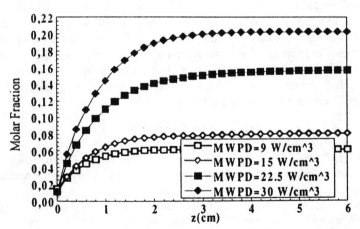

Figure 8. Evolution of H-atom mole fraction axial profile with the averaged microwave power density.

mole fraction which varies from 0.06 to 0.21 in the investigated domain of the averaged power density (Figure 8). The variation of H-atom mole fraction with the power density is also non-linear and is more important at high power density. At low MWP_{av} values the H-atom is mainly produced by electron-H_2 collision which is controlled by electron temperature. This is not the case at high MWP_{av}, where the dissociation of H_2 is mainly due to H_2-H_2 and H_2-H collisions which are controlled by gas temperature (thermal dissociation).

These results show that at high power density the energy transferred to the heavy species is dissipated in the activation of the H_2 thermal dissociation which leads to a higher increase of H-atom molar fraction and slower increase of T_g. They are in a good qualitative agreement with CARS and actinometry measurements. They also explain the high quality of the diamond films obtained under high power density discharge conditions, since the high H-atom concentration obtained for these conditions insures a high etching rate of the non-diamond phase.

5. CONCLUSIONS

The one-dimensional transport model presented here enables to estimate some key parameters for H_2 plasmas obtained under diamond deposition discharge conditions. The comparison of the model results and some experimental measurements shows that the gas and H_2 vibration temperatures are well predicted by the model. The discrepancy between the calculated and measured H-atom molar fraction show that the chemical kinetic modeling of the discharge must be improved by taking the non-Maxwellian behavior of the EEDF into account. The increase of the power density, while keeping constant the discharge volume, also strongly increases T_g and results in a higher H-atom molar fraction. The variations of T_g and [H] with this parameter are non-linear and the increase of H-atom mole fraction at high power is much more important. This may explain the high quality of the diamond films obtained at high power density.

REFERENCES

1. Scott, C. D., S. Farhat, A. Gicqual and K. Hassouni. 1996. "Determining Electron Temperature and Density in a Hydrogen Microwave Plasma," *Journal of Thermophysics and Heat Transfer*, 10:426.
2. Hassouni, K., C. D. Scott, S. Farhat and A. Gicquel. 1996. "Modeling of the Diffusional Transport of H_2 Plasma Obtained under Diamond Deposition Discharge Conditions," to be published in *Handbook for Industrial Diamonds and Diamond Films*, M. Prelas, G. Popovicci and K. Bigelow, eds.

3. Gicquel, A., E. Anger, M.-F. Ravet, D. Fabre, G. Scatena and Z.-Z. Wang. 1993. "Diamond Deposition in a Bell Jar Reactor: Influence of the Plasma and the Substrate Parameters on the Microstructure and the Growth Rate," *Diam. and Rel. Mat.*, 2:417–424.

4. Gicquel, A., K. Hassouni, S. Farhat, Y. Breton, C. D. Scott, M. Lefebvre and M. Péalat. 1994. "Spectroscopic Analysis and Chemical Kinetics Modeling of a Diamond Deposition Plasma Reactor," *Diamond and Related Materials,* 3:581–586.

5. Nishida, M. 1972. "Non Equilibrium Viscous Shock Layer in a Partially Ionized Gas," *Phys. Fluids,* 15:4.

6. Goodwin, D. G. 1993. "Scaling Laws for Diamond Chemical-Vapor Deposition. II. Atomic Hydrogen Transport," *J. Appl. Phys.,* 74(11):6895–6906.

On Estimation of Crystallization Temperature of Diamond Film

M. A. Prelas, S. K. Loyalka,* I. N. Ivchenko,
R. V. Tompson, G. Popovici and T. Sung
University of Missouri
Columbia, MO 65211, U.S.A.

B. V. Spitsyn, A. Alexenko and I. Galouchko
Institute of Physical Chemistry RAS
31 Leninsky Pr., 117915 Moscow, Russia

S. Khasawinah
Boston University
Boston, MA, U.S.A.

ABSTRACT: An analytical solution for the temperature difference between the substrate for diamond film regrowth and the support for the diamond CVD process has been formulated. The formula is valid for arbitrary gas pressure. Experimental measurements confirm the theoretical evaluation.

1. INTRODUCTION

The crystallization temperature of a growing diamond film is one of the most important parameters [1], because it specifies the growth rate of CVD diamond and quality of the film which depends on the presence of the intrinsic and impurity defects in the film.

Direct measurements of the crystallization temperature are difficult and moreover are connected with considerable errors. This problem

---

*Author to whom correspondence should be addressed.

originates from the very high temperature difference (more than one thousand degrees) of the gaseous crystallization medium relative to the surface of the crystallizing diamond film. That is why the temperature sensor should be located close to the growing diamond layer, inside the support of the substrate.

It is very important to note that there exists a gas gap between the planar surfaces of the support and substrate. The gas gap width depends on the specific construction of an experimental arrangement, and in some cases may be a controllable parameter. The temperature difference between the support and the diamond film may be evaluated if one considers the heat transfer process through the gas gap between the back side of the substrate for crystallization of the diamond film and the front side of support. Due to the comparatively high thermal conductivity of the substrate and of the diamond film relative to the gas, the back side substrate temperature can be nearly the same in comparison with the front side of the diamond film.

2. THE TEMPERATURE ESTIMATION

An analysis of the heat transfer problem between two planar and parallel surfaces has been carried out [2]. The analysis is based on solution of the Maxwell-Boltzmann moment system in the four moment approach. A new procedure to construct the moment system has been employed. For the first time it uses the Chapman-Enskog solution for thermal conductivity as a molecular property to construct the moment system that allows one to obtain analytical results for arbitrary gas pressure and arbitrary intermolecular potentials.

The temperature difference is found to be:

$$T_C - T_s = \frac{Q}{c \cdot P_o} \cdot \left(\frac{\pi \cdot m \cdot T_s}{2k} \right)^{1/2} \cdot (1 + 0.2716 \cdot c \cdot Kn^{-1}) \tag{1}$$

where $c = \alpha_T/(2 - \alpha_T)$.
Here the following notations are introduced:

T_C = diamond film crystallization temperature
T_s = support temperature
Q = net heat flux per unit time per unit surface area of the diamond film
α_T = energy accommodation coefficient
$Kn = \lambda/h$ = Knudsen number, where λ is the mean free path and h is the gap width

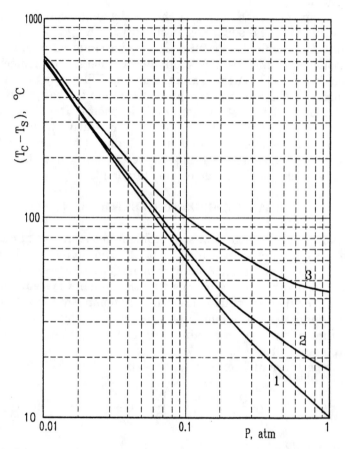

Figure 1. Substrate for diamond film crystallization (T_C) vs support (T_s) temperature difference in the dependence of gas (1% CH$_4$ plus H$_2$) pressure at different substrate to support flat gap values: 1–1 μm, 2–3 μm, 3–10 μm.

m = molecular mass of the gas

k = Boltzmann's constant

The CGS system is assumed to be used and molecules are assumed to be rigid spheres. The mean free path may be calculated as:

$$\lambda = \frac{k \cdot T_s}{P_o \cdot \pi\sqrt{2}\sigma^2}$$

where σ is the molecular diameter.

The heat flux density, Q, may be expressed in the form [1]:

$$Q = Q_{cond} + Q_{rad} + Q_{rec}$$

where Q_{cond} and Q_{rad} are thermal conduction and radiation energy fluxes from the activated crystallization medium (e.g., hot filament and surrounding gas phase) to the diamond film, and Q_{rec} is the energy flux owing to the recombination process of the atomic hydrogen. All of these energy fluxes may be evaluated in the specific experimental conditions. The analysis reported here allows one to control the diamond film temperature by means of very simple measurements of the support temperature.

Equation (1) permits one to calculate $T_C - T_S$, the temperature difference between the diamond film growing substrate and its support. For any diamond CVD, e.g., by HF technique, assuming, for example total heat flow $Q = 100$ W$\cdot$cm^{-2} and T_C of order 1200 K, one can obtain (Figure 1) the dependence of the temperature differences on gas phase (with contents of $\sim 1\%$ CH$_4$ plus $\sim 99\%$ H$_2$) pressure, at different substrate to support gaps (1, 3 and 10 μm).

With low gap values and high gas pressure the above mentioned temperature difference may be estimated to be on the order of several score degrees (Figure 1).

Experimental data using a hot filament diamond reactor demonstrates reasonable agreement of the formulation, Equation (1), with measured difference of $T_C - T_S$.

REFERENCES

1. Spitsyn, B. V. 1991. "The State of the Art in Studies of Diamond Synthesis from the Gaseous Phase and Some Unsolved Problems," in *Applications of Diamond Films and Related Materials*, Y. Tzeng, M. Yoshikawa, M. Murakawa, A. Feldman, eds., Elsevier Science Publishers B. V., 475–482.
2. Ivchenko, I. N., S. K. Loyalka and R. V. Tompson. In Press. *Fluid Dynamics*.

Diamond Deposition in Plasma Activated CVD Reactors. Two-Dimensional Modeling

Y. A. MANKELEVICH,* A. T. RAKHIMOV,
N. V. SUETIN AND S. V. KOSTYUK

Moscow State University
Nuclear Physics Institute
119899 Moscow, Russia

ABSTRACT: The general approaches to the elaboration of two-dimensional models of plasma activated chemical vapor deposition (PACVD) reactors are discussed. The models have been used to study the gas-phase and surface processes of diamond growth in continuous dc, pulsed dc and microwave discharge CVD reactors. The gas-phase and surface reaction mechanisms, the molecular diffusion and thermodiffusion, diffusional thermoeffects have been taken into account in transport equations that are numerically solved. The expressions for diamond growth and etching rates have been derived from surface kinetics.

Dependencies of the growth rate and species distribution from variations of dc discharge reactor parameters (total current, gas pressure, carbon content) are obtained. A possibility of diamond growth rate enhancement in a pulsed dc discharge CVD reactor is examined.

1. INTRODUCTION

A variety of gas discharges are used for diamond deposition by plasma activation of a reactive mixture [1,2,3]. For theoretical analysis of plasma-activated CVD reactors a correct description of coupled gas dynamics, electrodynamics and surface processes is necessary.

---

*Author to whom correspondence should be addressed.

Two-dimensional spatial modeling of existing reactors is required at least for reliable quantitative predictions.

In this report the description of the developed 2D models of continuous dc, pulsed dc and mw discharge CVD reactors is given and some calculated results are presented.

2. TWO-DIMENSIONAL MODELS OF PACVD REACTORS

We have used a general approach to development of 2D models for PACVD reactors [4,5]. The models consist of four blocks:

1. Transport block
2. Plasma-chemical kinetics block
3. Surface kinetics block
4. Electrodynamics block

that describe heat and mass transfer processes (block 1), neutral and plasma chemistry (block 2), gas-surface processes of diamond growth (block 3), and processes of plasma activation of a reactive mixture (block 4).

In the first and second block the set of conservation equations for mass, momentum, energy and species concentrations was numerically solved. The molecular diffusion and thermodiffusion (Soret effect), diffusional thermoeffects (Dufour effects) have been taken into account in these equations. Transport equations with the initial and boundary conditions, thermal and caloric equations of state are numerically integrated, until the steady state regime is reached. The detailed mathematical formulation of these equations in cylindrical coordinates is presented in Reference [4]. Neutral gas-phase chemistry and thermochemistry for a C/H/Ar mixture (15 components, 37 reversible reactions) are taken from the GRI-Mech 1.2 detailed reaction mechanism (see Reference 42 from [6]). Along with neutral chemistry electron-molecule, ion-molecule and electron-ion reactions are also considered. The ionization and dissociation rate coefficients, the H_2 vibrational excitation rate coefficients and other electron kinetic coefficients were determined by solving in the two-term approach the Boltzmann equation for the electron energy distribution function [5]. Analysis of plasma chemistry [7] shows that in the case under consideration main ions are CH_5^+ and H_3^+.

In the third block the reactions of hydrocarbons, atomic and molecular hydrogen with a solid [8–11] and surface migration of CH_2 group and H atom [10] are considered. Assuming that adsorption of CH_x ($x = 0,1,2,3$) and C_2H_2 on a dimer radical site [11] adjacent to a hydrogenated bridge site [11] is a limiting stage of diamond (100) growth, we have obtained the following formula for deposition rate G [μm/h]:

$$G = \frac{c1 \cdot \sqrt{T_{ns}} \cdot (CH_x + 0.05 \cdot C_2H_2)}{c2 \cdot e^{13.72/RT_s} + c3 \cdot e^{6.86/RT_s} + 1 + f(T_s, T_{ns}) \cdot H_2/H}$$

Here, CH_x, C_2H_2, H_2 and H are the concentration in cm^{-3} of the C_1-hydrocarbons, acetylene, atomic and molecular hydrogen just above the substrate, T_s and T_{ns} are the substrate temperature and the gas temperature near the substrate, $c1 = 7.68 \cdot 10^{-14}$, $c2 = 0.09$, $c3 = 0.6$, $c4 = 0.063$, $c5 = 0.21$, $R = 1.9873 \cdot 10^{-3}$ kcal/(mol·K), $f(T_s) = c4 \cdot \exp(-2/RT_s) + c5 \cdot \exp(-8.84/RT_s)$. To obtain the real growth rate the etching of C=CH_2 and CH_2 groups [12] should be taken into account. The estimation of the etching rate of CH_2 groups was derived from the mechanisms proposed in [11,12]: $G_{etch} = 6.5 \cdot 10^{16} \cdot \exp(-95.1/RT_s)$ [μm/h]. This rate becomes significant only for elevated substrate temperatures ($T_s > 1200$ K). According to the results from Reference [12], per-site rate of etching for C=CH_2 groups is one–two orders greater than etching rate for CH_2 groups.

The main effects of a mixture activation in a PACVD reactor assumed to be the gas heating by electric current (Joule heating) and the molecular hydrogen dissociation due to electron impact. In the electrodynamics block of a dc discharge reactor model the electron density and electric field distributions are determined from electron balance equation and the total current conservation condition [4]. The local field approximation approach is used and plasma electroneutrality is assumed. Electrode sheath effects (the gas excitation and ionization by cathode electron beam, beam relaxation) are considered singly using phase space particle motion scheme (PSPMS) for electron kinetic simulation [13]. It was found that for the conditions under study the cathode beam relaxation region is about 1 mm, and there is significant line radiation in this region.

As far as the mw discharge modeling is concerned, we have used the FDTD method [14] coupled with non-reflecting boundary conditions [15] to determine 2D distributions of electromagnetic fields inside the reactor with a fixed electron density profile. The self-consistent model, where Maxwell's equations are solved simultaneously with plasma-chemical kinetics and transport equations, is still under construction.

3. CALCULATION RESULTS FOR PACVD REACTORS

We have studied the dependencies of the deposition rate and species distribution from variations of the dc discharge reactor parameters (total current, gas pressure, carbon content). The modeled reactor consists of a chamber of 8 cm diameter with the cathode of 2.4 cm and the anode of 2.8 cm diameter with a gap of 2 cm between them. The range of the parameter values examined is as follows: total current 1.5–3 A, gas pressure 100–160 Torr, the methane mole fraction in the gas feed 1%–4%. The reactive mixture $1\%-4\%CH_4/1\%Ar/H_2$ is issuing from the hole system in the cathode. Argon, widely used as a tracer gas for actinometry, is included to study the effects of thermal diffusion. Possible effects of the heavy dilution of reactive mixture with noble gas are discussed in Reference [5].

The calculated results are summarized in Table 1, where the deposition rate (in $\mu m/h$) at the substrate center $G(0)$ and at the substrate periphery $G(R)$, the important species concentrations (in 10^{14} cm^{-3}) just above the substrate center ($r = 0$) and near the substrate edge ($r=R$, $R = 1.4$ cm is a substrate radius), and the maximal gas temperature (near the discharge center) are presented.

In the PACVD reactors, along with thermal dissociation of H_2 there are additional molecular hydrogen dissociation due to electron impact. The latter process, as our 2D calculations show, is an important source of H atoms for "soft" operating parameters when maximal gas temperature does not exceed 2900–3000 K. Thermal source of H atoms exceeds

Table 1. Table title.

Reactor Parameters	$G(0)$ $G(R)$	$H(0)$ $H(R)$	$CH_3(0)$ $CH_3(R)$	$C_2H_2(0)$ $C_2H_2(R)$	$CH_4(0)$ $CH_4(R)$	Maximal T_{gas}, K
P = 100 Torr 1% CH_4 I = 1.5A	1.45 0.48	4.20 1.95	0.44 0.26	1.40 1.00	31.7 47.4	2768
P = 100 Torr 1% CH_4 I = 3A	2.80 3.00	13.6 7.10	0.30 0.51	6.90 7.70	65.1 25.0	3291
P = 100 Torr 4% CH_4 I = 3A	6.52 6.49	13.0 6.50	0.52 1.02	19.8 21.0	11.2 48.1	3293
P = 160 Torr 1% CH_4 I = 1.5A	2.81 2.10	11.5 5.24	0.25 0.41	10.5 11.7	10.3 44.5	3072

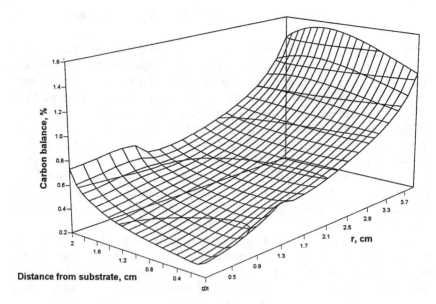

Figure 1. Carbon balance in a dc discharge CVD reactor.

plasma source in hot regions where the gas temperature $T_g > 2750–2850$ K. As it is seen from Table 1, the maximal gas temperature is increased with current and pressure that leads to the increase of atomic hydrogen concentration due to thermal dissociation. The higher is the H concentration, the deeper is the methane conversion in, mainly, CH_3 and C_2H_2. High level of the H, CH_3 and C_2H_2 concentrations results in the increase of the growth rate. It should be noted, however, that the extreme gas overheating leads to diminishing of the growth rates because of the thermodiffusion transfer of CH_x and C_2H_2 from hot discharge zone to the reactor walls. As seen in Figure 1, the calculated carbon balance (for the case P = 160 Torr, I = 1.5A, 1% CH_4) is significantly reduced in the hot region between the electrodes due to thermal diffusion to the cold walls and above the substrate due to carbon incorporation into the growing film. Similar reduction caused by thermodiffusion transfer is observed for the argon mole fraction. Minimal argon mole fraction is about 0.4% (50% of the inlet mole fraction) and is reached at the discharge center.

As far as gas-phase chemistry is concerned, the main chemical transformations occur in the hot reactive zone at the discharge center. For the case under study the most important reactions and reaction rates

$[10^{14}$ cm^{-3}s$^{-1}]$ in hot central discharge region and just above the substrate are the following:

1. Hydrogen dissociation and vibrational excitation

	At the Discharge Center		Near the Substrate	
	Forward	Reverse	Forward	Reverse
$H_2(v = 1) + H_2 \Rightarrow H_2 + H_2$	4.01e + 06	–	9.73E + 06	–
$H_2 + e \Rightarrow H_2(v = 1) + e$	4.01e + 06	–	9.94E + 06	–
$H_2 + e \Rightarrow \ldots \Rightarrow H + H + e$	3.29e + 04	–	6.09E + 04	–
$H + H + H_2 \Leftrightarrow H_2 + H_2$	9.00e + 03	4.06e + 05	4.05E + 01	8.57E – 02

2. Redistribution in CH_x $(x = 0$–$4)$ group: $CH_x \Leftrightarrow CH_y$

	At the Discharge Center		Near the Substrate	
	Forward	Reverse	Forward	Reverse
$H + CH_4 \Leftrightarrow CH_3 + H_2$	2.95e + 06	2.95e + 06	5.27E + 04	5.61E + 04
$CH + H_2 \Leftrightarrow H_3 + CH_2$	2.68e + 06	2.68e + 06	4.68E + 02	3.84E + 02
$CH_2 \text{ (singlet)} + H_2$ $\Leftrightarrow CH_3 + H$	8.20e + 05	8.21e + 05	8.97E + 02	7.22E + 02
$CH_2 + H_2 \Leftrightarrow CH_3 + H$	1.59e + 05	1.61e + 05	2.41E + 02	4.54E + 01
$H + CH \Leftrightarrow C + H_2$	1.32e + 05	1.32e + 05	1.86E + 00	6.88E + 00
$H + CH_2 \text{ (singlet)} \Leftrightarrow CH + H_2$	2.32e + 04	2.29e + 04	4.44E – 01	2.31E + 00
$H + CH_3 + M \Leftrightarrow CH_4 + M$	5.56e + 01	2.51e + 03	1.74E + 02	3.45E – 01

3. Redistribution in CH_2H_x $(x = 1$–$6)$ group: $C_2H_x \Leftrightarrow C_2H_y$

	At the Discharge Center		Near the Substrate	
	Forward	Reverse	Forward	Reverse
$C_2H + H_2 \Leftrightarrow H + C_2H_2$	6.99e + 06	6.99e + 06	8.37E + 02	7.46E + 02
$H + C_2H_4 \Leftrightarrow C_2H_3 + H_2$	6.67e + 04	6.56e + 04	3.19E + 02	4.36E + 02
$H + C_2H_3 \Leftrightarrow H_2 + C_2H_2$	1.27e + 04	1.15e + 04	2.76E + 01	3.28E – 01
$H + C_2H + M \Leftrightarrow C_2H_2 + M$	6.09e + 00	2.75e + 02	2.05E – 02	3.87E – 05
$H + C_2H_6 \Leftrightarrow C_2H_5 + H_2$	1.13e + 02	1.16e + 02	1.34E + 01	3.45E + 00
$H + C_2H_2 + M \Leftrightarrow C_2H_2 + M$	1.31e + 00	6.51e + 01	1.68E + 02	2.98E + 01
$C_2H_4 + M \Leftrightarrow H_2 + C_2H_2 + M$	4.44e + 01	8.76e – 01	1.98E – 01	1.52E + 00
$H + C_2H_3 + M \Leftrightarrow C_2H_4 + M$	2.71e – 01	1.24e + 01	8.91E – 02	1.38E – 04
$H + C_2H_5 \Leftrightarrow H_2 + C_2H_4$	9.12e + 00	1.57e + 00	1.61E – 01	1.89E – 04

4. Exchange between CH_x and C_2H_x groups: $CH_x <=> C_2H_y$

	At the Discharge Center		Near the Substrate	
	Forward	Reverse	Forward	Reverse
$CH_2 + CH_3 <=> H + C_2H_4$	1.10E + 03	1.89E + 02	2.31E + 00	1.87E − 03
$CH_3 + CH_3 <=> H + C_2H_5$	4.49E + 02	4.39E + 01	3.47E + 00	1.27E + 01
$CH + CH_4 <=> H + C_2H_4$	2.68E + 02	4.59E + 01	9.08E − 01	6.45E − 04
$C + CH_3 <=> H + C_2H_2$	1.92E + 02	2.93E + 01	6.09E − 02	1.78E − 07
$CH + CH_3 <=> H + C_2H_3$	1.37E + 02	2.30E + 01	1.13E − 02	1.03E − 05
$CH_2 + CH_2 <=> C_2H_2 + H_2$	1.06E + 02	1.64E + 01	9.69E − 03	2.41E − 08
$CH_3 + CH_3 + M <=> C_2H_6 + M$	6.60E − 02	2.87E + 00	1.16E + 01	3.50E − 01

As it is seen, for atomic hydrogen there is a significant local disbalance in loss/production terms that is compensated by diffusional transfer in both z and r directions.

Using the developed 2D model of a dc discharge reactor we have also studied diamond deposition processes in a pulsed dc discharge in a methane/hydrogen reactive mixture [5]. In a pulsed discharge, it is possible to reach higher electron temperatures than in a continuous dc discharge and, thereby, higher radical production rates. The time behavior of total current $I(t)$ is modeled as a series of rectangular pulses. During pulse period the gas temperature and hydrogen atom concentration in axial region are sharply increased. The high rate of H_2 dissociation by electron collision is established much faster than the rate of thermal dissociation. The increased concentration of H atoms induces the increase of CH_3 and diamond growth rate at the substrate center. The overall enhancement of growth rate depends on pulse width and the relationship between the pulse and pause period. As our calculations show, the optimal ratio of pulse period to pause time is about 0.2–0.4.

4. CONCLUSION

Self-consistent two-dimensional models based on transport equations of heat and mass transfer, neutral and plasma kinetics, coupled with electrodynamics block and surface kinetics are developed to study the diamond deposition processes in continuous dc and pulsed dc discharges. These models give the opportunity to describe adequately complex gas-phase and surface processes, to reveal key processes for optimization of the reactor parameters.

Simulation of a dc discharge CVD reactor with typical operational parameters (total current I = 1–3 A, gas pressure P = 100–160 Torr, E/N ~ 40 Td) shows:

- The growth rates are increased up to some level with total current and gas pressure because of the increase of the hydrogen thermal dissociation rate.
- Hydrogen dissociation due to electron impact is an important source of hydrogen atoms for "soft" discharge regimes when the maximal gas temperature is lower than 2900–3000 K.
- Significant reduction in the carbon balance and argon mole fractions, caused by thermal diffusion and carbon loss on the substrate (for carbon balance), was revealed in hot cylindrical region between electrodes. This effect reduces the growth rate, and, on the other hand, should be taken into account in actinometrical measurements with tracer gas.
- The most part of input power is absorbed by vibrational excitation of hydrogen molecules and then it is transferred to gas heating due to fast vibrational-translational relaxation.
- Diamond growth rate in a pulsed dc discharge CVD reactor is greater than in a continuous dc discharge for the same input energy. The overall enhancement depends on the relationship between the pulsed width and pause time.

ACKNOWLEDGEMENT

This work is supported, in part, by RFFI (Grant N-95-02-05184-a).

REFERENCES

1. Suzuki, K., A. Sawabe and T. Inuzuka. 1990. *Jpn. J. Appl. Phys.,* 29:153.
2. Satrapa, D., R. Haubner and B. Lux. 1991. *Surf. Coating Technol.,* 47:59.
3. McMaster, M. C., W. L. Hsu, M. E. Coltrin, D. S. Dandy and C. Fox. 1995. *Diamond and Rel. Mater.,* 4:1000.
4. Mankelevich, Y. A., A. T. Rakhimov and N. V. Suetin. 1995. *Diamond and Rel. Mater.,* 4:1065.
5. Mankelevich, Y. A., A. T. Rakhimov and N. V. Suetin. In press. *Diamond and Rel. Mater.,* 1996.
6. Yu, C.-L., C. Wang and M. Frenklach. 1995. *J. Phys. Chem.,* 99:14377.
7. Tahara, H., K. Minami and A. Murai, et al. 1995. *Jpn. J. Appl. Phys.,* 34:1972.
8. Harris, S. J. and D. G. Goodwin. 1993. *J. Phys. Chem.,* 97:23.

9. Krasnoperov, L. N., I. J. Kalinovski, H.-N. Chu and D. Gutman. 1993. *J. Phys. Chem.*, 97:11787.

10. Skokov, S., B. Weiner and M. Frenklach. 1994. *J. Phys. Chem.*, 98:7073.

11. Skokov, S., B. Weiner and M. Frenklach. 1995. *J. Phys. Chem.*, 99:5616.

12. Skokov, S., B. Weiner and M. Frenklach. 1995. *Elec. Soc. Proc.*, *95-4*, K. V. Ravi and J. D. Dismukes, eds., 41SDM Reno, p. 546.

13. Mankelevich, Y. A., A. T. Rakhimov and N. V. Suetin. 1991. *IEEE Trans. on Plasma Sci.*, 19:520.

14. Yee, K. S. 1966. *IEEE Trans. Antenas Propagat.*, AP-14:302.

15. Mur, G. 1981. *IEEE Trans. Electromagn. Compat.*, 23:377.

Detailed Modeling and Sensitivity Analysis of a Rotating Disk MOCVD Reactor

ALEXANDRE ERN*

CMAP-CNRS
Ecole Polytechnique
91128 Palaiseau cedex, France

(Received July 29, 1996)
(Accepted October 2, 1996)

ABSTRACT: We investigate computationally a rotating disk chemical vapor deposition reactor for the epitaxial growth of gallium arsenide. The model includes finite rate kinetics in the gas-phase and on the layer surface as well as detailed multicomponent transport algorithms. A converged numerical solution is obtained using an outer Newton's method for the gas-phase unknowns coupled to an inner Newton iteration yielding the surface species site fractions. A computationally efficient, Jacobian based method is used to perform a sensitivity analysis of both the gas-phase and the surface reaction mechanisms in order to investigate the rate limiting factors of the growth process.

KEY WORDS: sensitivity analysis, Newton's method, gas-phase chemistry, surface chemistry, multicomponent transport.

1. INTRODUCTION

The importance of metalorganic chemical vapor deposition (MOCVD) as a technique for growing thin epitaxial films of III-V compound semiconductors is widely appreciated. With the rapid progress in computer technology and numerical models achieved over the last decades, computational studies have become an extremely powerful tool for ana-

---

*Present address: CERMICS-ENPC, Cité Descartes, Champs sur Marne, 77455 Marne la Vallée cedex 2, France.

219

lyzing the growth process and also for reactor design purposes [1,2]. In the context of MOCVD modeling, two issues are particularly relevant. First, numerical models can provide valuable information on the link existing between the various operating parameters and product quality. Second, it is important to assess the dependence of the computed solution with respect to the wide range of parameters needed to describe the physical and chemical processes involved.

Sensitivity analysis can yield information about both questions in a mathematically elegant and computationally efficient way. In particular, when the problem is formulated as a set of coupled partial differential equations which are solved numerically using Newton's method, the Jacobian matrix appears as the fundamental link between model parameters and model predictions. This approach has already been applied to the analysis of model reaction-diffusion systems [3] and to laminar premixed flames [4]. For MOCVD systems which involve both gas-phase and surface reaction mechanisms, sensitivity analysis was only applied using direct techniques, where model parameters were varied one at a time and perturbed numerical solutions were sequentially computed in order to form the appropriate divided difference quotients [2,5]. When an extensive number of parameters has to be considered, as in MOCVD systems, this latter approach becomes extremely time consuming and the method based on Jacobian matrices is preferable [3]. It is thus one of the main goals of this work to generalize the methodology of References [3,4] to systems involving both gas-phase and surface reaction mechanisms and, subsequently, to perform a sensitivity analysis in order to identify the critical steps in the growth process.

In this paper we consider a rotating disk MOCVD reactor. We focus on the epitaxial growth of a gallium arsenide layer and consider the gas-phase and surface reaction mechanisms derived in Reference [5]. The rotating disk configuration is particularly attractive because it allows for high spatial uniformity of the grown layer [6]. In addition, the complexity of the governing conservation equations can be reduced substantially by seeking a similarity solution. In the next section we present the governing equations and the boundary conditions. The most relevant aspects of the gas-phase and surface reaction mechanisms are discussed in Section 3. The numerical solution techniques and the evaluation of sensitivity coefficients are subsequently presented in Sections 4 and 5, respectively. Finally, the numerical results are discussed in Section 6.

2. PROBLEM FORMULATION

We consider a MOCVD reactor in which a gallium arsenide substrate set on a heated susceptor is placed on a rotating disk. Since the growth rate of the epitaxial layer is of the order of a few micrometers per minute, the conservation equations for species mass, momentum and energy are written in steady-state form with the geometry of the reactor kept constant. In addition, the flow inside the reactor is assumed to be axisymmetric. Upon seeking a similarity solution, the set of conservation equations can be further reduced to a nonlinear two-point boundary value problem in the axial direction [6,7].

Let r and z denote the radial and axial coordinates, respectively, u, v, and w the radial, axial, and circumferential components of the velocity vector, respectively, φ the rotation rate, T the temperature, and Y_i the mass fraction of the ith species. Introducing a similarity solution of the form

$$u = rU(z), \quad v = v(z), \quad w = r\varphi W(z) \tag{1}$$

$$T = T(z), \quad Y_i = Y_i(z) \tag{2}$$

it is readily seen that the reduced pressure gradient in the radial direction is constant, i.e., we have

$$J = \frac{1}{r}\frac{\partial p}{\partial r} = \text{constant} \tag{3}$$

where p is the pressure. As a result, the equations expressing conservation of total mass, momentum, energy, and species mass take on the following form:

$$\frac{dV}{dz} + 2\rho U = 0 \tag{4}$$

$$\frac{d}{dz}\left(\eta\frac{dU}{dz}\right) - V\frac{dU}{dz} + \rho\varphi^2 W^2 - \rho U^2 - J = 0 \tag{5}$$

$$\frac{d}{dz}\left(\eta\frac{dW}{dz}\right) - V\frac{dW}{dz} - 2\rho UW = 0 \tag{6}$$

$$\frac{d}{dz}\left(\lambda\frac{dT}{dz}\right) - c_p V\frac{dT}{dz} - \sum_{i=1}^{n^{(g)}} \rho Y_i V_i c_{p,i}\frac{dT}{dz} - \sum_{i=1}^{n^{(g)}} h_i W_i \omega_i = 0 \tag{7}$$

$$-\frac{d}{dz}(\rho Y_i V_i) - V\frac{dY_i}{dz} + W_i\omega_i = 0, \quad 1 \le i \le n^{(g)} \tag{8}$$

with

$$\rho = \frac{pW}{RT}, \quad V = \rho v \tag{9}$$

In the above equations, ρ denotes the density, η the shear viscosity, λ the thermal conductivity, c_p the specific heat capacity of the mixture at constant pressure, $n^{(g)}$ the number of gas-phase species, V_i the diffusion velocity of the ith species in the axial direction, $c_{p,i}$ the specific heat capacity of the ith species at constant pressure, h_i the specific enthalpy of the ith species, W_i the molecular weight of the ith species, ω_i the molar production rate of the ith species due to gas-phase chemistry, W the mean molecular weight of the mixture, and R the ideal gas constant.

An important aspect in MOCVD modeling is to account for diffusion of precursor species in the gas-phase as a result of both temperature and species concentration gradients. From the kinetic theory of dilute polyatomic gas mixtures [8,9], the species diffusion velocities in the axial direction are given by

$$V_i = -\sum_{j=1}^{n^{(g)}} D_{ij}\frac{dX_j}{dz} - \theta_i\frac{1}{T}\frac{dT}{dz}, \quad 1 \le i \le n^{(g)} \tag{10}$$

Here, X_j, $1 \le j \le n^{(g)}$, denotes the mole fraction of the jth species, D_{ij}, $1 \le i,j \le n^{(g)}$, the species diffusion coefficients, and θ_i, $1 \le i \le n^{(g)}$, the thermal diffusion coefficients. A detailed investigation of the species diffusion velocities in MOCVD systems is presented in Reference [8] using the theoretical results derived in Reference [9]. For MOCVD systems involving reactant gases diluted in a carrier gas, the species diffusion coefficients can be evaluated accurately in the dilution limit discussed in Reference [8]. On the other hand, computationally efficient algorithms are used to evaluate the thermal diffusion coefficients [9]. Note, in particular, that the species diffusion velocities always satisfy the important mass conservation relation $\sum_{i=1}^{n^{(g)}} Y_i V_i = 0$.

The set of governing Equations (4)–(10) is closed with boundary conditions imposed at each end of the computational domain $0 \le z \le L$. More specifically, we have

Disk Surface ($z = 0$):

$$U = 0 \qquad (11)$$

$$V = 0 \qquad (12)$$

$$W = 1 \qquad (13)$$

$$\rho Y_i V_i = W_i \Omega_i^{(g)} - Y_i \sum_{j=1}^{n^{(g)}} W_i \Omega_j^{(g)}, \quad 1 \le i \le n^{(g)} \qquad (14)$$

$$T = T_s \qquad (15)$$

where $\Omega_i^{(g)}$ denotes the molar production rate of the ith gas-phase species due to surface chemistry and T_s the susceptor temperature.

Reactor Inlet ($z = L$):

$$U = 0 \qquad (16)$$

$$V = \rho_{in} v_{in} \qquad (17)$$

$$W = 0 \qquad (18)$$

$$Y_i = Y_{i,in}, \quad 1 \le i \le n^{(g)} \qquad (19)$$

$$T = T_{in} \qquad (20)$$

where the subscript *in* has been used to denote quantities specified at the reactor inlet. Since the axial velocity is specified at the reactor inlet, the conservation equations and boundary conditions form an eigenvalue problem for the constant quantity J. From a computational viewpoint, it is then convenient [7] to introduce the additional dummy equation

$$\frac{dJ}{dz} = 0 \qquad (21)$$

and to treat J as an additional unknown in the axial direction.

In order to evaluate the quantities ω_i and $\Omega_i^{(g)}$, we need to specify gas-phase and surface reaction mechanisms. This aspect of the model is discussed in the next section.

3. GAS-PHASE AND SURFACE CHEMISTRY

We consider a rotating disk reactor fed with an inlet mixture of trimethyl-gallium and arsine diluted in a hydrogen carrier gas. The goal of our model is twofold: first, to predict the growth rate of the gallium arsenide layer and, second, to predict carbon incorporation levels into the grown layer. To this purpose we consider gas-phase and surface reaction mechanisms discussed in Reference [5]. Model predictions obtained with these mechanisms have been compared to experimental results in References [5] and [2].

The gas-phase reaction mechanism includes $n^{(g)} = 15$ gas-phase species participating in the $m^{(g)} = 17$ reactions presented in Table 1. Reactions G1 and G2 involve methylated gallium species $Ga(CH_3)_x$, $x = 1$, 2, or 3, and correspond to the pyrolitic decomposition of trimethyl-gallium into dimethyl- and monomethyl-gallium. The released methyl radicals subsequently attack arsine and hydrogen molecules (reactions G3 and G4), leading, in particular, to formation of atomic hydrogen. Atomic hydrogen can abstract methyl radicals from the methylated gallium species (reactions G8 and G9). The last part of the mechanism (reactions G10–G17) involves the gallium-carbene species $Ga(CH_3)_2CH_2$, $GaCH_3CH_2$, and $GaCH_2$. These species are present in the reactor in very small concentra-

Table 1. Gas-phase reaction mechanism (gas-phase reaction rates are in $[mol/s/cm^3]$, E is in $[kcal/mol]$).

	Reaction	A	β	E
G1.	$Ga(CH_3)_3 \rightleftharpoons Ga(CH_3)_2 + CH_3$	3.5×10^{15}	0.0	59.50
G2.	$Ga(CH_3)_2 \rightleftharpoons GaCH_3 + CH_3$	8.7×10^{07}	0.0	35.40
G3.	$CH_3 + AsH_3 \rightleftharpoons AsH_2 + CH_4$	3.9×10^{10}	0.0	1.65
G4.	$CH_3 + H_2 \rightleftharpoons CH_4 + H$	2.9×10^{12}	3.1	8.71
G5.	$H + H + M \rightleftharpoons H_2 + M$	1.0×10^{16}	0.0	0.00
G6.	$CH_3 + H + M \rightleftharpoons CH_4 + M$	2.4×10^{22}	−1.0	0.00
G7.	$CH_3 + CH_3 \rightleftharpoons C_2H_6$	2.0×10^{13}	0.0	0.00
G8.	$Ga(CH_3)_3 + H \rightleftharpoons Ga(CH_3) + CH_4$	5.0×10^{13}	0.0	10.00
G9.	$Ga(CH_3)_2 + H \rightleftharpoons GaCH_3 + CH_4$	5.0×10^{13}	0.0	10.00
G10.	$Ga(CH_3)_3 + CH_3 \rightleftharpoons Ga(CH_3)_2CH_2 + CH_4$	2.0×10^{11}	0.0	10.00
G11.	$Ga(CH_3)_2CH_2 + H \rightleftharpoons Ga(CH_3)_3$	1.0×10^{14}	0.0	0.00
G12.	$Ga(CH_3)_2 + CH_3 \rightleftharpoons GaCH_3CH_2 + CH_4$	2.0×10^{11}	0.0	10.00
G13.	$GaCH_3CH_2 + H \rightleftharpoons Ga(CH_3)_2$	1.0×10^{14}	0.0	0.00
G14.	$GaCH_3 + CH_3 \rightleftharpoons GaCH_2 + CH_4$	2.0×10^{11}	0.0	10.00
G15.	$GaCH_2 + H \rightleftharpoons GaCH_3$	1.0×10^{14}	0.0	0.00
G16.	$Ga(CH_3)_2CH_2 \rightleftharpoons GaCH_3CH_2 + CH_3$	3.5×10^{15}	0.0	59.50
G17.	$GaCH_3CH_2 \rightleftharpoons GaCH_2 + CH_3$	8.7×10^{07}	0.0	35.40

tions and can be neglected without any effect on predicted growth rates. These species, however, are those responsible for carbon incorporation, as discussed below.

The molar production rates due to gas-phase chemistry are expressed as

$$\omega_i = \sum_{j=1}^{m^{(g)}} (\nu''_{ij} - \nu'_{ij}) q_j \tag{22}$$

where ν'_{ij} and ν''_{ij} are stoichimetric coefficients and where q_j is the rate of progress for the jth reaction given by

$$q_j = k_j(T) \left(\prod_{i=1}^{n^{(g)}} [\mathcal{X}_i]^{\nu_{ij}} - \frac{1}{K^e_j(T)} \prod_{i=1}^{n^{(g)}} [\mathcal{X}_i]^{\nu_{ij}} \right) \tag{23}$$

Here, $[\mathcal{X}_i] = \rho Y_i / W_i$ is the concentration of the ith gas-phase species with symbol $\mathcal{X}_i$. The equilibrium constant $K^e_j(T)$ is obtained from a thermo-dynamical data base [10] while the forward rate constant $k_j(T)$ is expressed as

$$k_j(T) = A_j T^{\beta_j} \exp(-E_j / RT), \quad 1 \le j \le m^{(g)} \tag{24}$$

with the parameters A_j, β_j, and E_j tabulated in Table 1. We have used the parameters given in Reference [5] except for reaction G4 where we have used the parameters given in Reference [11].

The present surface reaction mechanism has been derived for a (110) surface orientation, i.e., with an equal number of gallium and arsenic surface atoms at the surface [5]. We consider a total of $n^{(s)}=7$ surface species. The first three ($H^{(G)}$, $CH_3^{(G)}$, and $GaCH_3^{(G)}$) occupy gallium sites at the surface and the last four ($H^{(A)}$, $CH_3^{(A)}$, $AsH^{(A)}$, and $As^{(A)}$) occupy arsenic sites. In addition, we denote by S_G and S_A, respectively, a free gallium or arsenic site. Finally, two bulk species are included in the model, $GaAs^{(b)}$ and $GaC^{(b)}$, the latter being responsible for carbon incorporation into the epitaxial layer. The concentration of a surface species S_i is expressed as

$$[S_i] = (S_0 / \mathcal{N}) \sigma_i \tag{25}$$

where $\mathcal{N}$ is the Avogadro constant, $S_0 = 4.425 \times 10^{14}$ atom/cm^2 the concentration of surface atoms, and σ_i, $1 \le i \le n^{(s)}$, the site fraction of the ith

Table 2. Surface reaction mechanism (surface reaction rates are in
$[mol/s/cm^2]$, E is in $[kcal/mol]$).

	Reaction	A	E
S1.	$H + S_G \rightarrow H^{(G)}$	—	0.0
S2.	$H + S_A \rightarrow H^{(A)}$	—	0.0
S3.	$CH_3 + S_G \rightarrow CH_3^{(G)}$	—	0.0
S4.	$CH_3 + S_A \rightarrow CH_3^{(A)}$	—	0.0
S5.	$GaCH_3 + S_G \rightarrow GaCH_3^{(G)}$	—	0.0
S6.	$Ga(CH_3)_2 + S_G \rightarrow GaCH_3^{(G)}$	—	0.0
S7.	$Ga(CH_3)_3 + S_G \rightarrow GaCH_3^{(G)} + 2CH_3$	—	0.0
S8.	$GaCH_2 + S_G + S_A \rightarrow GaC^{(b)} + H_2$	—	0.0
S9.	$GaCH_3CH_2 + S_G + S_A \rightarrow GaC^{(b)} + CH_3 + H_2$	—	0.0
S10.	$Ga(CH_3)_2CH_2 + S_G + S_A \rightarrow GaC^{(b)} + 2CH_3 + H_2$	—	0.0
S11.	$AsH + S_A \rightarrow AsH^{(A)}$	—	0.0
S12.	$AsH_2 + S_A \rightarrow AsH^{(A)} + H$	—	0.0
S13.	$AsH_3 + S_A \rightarrow AsH^{(A)} + H_2$	—	0.0
S14.	$CH_3 + H^{(G)} \rightarrow CH_4 + S_G$	—	0.0
S15.	$CH_3 + H^{(A)} \rightarrow CH_4 + S_A$	—	0.0
S16.	$H + CH_3^{(G)} \rightarrow CH_4 + S_G$	—	0.0
S17.	$H + CH_3^{(A)} \rightarrow CH_4 + S_A$	—	0.0
S18.	$CH_3 + AsH^{(A)} \rightarrow A_s^{(A)} + CH_4$	—	20.0
S19.	$H^{(G)} + CH_3^{(A)} \rightarrow CH_4 + S_G + S_A$	1.0×10^{17}	10.0
S20.	$H^{(A)} + CH_3^{(G)} \rightarrow CH_4 + S_G + S_A$	1.0×10^{17}	10.0
S21.	$H^{(G)} + H^{(A)} \rightarrow H_2 + S_G + S_A$	1.2×10^{17}	20.0
S22.	$CH_3^{(G)} + CH_3^{(A)} \rightarrow C_2H_6 + S_G + S_A$	1.0×10^{17}	20.0
S23.	$GaCH_3^{(G)} + AsH^{(A)} \rightarrow GaAs^{(b)} + CH_4 + S_G + S_A$	5.0×10^{18}	29.3
S24.	$AsH^{(A)} + AsH^{(A)} \rightarrow As_2 + H_2 + 2S_A$	1.0×10^{17}	35.0
S25.	$As^{(A)} + As^{(A)} \rightarrow As_2 + 2S_A$	1.0×10^{18}	30.0
S26.	$GaCH_3^{(G)} + As^{(A)} \rightarrow GaAs^{(b)} + CH_3 + S_G + S_A$	5.0×10^{18}	20.0
S27.	$CH_3^{(G)} \rightarrow CH_3 + S_G$	1.0×10^{12}	20.0
S28.	$CH_3^{(A)} \rightarrow CH_3 + S_A$	1.0×10^{12}	20.0
S29.	$GaCH_3^{(G)} \rightarrow GaCH_3 + S_G$	1.0×10^{14}	40.0
S30.	$AsH^{(A)} \rightarrow AsH + S_A$	1.0×10^{14}	40.0

Table 3. Rate of progress q_k for the kth surface reaction.

Type	Form of Reaction	Rate of Progress
Adsorption	$\mathcal{X}_i$ + free site(s) → products	$f_i \exp(-E_k/RT)\mathcal{X}_i g(\sigma)$
Abstraction	$\mathcal{X}_i + S_j$ → products	$f_i \exp(-E_k/RT)\mathcal{X}_i \sigma_j$
Recombination	$S_i + S_j$ → products	$A_k \exp(-E_k/RT)[S_i][S_j]$
Desorption	S_i → products	$A_k \exp(-E_k/RT)[S_i]$

surface species. In addition, the site fraction of free sites directly results from the overall site balance equations

$$\sigma_G = 1 - \sum_{i=1}^{3}\sigma_i, \quad \sigma_A = 1 - \sum_{i=4}^{7}\sigma_i \qquad (26)$$

The surface mechanism presented in Table 2 consists of $m^{(s)} = 30$ elementary irreversible reactions, which can be split into adsorption (S1–S13), abstraction (S14–S18), surface recombination (S19–S26), and desorption (S27–S30) reactions. The molar production rates due to surface chemistry may be written

$$\Omega_i = \sum_{k=1}^{m^{(s)}} v_{ik}^{(s)} q_k \qquad (27)$$

where $v_{ik}^{(s)}$ is the stoichiometric coefficient of the ith chemical species in the kth surface reaction. The rate of progress q_k is given in Table 3, where we have set $f_i = p/(2\pi W_i RT)^{1/2}$. In addition, for single-site adsorption, we have either $g(\sigma) = \sigma_G$ or $g(\sigma) = \sigma_A$, depending on the site type involved, and for two-site adsorption, we have $g(\sigma) = \sigma_G \sigma_A$. Finally, the pre-exponential factor A_k and the activation energy E_k are given in Reference [5] and are tabulated for completeness in Table 2. For later convenience, we also introduce the vectors $\Omega^{(g)}$, $\Omega^{(s)}$, $\Omega^{(b)}$ denoting, respectively, the gas-phase, surface, and bulk species surface production rates. Thus, $\Omega^{(g)}$ has 15 components, $\Omega^{(s)}$ has 7 components, and $\Omega^{(b)}$ has 2 components.

Upon neglecting diffusion processes along the surface, mass conservation for the surface species yields

$$\Omega_i^{(s)} = 0, \quad 1 \le i \le n^{(s)} \qquad (28)$$

thus providing a system of $n^{(s)}$ nonlinear equations from which the site fractions can be obtained as implicit functions of the gas-phase mole fractions.

4. METHOD OF SOLUTION

Discretization of the governing equations and boundary conditions leads to a system of algebraic equations in the form

$$F(Z,\sigma) = 0 \qquad (29)$$

$$\Omega^{(s)}(Z,\sigma) = 0 \qquad (30)$$

where $Z = (Z_i)_{1 \leq i \leq n_Z}$, n_Z is the number of grid nodes in the axial direction, $Z_i = (U_i, V_i, W_i, T_i, J_i, Y_{i,1}, \ldots, Y_{i,n(g)})$ denotes the gas-phase unknowns at the ith node, and $\sigma = (\sigma_1, \ldots, \sigma_{n(s)})$ denotes the site fractions of the surface species. The residual F results from finite difference approximations of the conservation equations and boundary conditions, while $\Omega^{(s)}$ denotes the molar production rates of the surface species. Optimized fortran libraries are used in order to evaluate the terms associated with gas-phase chemistry and multicomponent transport in the governing equations [12,13].

A numerical solution of Equations (29)–(30) is sought using Newton's method. At each iteration the site fractions are first recovered from Equation (30) as an implicit function of the gas-phase unknowns. This step is performed numerically using an inner standard Newton iteration. Three to four iterations typically yield convergence to machine accuracy. As a result, for the gas-phase unknowns, we form a sequence of iterates Z^n given by the following damped Newton iteration:

$$J^n(Z^{n+1} - Z^n) = -\lambda^n F(Z^n, \sigma^n) \qquad (31)$$

with the site fractions σ^n satisfying

$$\Omega^{(s)}(Z^n, \sigma^n) = 0 \qquad (32)$$

In addition, λ_n is the nth damping parameter and J^n is the Schur complement of the block $\partial F/\partial Z$ in the full Jacobian matrix of Equations (29)–(30). Specifically, we thus have

$$J^n = \left[\frac{\partial F}{\partial Z} - \frac{\partial F}{\partial \sigma} \left(\frac{\partial \Omega^{(s)}}{\partial \sigma} \right)^{-1} \frac{\partial \Omega^{(s)}}{\partial Z} \right] (Z^n, \sigma^n) \qquad (33)$$

Convergence of the outer iteration (31) is achieved when the scaled norm of the update vector $Z^{n+1} - Z^n$ is reduced below a given tolerance, typically 10^{-9}.

Several numerical techniques are used in order to optimize the evaluation of the Jacobian matrix J^n. In particular, this matrix is evaluated numerically using vector function evaluations with transport and chemistry properties re-evaluated at the perturbed nodes only. Once the Jacobian is formed, Equation (31) is solved directly using an efficient block tridiagonal solver.

Adaptive gridding techniques are used in order to cluster grid nodes where the numerical solution presents steep variations. Weight functions based on the local gradient and curvature of a given numerical solution are equidistributed over the computational domain in order to generate a new grid. With this procedure, a sequence of coarse to fine grids is obtained, until the finest grid resolves the solution profiles adequately. In the present computations, the finest grid contains 125 nodes, since subsequent grid refinement has confirmed the grid independence of the numerical solution.

5. SENSITIVITY ANALYSIS

The use of Newton's method for the numerical solution of the governing equations provides a very efficient and elegant means to perform sensitivity analysis [3,4]. At the most basic level, first-order sensitivity coefficients are partial derivatives of the numerical solution with respect to some parameter. The nonlinear residual Equation (29) and the mass conservation relations for the surface species, Equation (30), are rewritten in the form

$$F(Z, \sigma; \alpha) = 0 \tag{34}$$

$$\Omega^{(s)}(Z, \sigma; \alpha) = 0 \tag{35}$$

where α denotes a given parameter. Differentiation of Equations (34)–(35) with respect to the parameter α yields

$$\frac{\partial F}{\partial Z}\frac{\partial Z}{\partial \alpha} + \frac{\partial F}{\partial \sigma}\frac{\partial \sigma}{\partial \alpha} + \frac{\partial F}{\partial \alpha} = 0 \tag{36}$$

$$\frac{\partial \Omega^{(s)}}{\partial Z}\frac{\partial Z}{\partial \alpha} + \frac{\partial \Omega^{(s)}}{\partial \sigma}\frac{\partial \sigma}{\partial \alpha} + \frac{\partial \Omega^{(s)}}{\partial \alpha} = 0 \tag{37}$$

Upon eliminating $\partial\sigma/\partial\alpha$, it is possible to express the quantity $\partial Z/\partial\alpha$ as the solution of a linear system

$$A \frac{\partial Z}{\partial \alpha} = -b \tag{38}$$

where the matrix A reads

$$A = \frac{\partial F}{\partial Z} - \frac{\partial F}{\partial \sigma} \left(\frac{\partial \Omega^{(s)}}{\partial \sigma} \right)^{-1} \frac{\partial \Omega^{(s)}}{\partial Z} \tag{39}$$

and the right member b

$$b = \frac{\partial F}{\partial \alpha} - \frac{\partial F}{\partial \sigma} \left(\frac{\partial \Omega^{(s)}}{\partial \sigma} \right)^{-1} \frac{\partial \Omega^{(s)}}{\partial \alpha} \tag{40}$$

It is readily seen that the matrix A is actually the Schur complement of the upper left block in the full Jacobian matrix, already considered in Equation (33). The first-order sensitivities $\partial Z/\partial \alpha$ are formed only when a converged numerical solution has been obtained on the finest grid. In practice, the last numerical Jacobian evaluated in Newton's method is used in Equation (38). For a perturbation analysis resulting from the use of an approximate Jacobian matrix, we refer to Reference [3].

In this work, we are interested in predicting the effects of the variation of the parameter α on the surface production rates of the bulk species $\Omega^{(b)}$. This leads to the following derived sensitivity coefficients:

$$\frac{d}{d\alpha} \Omega^{(b)}(Z, \sigma; \alpha) = \frac{\partial \Omega^{(b)}}{\partial Z} \frac{\partial Z}{\partial \alpha} + \frac{\partial \Omega^{(b)}}{\partial \sigma} \frac{\partial \sigma}{\partial \alpha} + \frac{\partial \Omega^{(b)}}{\partial \alpha}$$

$$= S_1(\alpha) \frac{\partial Z}{\partial \alpha} + S_2(\alpha) \tag{41}$$

Here, we have introduced the quantities

$$S_1(\alpha) = \frac{\partial \Omega^{(b)}}{\partial Z} - \frac{\partial \Omega^{(b)}}{\partial \sigma} \left(\frac{\partial \Omega^{(s)}}{\partial \sigma} \right)^{-1} \frac{\partial \Omega^{(s)}}{\partial Z} \tag{42}$$

and

$$S_2(\alpha) = \frac{\partial \Omega^{(b)}}{\partial \alpha} - \frac{\partial \Omega^{(b)}}{\partial \sigma} \left(\frac{\partial \Omega^{(s)}}{\partial \sigma} \right)^{-1} \frac{\partial \Omega^{(s)}}{\partial \alpha} \tag{43}$$

For one-dimensional problems, the Jacobian matrix is block tridiagonal. Its LU decomposition can thus be formed once and for all in a preprocessing stage before computing any sensitivity coefficients. In a second step, various parameters α are considered thus leading to various right members b and the linear systems (38) are then solved to obtain the various sensitivity coefficients. This approach is termed here the "differential" approach. Another approach in sensitivity analysis consists in first perturbing the parameters, then obtaining a new converged solution, and finally forming the appropriate divided difference quotients to evaluate the sensitivities. This approach is termed here the "direct" method. From the above discussion, it is clear that when a large number of sensitivity coefficients has to be evaluated, the differential method becomes much more computationally efficient. A detailed comparison of the computational costs of both methods is presented in Reference [3]. On the other hand, we point out that the differential method provides local information on the system since it evaluates partial derivatives, while the direct method may yield information of a more global type since significant changes in the parameters can be considered when forming the perturbed numerical solutions.

6. RESULTS AND DISCUSSION

In this section we investigate numerically the rotating disk MOCVD reactor described in the previous sections. The numerical results have been obtained on HP750 workstations with typical CPU times varying between one and ten minutes. The reactor is operated at a total pressure of one atmosphere and fed with trimethyl-gallium and arsine with partial pressure set to 1.8×10^{-4} atm and 3.3×10^{-3} atm, respectively. Carrier gas is hydrogen. The susceptor is set at a temperature of 1000 K and rotates at $\varphi = 600$ rpm. The inlet flow velocity is 25 cm/s and the inlet temperature of the fresh gases is 300 K.

In Figure 1 we first consider fluid flow and temperature predictions. The radial velocity peaks at a distance of about 0.3 cm away from the susceptor and then decreases to zero. The circumferential velocity goes to zero away from the susceptor at a much higher rate than the radial velocity. Finally, the temperature profile is almost linear, the thermal boundary layer thickness being on the order of 1 cm. Similar results have been obtained for other rotating disk reactors [6].

We next verify the validity of a widely used assumption in CVD modeling based upon neglecting the source terms $\Sigma_{i=1}^{n(g)} \rho Y_i V_i c_{p,i}(dT/dz)$ and $\Sigma_{i=1}^{n(g)} h_i W_i \omega_i$ in the energy Equation (7). This assumption along with the dilute approximation is often invoked in multidimensional models

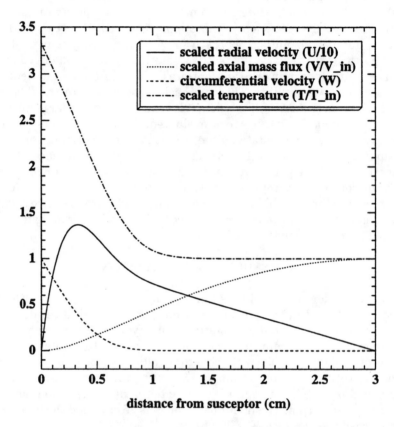

Figure 1. Scaled velocities (radial velocity: $U/10$; axial mass flux: V/V_{in}; circumferential velocity: W) and reduced temperature (T/T_{in}) profiles as a function of height above the susceptor. Susceptor temperature: 1000 K; rotation rate: 600 rpm; inlet axial velocity: 25 cm/s; inlet temperature: 300 K; partial pressure of trimethyl-gallium: 1.8×10^{-4} atm; partial pressure of arsine: 3.3×10^{-3} atm; total pressure: 1 atm.

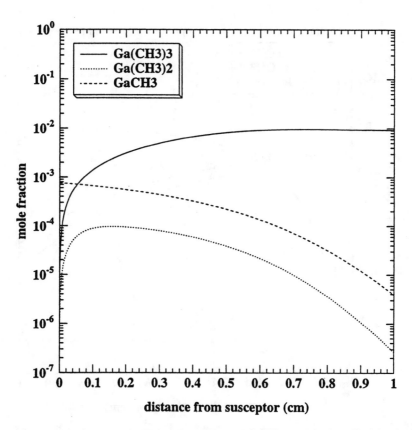

Figure 2. Mole fraction of Ga(CH₃)₃, Ga(CH₃)₂, and GaCH₃ as a function of height above the susceptor. Growth conditions as in Figure 1.

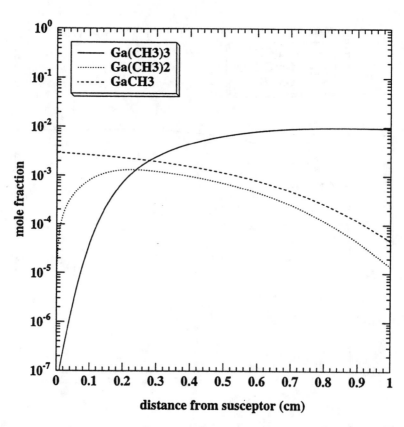

Figure 3. Mole fraction of Ga(CH₃)₃, Ga(CH₃)₂, and GaCH₃ as a function of height above the susceptor. Susceptor temperature: 1200 K. Other growth conditions as in Figure 1.

since it provides a means of uncoupling the fluid flow and heat transfer problem from chemistry aspects and thus reduces considerably the computational costs. For the present reactor, omitting the source terms in the energy equation yields relative errors on the predicted growth rates of less than one percent. The relative errors for carbon incorporation levels are even smaller.

The axial profile of the methylated gallium species is plotted in Figures 2 and 3 for susceptor temperatures of 1000 K and 1200 K, respectively. Larger amounts of monomethyl- and dimethyl-gallium are formed at higher temperatures due to the enhanced pyrolitic decomposition of trimethyl-gallium. Moreover, at high temperatures, much smaller amounts of trimethyl-gallium are present near the susceptor. Indeed, in a thin zone extending up to 1 mm above the susceptor, methylated gallium species are mostly present through monomethyl-gallium. This phenomenon is due to desorption of surface species $GaCH_3^{(G)}$ from the layer surface and becomes more significant at high temperatures. In Figure 4 we next consider the axial variations of the mole fraction for species AsH_3 and AsH at susceptor temperatures of 1000 K and 1200 K. The two other arsenic containing species, AsH_2 and As_2, are only present in trace amounts in the reactor. For both temperatures, AsH is the predominant arsenic containing species within a distance of 2.5 mm away from the susceptor. This phenomenon results from the fact that in the present model arsine adsorbs at the layer surface producing surface species $AsH^{(A)}$ (reaction S13) which, in turn, desorbs under the form of AsH (reaction S30).

In Figure 5 we consider the variations of the growth rate as a function of susceptor temperature. As a comparison we have also included the growth rate obtained by neglecting thermal diffusion. This comparison clearly shows the critical importance of detailed transport modeling for accurate growth rate predictions. Relative errors of up to 30 percent are obtained when omitting thermal diffusion in the numerical model. Predicted carbon incorporation levels are also sensitive to thermal diffusion. In this case, omitting thermal diffusion yields relative errors on the order of 10 percent. Figure 5 also illustrates the three different growth regimes arising in function of susceptor temperature: a regime controlled by surface kinetics at low temperatures (800–850 K), a regime controlled by diffusion of precursor species in the gas-phase at intermediate temperatures (850–1000 K) and a regime controlled by desorption of surface species at high temperatures (1000–1200 K). These observations are further supported by Figure 6 where the four largest site fractions at the layer surface are plotted as a function of susceptor temperature. It is clearly seen that at high temperatures most of the arsenic and gallium

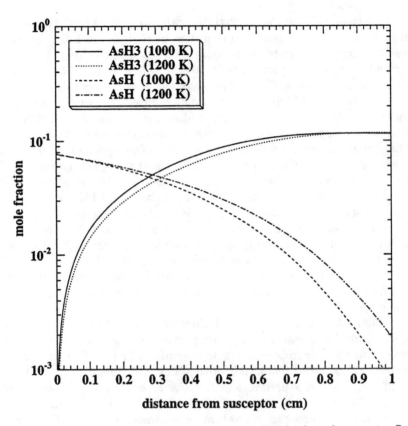

Figure 4. Mole fraction of AsH$_3$ and AsH as a function of height above the susceptor. Susceptor temperature: 1000 K and 1200 K. Other growth conditions as in Figure 1.

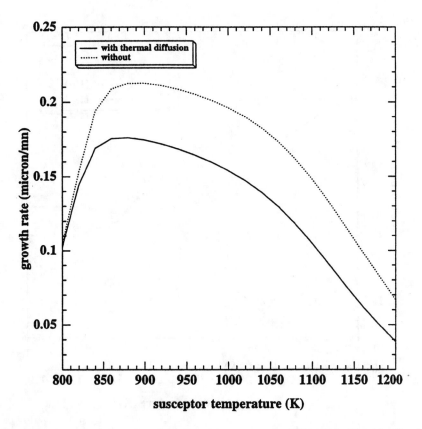

Figure 5. Growth rate as a function of susceptor temperature obtained with and without thermal diffusion included in the model.

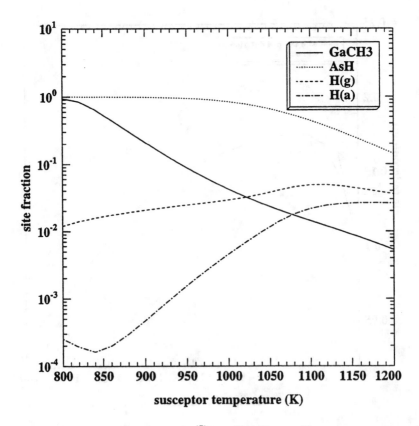

Figure 6. Site fraction for species $GaCH_3^{(G)}$, $AsH^{(A)}$, $H^{(G)}$ and $H^{(A)}$ as a function of susceptor temperature.

sites are actually free. The site fraction of the major surface species, $GaCH_3^{(G)}$ and $AsH^{(A)}$, decreases, respectively, by two orders and one order of magnitude between 800 and 1200 K.

In Figures 7 and 8 we consider the axial variation of the mole fraction of the gallium-carbene species. The susceptor temperature is 1000 K in Figure 7 and 1200 K in Figure 8. We restate that the gallium-carbene species, specifically $Ga(CH_3)_2CH_2$, $GaCH_3CH_2$ and $GaCH_2$, are those responsible for carbon incorporation into the epitaxial layer through two-site adsorption (reactions S8–S10 in Table 2). At a temperature of 1000 K, gallium-carbene species are present as $Ga(CH_3)_2CH_2$ and $GaCH_2$, the mole fraction of $GaCH_3CH_2$ being one order of magnitude smaller. At this temperature, $GaCH_2$ is mainly produced from species $GaCH_3$ through chemical reactions in the gas-phase (reactions G14 and G15). Species $GaCH_3$, in turn, results from desorption of $GaCH_3^{(G)}$ from the surface. At a temperature of 1200 K, $GaCH_2$ is the gallium-carbene species with the highest mole fraction in a region extending up to 3 mm above the susceptor, whereas $Ga(CH_3)_2CH_2$ is now present in smaller amounts.

We next investigate the influence of gas-phase and surface chemistry on model predictions by performing a sensitivity analysis. We first consider the gas-phase reaction mechanism presented in Table 1. A dummy multiplicative factor $\alpha_j = 1$ is introduced in the right member of Equation (23). Using the differential approach discussed in Section 5, we then evaluate the normalized sensitivity coefficients

$$s_j^{(g)} = \frac{\partial \ln(\Omega^{(b)})}{\partial \alpha_j} \tag{44}$$

These sensitivity coefficients have two components, one associated with $GaAs^{(b)}$ and one with $GaC^{(b)}$. For $GaAs^{(b)}$, the four most significant sensitivities are presented in Figure 9 as a function of susceptor temperature. We first notice that the sensitivities are rather small in magnitude except at high temperatures, where gas-phase chemistry has a stronger impact on the growth process. In particular, at high temperatures, the sensitivity associated with reactions G1, G8, and G9, which correspond to demethylation of trimethyl- and dimethyl-gallium, becomes insignificant.

For $GaC^{(b)}$, the five most significant sensitivities are presented in Figure 10 as a function of susceptor temperature. In this case, sensitivities tend to decrease in magnitude as the susceptor temperature is increased. Note also that the sensitivities are two orders of magnitude higher than those reported in Figure 9. Among the gas-phase reactions involving gallium-carbene species, the most sensitive ones are reactions G10, G14, and G15.

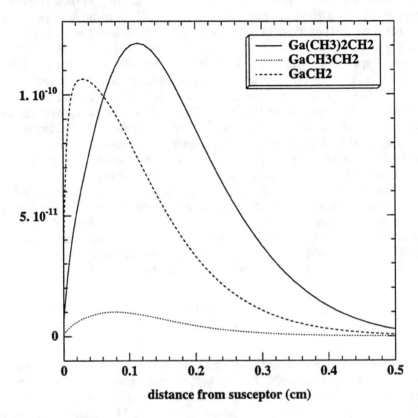

Figure 7. Mole fraction of Ga(CH$_3$)$_2$CH$_2$, GaCH$_3$CH$_2$, and GaCH$_2$ as a function of height above the susceptor. Growth conditions as in Figure 1.

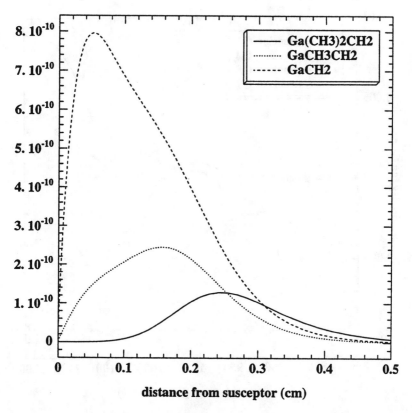

FIGURE 8. Mole fraction of $Ga(CH_3)_2CH_2$, $GaCH_3CH_2$, and $GaCH_2$ as a function of height above the susceptor. Susceptor temperature: 1200 K. Other growth conditions as in Figure 1.

241

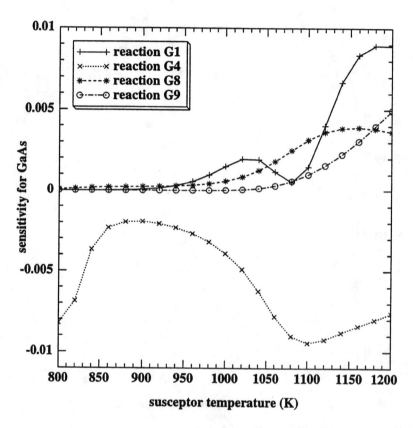

Figure 9. Normalized sensitivity coefficients for GaAs[b] with respect to gas-phase chemical reactions as a function of susceptor temperature.

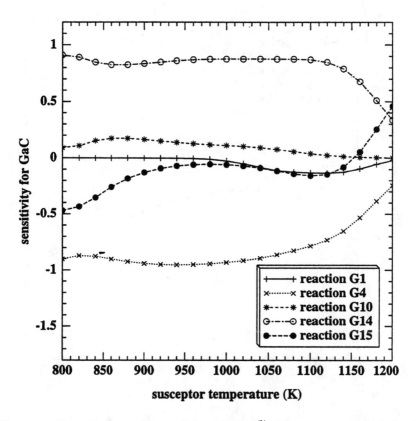

Figure 10. Normalized sensitivity coefficients for GaC[(b)] with respect to gas-phase chemical reactions as a function of susceptor temperature.

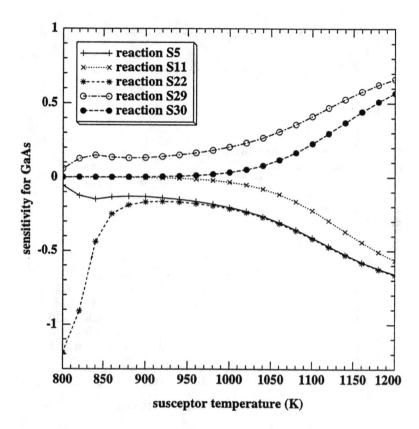

Figure 11. Normalized sensitivity coefficients for GaAs[b] with respect to surface chemical reactions as a function of susceptor temperature.

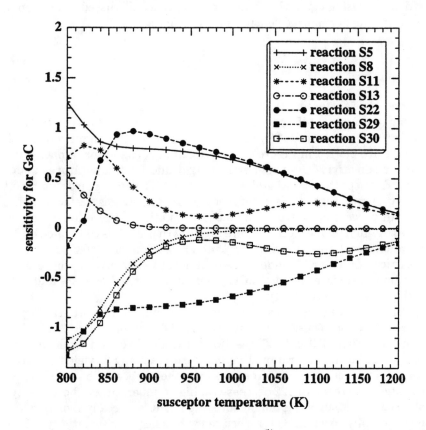

Figure 12. Normalized sensitivity coefficients for GaC[(b)] with respect to surface chemical reactions as a function of susceptor temperature.

The most significant is reaction G14 leading to formation of GaCH$_2$ from GaCH$_3$. The corresponding sensitivity is clearly positive, since reaction G14 increases the gallium-carbene pool near the surface.

We next consider the surface reaction mechanism presented in Table 2. We use the activation energy as a parameter to evaluate the sensitivity coefficients. Using again the differential approach discussed in Section 5, we form the normalized sensitivity coefficients

$$s_j^{(s)} = RT_{ref} \frac{\partial \ln(\Omega^{(b)})}{\partial E_j} \tag{45}$$

where T_{ref} is a reference temperature set to 1000 K. The sensitivity coefficients in Equation (45) have again two components, one associated with GaAs$^{(b)}$ and one with GaC$^{(b)}$. For GaAs$^{(b)}$, the five most significant sensitivities are plotted in Figure 11 as a function of susceptor temperature. All the sensitivities are higher in magnitude for temperatures above 1100 K except that associated with reaction S22 which is also significant at low temperatures. This latter result clearly illustrates the fact that the growth process is controlled by surface kinetics at these low temperatures. At high temperatures where the growth process is controlled by desorption reactions, we observe that adsorption of GaCH$_3$ and AsH (reactions S5 and S11) indeed yields a negative sensitivity whereas desorption of surface species GaCH$_3^{(G)}$ and AsH$^{(A)}$ (reactions S29 and S30) yields a positive sensitivity.

For GaC$^{(b)}$, the seven most significant sensitivities are plotted in Figure 12 as a function of susceptor temperature. Adsorption (S5, S11, and S13) and desorption (S29 and S30) reactions are found to have a significant impact on carbon incorporation. Furthermore, reaction S22 yields a positive sensitivity since it removes free sites from the surface. The sensitivity associated with reaction S22 decreases at high temperatures where the free site pool at the surface is much more important, as already shown in Figure 6. Finally, reaction S8 is associated with a negative sensitivity as is to be expected. Reaction S8 plays a more important role than reactions S9 and S10 due to the enhanced availability of species GaCH$_2$ near the surface.

7. CONCLUSIONS

In this paper we have presented a detailed computational model of a gallium arsenide CVD reactor with a rotating disk. Numerical results have been obtained over a wide range of susceptor temperatures covering three different growth regimes, controlled primarily by surface kinetics, diffusion of precursor species and desorption reactions from the surface, respectively. Our numerical results show the critical importance of accurate transport algorithms for model predictions. In addition, we have generalized to MOCVD systems the approach of References [3,4] in order to compute efficiently sensitivity coefficients. The new methodology has been applied to investigate the impact of both gas-phase and surface chemistry on the growth process of the gallium arsenide layer.

REFERENCES

1. Jensen, K. F., Fotiadis, D. I., and Mountziaris, T. J. 1991. "Detailed models of the MOVPE process," *J. Cryst. Growth*, 20:1–11.

2. Ern, A., Giovangigli, V., and Smooke, M. D. 1996. "Numerical study of a three-dimensional chemical vapor deposition reactor with detailed chemistry," *J. Comput. Phys.*, 126:21–39.

3. Reuven, Y., Smooke, M. D., and Rabitz, H. 1986. "Sensitivity analysis of boundary value problems: Application to nonlinear reaction-diffusion systems," *J. Comput. Phys.*, 64:27–55.

4. Smooke, M. D., Rabitz, H., Reuven, Y., and Dryer, F. L. 1988. "Application of sensitivity analysis to premixed hydrogen-air flames," *Combust. Sci. and Tech.*, 59:295–319.

5. Mountziaris, T. J. and Jensen, K. F. 1991. "Gas-phase and surface reaction mechanisms in MOCVD of GaAs with trimethyl-gallium and arsine," *J. Electrochem. Soc.*, 138:2426–2439.

6. Coltrin, M. E., Kee, R.J., and Evans, G. H. 1989. "A mathematical model of the fluid mechanics and gas-phase chemistry in a rotating disk chemical vapor deposition reactor," *J. Electrochem. Soc.*, 136:819–829.

7. Smooke, M. D. and Giovangigli, V. 1992. "A comparison between experimental measurements and numerical calculations of the structure of premixed rotating counterflow methane-air flames," 24th Symposium (International) on Combustion, The Combustion Institute, 161–168.

8. Ern, A. and Giovangigli, V. 1994. "On the evaluation of thermal diffusion coefficients in chemical vapor deposition processes," *J. Chem. Vapor Depos.*, 3:3–31.

9. Ern, A. and Giovangigli, V. 1994. *Multicomponent Transport Algorithms.* Heidelberg, Springer-Verlag.

10. Kee, R. J., Rupley, F. M., and Miller, J. A. 1989. "Chemkin-II: A fortran chemical kinetics package for the analysis of gas-phase chemical kinetics," Sandia Report, SAND89-8009.

11. Coltrin, M. E. and Kee, R. J. 1989. "A mathematical study of the gas-phase and surface chemistry in GaAs MOCVD," *Mat. Res. Symp. Proc.*, 145:119–124.

12. Giovangigli, V. and Darabiha, N. 1988. "Vector computers and complex chemistry combustion," in *Mathematical Modeling in Combustion and Related Topics*, C. Brauner and C. Schmidt-Lainé, eds., NATO Adv. Sci. Inst. Ser. E Vol. 140, Martinus Nijhoff Pub., Dordrecht.

13. Ern, A. and Giovangigli, V. 1996. "Optimized transport algorithms for flame codes," *Combust. Sci. and Tech.*, 118:387–395.

Calculation of Diamond Chemical Vapor Deposition Region in C-H-O, C-H-O-F and C-S-F Phase Diagrams

E. G. Rakov, A. V. Modin and A. V. Grunsky

D. I. Mendeleev University of Chemical Technology of Russia
Moscow A-47, 125047
Russia

INTRODUCTION

The composition of starting mixtures for diamond chemical vapor deposition, as well as the overall pressure and the temperature of gas and substrate plays an important role in the optimization of the deposition process. Bachman et al. have compiled a significant part of the presently known data in the literature and have developed the ternaries C-H-O [1], C-H-Cl [2] and C-H-F [2] indicating the boundaries of the diamond deposition region. Such diagrams remain the main tool for estimation of the concentration limits for diamond chemical vapor deposition of any mixtures containing hydrocarbons, oxygen, chlorinated or fluorinated hydrocarbons and halogen, as well as of mixtures containing any other agents of proper composition.

The extension of a calculation method for determination of deposition region boundaries is therefore of great importance. A constrained chemical equilibrium calculation was performed to examine the nature of the diamond-growing region on a ternary C-H-O diagram [3]. It could be shown that the calculated lower boundary of the diamond deposition region is close to the experimental boundary. Small deviations in elemental composition from the H_2-CO tie line gave rise to very large differences in the concentration of secondary species (C, C_n, C_nH_m) in the gas phase.

Another method of computation based on the estimation of the equilibrium yield of graphite and the assumption that the diamond deposition region lies within the graphite deposition region, is offered in Reference [4]. The calculation allows one to analyze the influence of substrate temperature, overall pressure, concentration of inert or active additives on the location of the lower boundary of the deposition region. The calculated data for the C-H-O system [4] are in satisfactory agreement with the most precise experimental values [5,6]. It was demonstrated that for some systems the mechanism of diamond deposition might include spontaneous growth-etch cycles due to fluctuation of gas component concentrations.

Calculations for the C-H-F and C-S-F systems by means of method [4] are reported in this work.

COMPARISION OF THE DATA IN THE C-H-F SYSTEM

According to the calculated data the lower boundary of the diamond deposition region approximately coincides with the HF-CF_4 tie line (point 0.5 on H-F side of the triangle and point 0.2 on F-C side). For example, if the initial H/C atomic ratio is equal to 1.0, the calculated F/C ratio is near 5.0; if H/C = 2.0 F/C = 6.0; if H/C = 4, F/C = 8.0; if H/C = 20, F/C = 24, etc. This means that the main products, apart from graphite, on the boundary of the deposition region are HF and CF_4, and that the final HF/CF_4 ratio strongly depends on the composition of the initial mixture.

It must be noted that changing the overall pressure by a factor of 100 (at an interval 100,000–1000 Pa) has no influence on the lower boundary of the deposition region.

Experimental data on diamond deposition in the ternary C-H-F system Reference [2] does not contradict the calculated values. However, the gas composition used in Reference [2] is far from the lower boundary of deposition region and lies along the H-CF_4 tie line.

A POSSIBLE MECHANISM OF DIAMOND DEPOSITION

The mechanism of diamond formation in the process of chemical vapor deposition, as was shown for the C-H-O system [4], may include growth-etch cycles. These cycles are the result of spontaneous fluctuation of gas component concentration. The existence of growth-etch cycles in other deposition systems can be proven by the appearance of marked shifts of

Table 1. The dependence of graphite yield (%) in the C-H-F system on the atomic ratio F/C and the substrate temperature (P = 101,325 Pa, H/C = 1).

F/C Ratio	T, K							
	900	1000	1200	1500	1700	1800	1900	2000
5.00	0	0	0	0	0	0	0	0
4.99	0.25	0.25	0.25	0.24	0.17	0.02	0	0
4.95	1.25	1.25	1.25	1.25	1.18	1.03	0.06	0

the temperature deposition boundary as a result of a small change of gas composition. The calculated data for the C-H-F system are presented in Table 1. One notes that a minor deviation of the F/C ratio results in a considerable change of the solid phase yield and in the shift of the deposition boundary to the higher temperatures. That means that local fluctuations of gas composition near the substrate play an important role during diamond formation.

THE INFLUENCE OF COMPOSITION DEVIATION FROM THE BOUNDARY ON CONCENTRATION OF SECONDARY SPECIES

Our method allows one to examine the equilibrium yield of graphite and yields of different gas species simultaneously. It was shown that the lower boundary of the diamond deposition region in the ternary diagram C-H-F at a C/H atomic ratio of 1.0, atmospheric pressure and the substrate temperature of 1500 K, lays at an F/C atomic ratio of 5.0. The main products in the vicinity of deposition on the boundary are CF_4 and HF

Table 2. Species yields vs. F/C atomic ratio (H:C = 1.0, T = 1500 K, P = 101,325 Pa)

F/C	Yield, mole/kg				
	H_2	F	CF_2	CF_3	C_2F_6
4.50	4.3×10^{-7}	3.1×10^{-5}	4.7×10^{-4}	3.5×10^{-4}	9.5×10^{-5}
4.80	3.8×10^{-7}	3.0×10^{-5}	4.7×10^{-4}	3.6×10^{-4}	9.9×10^{-5}
4.90	3.7×10^{-7}	3.0×10^{-5}	4.7×10^{-4}	3.6×10^{-4}	1.0×10^{-5}
4.95	3.6×10^{-7}	3.0×10^{-5}	4.7×10^{-4}	3.6×10^{-4}	1.0×10^{-5}
5.00	1.4×10^{-8}	1.5×10^{-4}	1.9×10^{-5}	7.3×10^{-5}	4.2×10^{-6}
5.05	1.6×10^{-15}	4.5×10^{-1}	2.1×10^{-12}	2.4×10^{-8}	4.5×10^{-13}
5.10	4.0×10^{-16}	8.9×10^{-1}	5.6×10^{-13}	1.2×10^{-8}	1.2×10^{-13}
5.20	1.0×10^{-16}	1.7	1.5×10^{-13}	6.5×10^{-9}	3.0×10^{-14}
5.50	1.9×10^{-17}	4.0	3.2×10^{-14}	2.9×10^{-9}	5.5×10^{-15}

(each 9.26 mole/kg of mixture). Changing the F/C ratio to 4.5 or to 5.5 has no significant influence on the yield of the main products.

At the same time the distribution of other (minor) gas species dramatically alters near the deposition boundary (Table 2). We can conclude that the behavior of the C-H-O and C-H-F systems is similar.

THE C-S-F SYSTEM

The possibility of diamond deposition from the reaction between CS_2 and F_2, without hydrogen, have been demonstrated earlier [7]. Our calculation shows that some new reactions may be used for diamond chemical vapor deposition in this system, e.g., CS_2 + SF_4 and SC_2 + SF_6. The lower boundary of the diamond deposition region at 750 K crosses the C_2S_2-F_2, CS_2-F_2, CS_2-SF_6 and CS_2-SF_4 tie lines at points corresponding to F/(F + C) ratios of 0.725, 0.65, 0.63 and 0.62, respectively.

CALCULATION OF COMBUSTION TEMPERATURES

In order to evaluate the possibility of the simplest way to produce diamond deposits, e.g., combustion flame synthesis, it is necessary to calculate the combustion temperatures. The program used in this work along with some additional operations, allows one to perform such calculations. The combustion temperatures for different mixtures on the C-H-O, C-H-F and C-S-F systems have been estimated. The preliminary results show that the adiabatic combustion temperature for CH_4-F_2 mixtures with F_2/CH_4 ratio of 4.0 is equal to 3557 K, and for CS_2-F_2 mixture with F_2/CS_2 ratio of 1.5 reach 3557 K, which is equal to 2569 K at an initial gas temperature of 500 K. These temperatures exceed the corresponding values for CH_4-O_2 and CS_2-O_2 mixtures. Due to a large consumption of elemental F for diamond deposition in the C-H-F system, the C-H-O-F system appears more appropriate.

ATTEMPTS AT ESTIMATING THE UPPER CONCENTRATION BOUNDARY OF THE DIAMOND DEPOSITION REGION

The upper boundary of the diamond stability region, as far as we know, has not been calculated before. We did try to estimate the position of this boundary in two ways:

1. By calculation of O/C or F/C ratio corresponding to the fixed non-zero yield of graphite

2. By the supposition that the gas composition fluctuations are equivalent to the fluctuations of temperature (in this case the upper boundaries are at an average temperature T_a coinciding with the lower boundary at the temperature $T_a + \Delta T$).

The first method did not work; the calculation using experimental data [5,6] shows that the graphite yield at the upper diamond deposition boundary diminishes from 12.8% for a CH_4-CO_2 mixture to 1.7% for a C_2H_2-CO_2 mixture.

The second method seems to be the more realistic one. Taking an experimental value for $T_a = 1173$ K [5,6] and $\Delta T = 200$ K (arbitrary value) we achieved satisfactory agreement of experimental and estimated values. The hydrocarbon content in the no-growth region, at the upper boundary in the diamond deposition region at the lower boundary, and in the non-diamond growth region for CH_4-CO_2 mixtures are equal to 47.7, 47.57 (calc.), 49.1, 49.99 (calc.), 56.3%; and for C_2H_2-CO_2 mixtures are equal to 32.0, 30.85 (calc.), 32.7, 33.35 (calc.), 33.4%, respectively. The uncertainty in the ΔT value does not permit the use of the method without experimental checking.

ACKNOWLEDGEMENT

This study was supported, in part, by the Russian Foundation of Fundamental Research, Grant No. 95-03-09176a.

REFERENCES

1. Bachmann, P. K., D. Leers and H. Lydtin.1991. *Diamond Relat. Mater.*, 1:1.
2. Bachmann, P. K., H.-J. Hagemann,H. Lade, et al. 1994. *Mater. Res. Soc. Symp. Proc.*, p. 339, *Mater. Res. Soc.*, p. 267.
3. Prijaya, N. A., J. C. Angus and P. K. Bachmann. 1993. *Diamond Relat. Mater.*, 3:129.
4. Rakov, E. G. 1996. Dokladi, Russian Academy of Sciences, in press.
5. Marinelli, M., E. Milani, M. Montouri, et al. 1994. *J. Appl. Phys.*, 76:5702.
6. Marinelli, M., E. Milani, M. Montouri, et al. 1994. *Appl. Phys. Lett.*, 65:2839.
7. Patterson, D. E. et al. 1991. In: "Applications of Diamond Films and Related Materials," Ed., B. Y. Tzeng, E. A. Elsevier, Amsterdam, Netherlands. p. 569.

Numerical Simulation of Generalities of Gas Dynamics and Chemical Species Diffusion in Plasma Jet CVD Reactor

G. G. GLADUSH AND N. B. RODIONOV
TRINITI
142092 Troitsk
Moscow Reg., Russia

INTRODUCTION

The gas dynamics and chemical species diffusion are general processes in plasma jet CVD reactors [1]. Therefore, the consideration of these processes can be of independent interest, as all chemical reactions occur in the background of these processes.

Specifically, the deposition rate is determined by carrier gas flux and chemical species diffusion on a substrate. In some reactors the gas jet impinges normally to substrate surface. It is the so-called stagnation point flow. This flow may be perspective for the enhancement of film homogeneity as the ideal stagnation point flow gives uniform boundary layer [2] and, consequently, uniform chemical species flux on the substrate if the chemical species density is uniform in carrier gas jet. In reality, the gas flow is usually far from being the ideal one, and it is interesting to know in what degree the useful property is conserved.

Gas dynamics and diffusion are important not only from the point of view of direct influence on deposition rate and uniformity of films. In

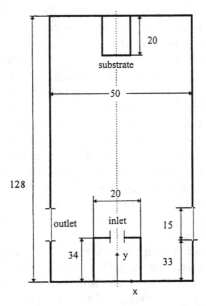

Figure 1. The outline of reactor for plasma deposition [1,3]. All sizes are in cm.

some cases it may use this influence not by direct manner but by the interaction with the lateral wall of the chamber. For example, Otorbaev et al. [3] assumed that quick recombination of expanding argon-hydrogen plasma jet can be explained by the peculiarities of gas dynamics in such reactor. The point is that in a reactor [3] the jet turns around near the substrate surface and goes out through the outlet holes on lateral walls (Figure 1). The hydrogen atoms associate in molecules on the chamber walls. These molecules may return in a jet due to circular flows and initiate there a quick dissociation recombination.

So, the goal of our work is to study the generalities of this gas flow, a mass transfer in the simplest case, when only two processes are taken into account: species transfer by gas flow, and diffusion. This research will be carried out for the reported experimental [3] conditions, where volume chemical reactions are not so important since the gas velocity is large and the pressure is not so large.

BASIC EQUATIONS

Reynolds and Peclet numbers are large in the conditions of these experiments [1]. Therefore, the viscosity and heat diffusivity must be important in thin boundary layers only. According to Prandtl hypothesis (see for example Reference [4]) the circulation flow, which arises due to viscosity, can be considered in the approximation of nonviscous fluid. The viscosity and heat conduction must not be important for mass transfer on the target and chamber walls too, as we consider the condition where Schmidt and Lewis number are small. We studied the subsonic flows. It is known that the flows are supersonic [1]. We assume that general velocity pattern depends weakly on shock wave structure, as shock wave sizes are very small. So, we solved the usual ideal gas 2D-system of equations with 2D-chemical species diffusion equation taking into account of convective mass transfer. The plane system of co-ordinate is used. The boundary conditions are the usual for these tasks. The velocity and chemical species concentration are zero on the hard walls. The velocity and chemical species profiles are determined at inlet and gas pressure is determined at outlet.

Our investigations were carried out in a wide region of parameters, extending that of other experiments [1,3]. It will allow the opportunity to look on the experimental situation from a more general point of view. The gas flow varied in a large range of values of gas velocity (1 m/s < V < 500 m/s) and gas pressure (1–100 mbar). The gas argon is hot, the sound velocity is 1000 m/s. Most of the calculations were carried out for two values of substrate width d_s = 10 and 20 cm, but in general d_s was changed from 5 to 40 cm.

FEATURES OF GAS DYNAMICS

The calculations showed that the global structure of gas flow is sensitive to chamber geometry: the place of outlet holes, the relation of chamber width d_c to its length L_c. Typical velocity pattern is shown in Figure 2 for geometry [1,3] (see Figure 1 d_s = 10 cm, jet width d_j = 8 cm, V_0 = 300 m/s). As is seen, the large scale curls arise in the chamber, and weak curls arise additionally in the regions behind the substrate. There are some weak curls inside large curls, which disappear at decreasing velocity or pressure.

Let us consider the velocity pattern in detail (Figure 3). It is seen, that the flow is similar ideal flow with stagnation point: region L_x, with the lateral velocity increasing linearly with the distance from the substrate center, exceeds the substrate size. Lateral velocity and velocity of return stream are about the same as inlet velocity. The velocity behind the substrate is ten times less than inlet velocity. It is worth noting that when the target size d_s decreases to 5 cm (i.e., when jet width is twice as much as target size), the inlet jet does not penetrate behind the substrate but drops here partially.

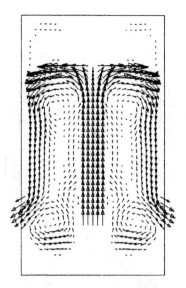

Figure 2. Field of velocities.

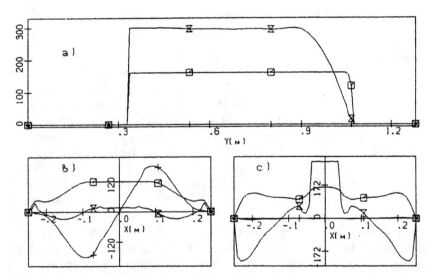

Figure 3. The profiles of velocity and chemical species density in reactor. The species profile in inlet is the rectangle, (a) at section $x = 0$, (b) at section near substrate, (c) at section $y = 0.06$ + lateral velocity (m/s), X—longitudinal velocity (m/s), □—chemical species density (dimensionless).

257

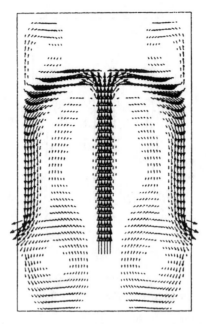

Figure 4. Field of velocities in wide chamber.

Interestingly, what is the size of region L_x, where $\partial V_y/\partial x \approx$ const., is determined? Our calculations showed that L_x depends very little on substrate width. The increase of width d_c to 1 m changes gas flow near the substrate not essentially, but the global gas flow changes qualitatively. In the wide chamber the additional curls arise with opposite circulation direction (Figure 4). The increase of jet width d_j yields increasing L_x linearly in such a wide chamber. The characteristic size of changing of longitudinal velocity L_y is proportional to the jet width also [Figure 3(a)]. Note that the theory of ideal flow with stagnation point yields $L_x = L_y$ [2]. In our conditions $L_x \approx L_y \approx 1.5\,d_j$. So, from the point of view of gas dynamics it is possible to obtain the thin film size, which is three times as much as jet width. Further, we will see whether or not it is possible when the chemical species diffusion is taken into account. At decreasing chamber width d_c the position of max(V_x) is weakly displaced to the substrate center. When $d_c <$ 33 cm gas flow is changed qualitatively. The main part of inlet jet turns around far from the substrate (Figure 5). Only a small part of gas jet impinges on the substrate surface with velocity value four times less than the inlet velocity. There is the same effect when increasing jet width. It is seen from Figure 5 that this effect takes place far from the direct interaction of inlet jet with return stream. It happens when jet width does not exceed the quarter of chamber width. So the lateral walls (or more exactly, the return stream) may influence the gas flow near the substrate.

DIFFUSION

A uniform distribution of chemical species in an ideal model of gas flow with a stagnation point solution of diffusion equation gives uniform flux on the substrate surface J [5]:

$$J = nV_0/\sqrt{Pe_d}, \qquad Pe_d = L_y V_0/D \gg 1 \qquad (1)$$

Applicability of Equation (1) to our more complex situation is seen from Figure 6. In Figure 6 the value of concentration n is chosen at the distance L_y from substrate. The simulations were carried out for different values of velocity, diffusion coefficient D ($0.003 < D < 0.3$ m²/s), jet width ($0.08 < d_j < 0.16$ m), chemical species distribution width ($0.04 < d_{SP} < 0.16$ m), substrate width, chamber width and stand off distance between the substrate and gas inlet. Our calculations show that maximum flux on substrate is independent of the four last parameters, if the chemical species density does not decrease by diffusion during the transport to a substrate. Here it is useful to discuss the following effect which seemed

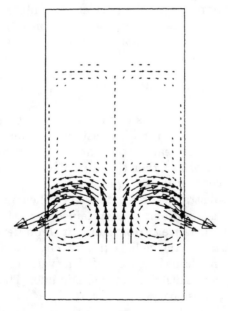

Figure 5. Field of velocities in narrow chamber.

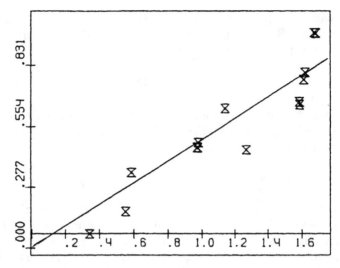

Figure 6. The maximum of the substrate flux vs. complex $nV_0/\sqrt{Pe_d}$.

to be strange. At increasing jet and species distribution widths the substrate flux width increases but maximum value of flux decreases. This decrease may be explained by Equation (1), where the value of L_y increases with increasing jet width.

It is more difficult to explain the generalities of flux distribution on the substrate surface. The theoretical consideration in framework of the ideal model gives for degree of nonuniformity $\Delta J/J = [J(0) - J(x)]/J(0)$ the following expression [6]:

$$\Delta J/J \sim xM/d_{SP}Pe_d, \qquad Pe_d \gg 1 \qquad (2)$$

where M is Mach number, x is distance from substrate center, d_{SP} is chemical species distribution width at distance L_y from substrate surface. When substrate is placed near inlet (distance between inlet and substrate is equal to the outlet width), then the value of $\Delta J/J(0)$ increases with decreasing d_{SP}, in accordance with Equation (2).

But, when the substrate is far from the inlet, then $\Delta J/J$ is constant for different species distributions. It is explained by the uniformity of species distribution at distance L_y from substrate for different values of d_{SP}. The uniformity is due to the global curls, which smooth the species density in the chamber volume [Figure 3(c)]. The value of this density is determined by the species distribution width in inlet. Therefore the maximum flux $J(0)$ depends on d_{SP}.

The substrate influences the target flux uniformity when its width is equal to or less than jet width. Although we remarked that target does not influence the global velocity pattern, in this case the boundary effect is present as the jet drops partly behind the substrate. For this chemical species flux increases slightly from center to the edge ($\approx 10\%$).

It is interesting to consider the target flux properties at increasing jet width, when the jet interacts with return stream. The gas velocity decreases four times and target flux decreases two times in correspondence with Formula (1). The degree of flux nonuniformity $\Delta J/J$ increases more than three times. This is explained by qualitative rebuilding of gas flow in the chamber.

ACKNOWLEDGEMENT

This work was supported by INTAS (ref. number INTAS-95-2922).

REFERENCES

1. Kroesen, G. M. M., D. C. Schram and J. C. M. de Hoag. 1990. *Plasma Chem., and Plasma Pros.*, 10:531.
2. Schlichting, H. 1968. *Boundary Layer Theory*, 6th Ed.,NY: McGraw-Hill.
3. Otorbaev, D. K., M. C. M. van de Sanden and D. C. Schram. 1995. *Plasma Sources Sci., Tech.*, 4:293.
4. Goldshtic, M. A. 1981. *Curl Flows*, Nauka, Novosibirsk.
5. Halstein, W. L. 1992. *Proc. Crystal Growth and Charact.*, V24:111–211.
6. Gladush, G. G. and A. A. Samokhin. 1994. ESCAMPIG, XIII, the Netherlands, August 23–26.

Keating-Harrison Model for Wide Band Gap Semiconductor Elastic Properties Description

S. YU. DAVYDOV* AND S. K. TIKHONOV

A. F. Ioffe Physico-Technical Institute
Russian Academy of Sciences
194021, Polytechnicheskaya str. 26
St. Petersburg, Russia

ABSTRACT: On the basis of the direct comparison of deformation schemes in Keating and Harrison models the expressions relating force constants to crystal covalence, covalent energy and nearest neighbor distance have been obtained. Also the numerical calculations of elastic coefficients for the cubic and hexagonal modifications of diamond, silicon carbide and nitrides of B, Al, and Ga have been performed.

Investigations of elastic characteristics of diamond and other wide band gap semiconductors are of great importance for growth processes as well as for the device design and fabrication. These investigations stimulate the creation of theoretical models that permit one not only to describe the experimental data available but also estimate in a simple straightforward manner the presently unknown values of different physical characteristics. Here we put forward the model which describes satisfactorily the elastic constants C_{ij} for diamond and silicone carbide and enables us to evaluate C_{ij} for the nitrides of B, Al, and Ga.

It is well known that the best results in theoretical studies of semiconductor elastic properties within the traditional force constants

---

*Author to whom correspondence should be addressed.

method are obtained with the Keating model [1,2]. In this model the elastic constants of cubic crystals are determined by the force constants of central α and non-central β nearest neighbor (n.n.) interaction:

$$C_{11} = \frac{\alpha + 3\beta}{4a_0}, \quad C_{12} = \frac{\alpha - \beta}{4a_0}, \quad C_{44} = \frac{\alpha\beta}{a_0(\alpha + \beta)} \tag{1}$$

where $4a_0$ is a lattice constant. It follows from Equation (1) that the parameter

$$R = \frac{C_{44}(C_{11} + C_{12})}{C_S(C_{11} + 3C_{12})} \tag{2}$$

is equal to unity which is just the case for the experimental values of C_{ij} for diamond and SiC (see Reference [3]). However, to obtain α and β one has to use the experimental values of C_{ij} for corresponding crystal. But for prediction of the elastic properties we must use a model based on general principles. The simple quantum-mechanical approach to this problem was proposed by Harrison [4] in terms of bond orbital model (BOM). In our previous paper [5] we have used BOM for the calculation of wide band gap semiconductors elastic moduli. Our results (see Table 1) show, however, that the equality $R = 1$ is satisfied only with the accuracy of the order of 15%. For this reason it seems useful to combine BOM with the Keating model.

By the direct comparison of deformation schemes of Keating and Harrison one can find

$$\alpha = \frac{8}{3}\frac{\alpha_c^3}{d^2} |V_2| \tag{3}$$

$$\beta = \lambda \frac{\alpha_c^3}{d^2} |V_2| \tag{4}$$

Here α_c is a covalency, V_2, covalent energy, $\lambda = 0.85$ is the constant which describes the dependence of V_2 on the angle between $|sp^3\rangle$ orbitals, $d = a_0/\sqrt{3}$ is the n.n. distance. We must emphasize here that the non-central interactions in Keating and Harrison models differs to some extent. In the former one β is related with the variation of the angle between $|sp^3\rangle$ orbitals which are centered on the same atom. In the Harrison model the non-central interaction is associated with the mutual arrangement of $|sp^3\rangle$ orbitals centered on adjacent atoms.

Table 1. Elastic constants C_{ij} of cubic modification (in 10^{10} Pa).

		C	SiC	BN	AlN	GaN
Harrison model	C_{11}	96.9	32.7	72.7	18.7	15.6
	C_{12}	30.1	10.1	24.5	5.7	4.8
	C_{44}	42.6	14.3	31.8	8.1	6.8
	R	0.87	0.86	0.84	0.85	0.87
Keating-Harrison	C_{11}	102.7	34.6	78.0	19.8	16.5
model	C_{12}	35.9	12.0	26.8	6.8	5.7
	C_{44}	50.7	17.0	37.8	9.6	8.1
	R	1.00	1.00	1.00	1.00	1.00
Experiment	C_{11}	107.9	41.1	—	—	—
	C_{12}	12.4	16.4	—	—	—
	C_{44}	57.8	19.4	—	—	—
	R	1.00	1.00	—	—	—

Note also, that the Keating theory describes only homopolar semiconductors. The model for heteropolar compounds $A_N B_{8-N}$ was elaborated in [6]. As in Reference [6] we use the same force constant β both for A and B atoms but contrary to Reference [6] we omit the interionic Coulomb interaction due to very high covalency of wide band gap semiconductors.

Substituting Equations (3) and (4) in Equation (1) we obtain the Keating-Harrison (KH) expressions for the bulk modulus $B = (C_{11} + 2C_{12})/3$ and shear moduli $C_S = (C_{11} - C_{12})/2$ and C_{44}:

Table 2. Elastic constants C_{ij} of hexagonal modification, Keating-Harrison model.

		C_{11}	C_{33}	C_{12}	C_{13}	C_{44}	C_{66}
SiC	Theory	38.8	42.2	11.2	8.2	12.7	14.6
	Experiment	50.0	56.4	9.2	—	16.8	—
BN	Theory	89.1	94.3	23.8	18.7	28.7	32.6
	Experiment	—	—	—	—	—	—
AlN	Theory	22.6	23.9	6.1	4.7	7.3	8.3
	Experiment	34.5	39.5	12.5	12	11.8	—
GaN	Theory	18.9	20.1	5.1	3.9	6.1	6.9
	Experiment	—	—	—	—	—	—

$$B = \frac{(8 + \lambda)\sqrt{3}}{12} \frac{\alpha_c^3}{d^3} \, |V_2|$$

$$C_S = \frac{\lambda\sqrt{3}}{2} \frac{\alpha_c^3}{d^3} \, |V_2| \tag{5}$$

$$C_{44} = \frac{8\lambda\sqrt{3}}{8 + 3\lambda} \frac{\alpha_c^3}{d^3} \, |V_2|$$

The results of calculations based on such KH model for cubic modifications are listed in the second row of Table 1. From the comparison with the experiment for diamond [3] one can see that the KH model gives better results for C_{11} and C_{44} but less appropriate for C_{12}. As for silicon carbide, our approach is in better agreement with all moduli. Thus we can hope that the predicted KH values for the nitrides are more correct compared with the BOM results [5].

Table 2 demonstrates the theoretical values for the hexagonal modifications. We have used here our previous results [7]. Again one can see the advantages of KH model. Moreover, using these results it is easy to calculate the elastic constants for all other crystal politypes.

This work was supported in part by the University of Arizona.

REFERENCES

1. Keating, P. N. 1966. *Phys. Rev.,* 145:637.
2. Keating, P. N. 1966. *Phys. Rev.,* 149:674.
3. Nikanorov, S. P. and B. K. Kardaschov. 1985. *Elasticity and Dislocation Non-elasticity of Crystals,* In Russian, Nauka, Moscow.
4. Harrison, W. A. 1980. *Electronic Structure and the Properties of Solids,* Freeman, San Francisco.
5. Davydov, S. Yu. and S. K. Tikhonov. 1996. "Elastic Constants and Phonon Frequences of Wide Band Gap Semiconductors," *Semiconductors,* 30, to be published.
6. Martin, R. M. 1970. *Phys. Rev.,* B1:4005.
7. Davydov, S. Yu. and S. K. Tikhonov. 1995. *Fiz. Tverd. Telas,* St. Petersburg, 37:221.

Quantum Technology Applied to Alpha-C:H Film Modeling Aspects

E. F. SHEKA*
Russian Peoples' Friendship University
ul. Ordgonikidze, 3
117302 Moscow
Russia

V. D. KHAVRYUTCHENKO
Institute of Surface Chemistry
Nat. Ac. Sci. of the Ukraine
pr. Nauki, 31
252028 Kiev
Ukraine

E. A. NIKITINA
Institute of Applied Mechanics
Russian Ac. Sci.
Leninsky pr.
31a, Moscow
Russia

INTRODUCTION

The possibility of atomic-level manipulation on solid surfaces in the field of the tip of a scanning tunnelling microscope (STM) was first demonstrated by Becker et al. in 1987 [1] as a formation of nanosize protuberances on the Ge (111) surface. In the following years, the fundamental investigations of the atomic-manipulation processes have dealt

---

*Author to whom correspondence should be addressed.

266

mainly with pure surfaces of some solids, such as gold, silver, and platinum, tungsten and molybdenum chalcogenides. Silicon surfaces in ultrahigh vacuum were studied most thoroughly (see References [2,3] and references therein). At the same time, a wish to simplify the technological conditions made it necessary to search for a possibility of performing atomic manipulation on a surface in gaseous or liquid media. This lead to papers demonstrating the first successful attempts to produce single nanosize growths on the surface of highly oriented pyrolytic graphite from liquid [4] and gas [5–7] media. It was shown that in all cases the growths are formed in the tunnelling gap when the voltage on the STM tip reaches a threshold value in the range of 3.5–4.5 V (depending on the system). However, the formation mechanism itself and the atomic composition of the structures formed have not been determined.

DESCRIPTION OF THE SYSTEM TO BE MODELLED

Two fundamental properties provide the choice of graphite as a working surface in the above experiments: high electrical conductance and adequate chemical inertness with respect to many gaseous and liquid media. A serious drawback, however, is its low mechanical strength which accounts for the peeling off of the surface layers as a result of an electric pulse application [6,7]. The intention to preserve the positive qualities of graphite as well as to improve its mechanical properties, led to the natural choice of the alpha-C:H films as a reference base for nanostructure building [8]. This X-Ray-amorphous solid carbon maintains a good electron conductivity, being a semiconductor, it retains chemical inertness and its mechanical strength is close to that of diamond. In contrast to the homogeneous surface of pyrolytic graphite, however, the surface of the film is heterogeneous, as determined by many investigations, see review [9] and references therein. It consists of a complex diamond-like solid matrix structure formed by the sp3-hybridized carbon atoms disordered by embedded clusters of a graphite-like structure (better, the skeletons of the "aromatic" sp2-hybridized carbon atom configurations) with linear sizes ranging from 0.3 to 0.9 nm. As a result, the surface of the film is substructured by the fragments of the "quasidiamond" and "quasigraphite" phases.

Many experimental attempts in creating atomic-scale structures on the film surface when applying pulsed electric field to the STM tip in the presence of various adsorbates, have not resulted in any controllable technological process [8]. That is why we have tried to approach the problem from the opposite side and have performed a computational experiment aiming at choosing an appropriate substance for both surrounding medium and adsorbate suitable for construction of atomic "buildings" on the film surface. The following criterion for the stable nanostructure formation was suggested: the structures must be bonded to the surface by strong chemical bonds and they must incorporate a possibility of mass growth, again via new chemical bonds.

As for the bonding to the substrate, it should clearly be provided by the formation of strong C—C sigma-bonds between the substrate and adsorbate. Generally speaking, both film components can be involved in the process, although with different effectiveness. Actually, the formation of the C—C bond with adsorbate means a breaking of the basic C—C pi-bond for the sp2-component while the peripheral dangling bond of the carbon atom saturated by hydrogen should be introduced into the reaction from the sp3-component side.

As for the adsorbate, the main criterion for the choice is dictated by its role as the main block in the building on the surface. Therefore, it should be quite simple in structure, easily involved in the reaction of the C—C bond formation, and well suited to the technological process. Among candidates meeting these requirements, there is trichloroethylene C_2HCl_3 (TCE) which was chosen for the present study. Therefore, the computational experiment involved examining quantum-chemically the possible chemical transformations happening in the system TCE-alpha-C:H film when applying an electric field.

CHARACTERISTICS OF MODELING PROCESS
AND MODEL CLUSTERS

The calculations have been performed in the cluster approximation by using a computational software DYQUAFIELD [11], based on modern efficient quantum-chemical semiempirical methods (AM1, PM3, NDDO), specially elaborated for determining the structure and properties of nanosize atomic system in the electric field of arbitrary configurations. The electric field is configured by a set of point charges whose total number approaches 201. A successful application of the software to the solution of a similar problem has been recently shown [12]. A personal computer with a Pentium90 processor has been used for the computational performance.

The alpha-C:H film was modelled by a cluster consisting of 207 atoms (Figure 1) and involving both sp3- and sp2-components. As shown in the figure, the latter was presented by two configurations, linear and squared ones, consisting of four benzene rings that correspond to a band gap energy of 2.5–3 eV as follows from Reference [10]. After completing the opti-

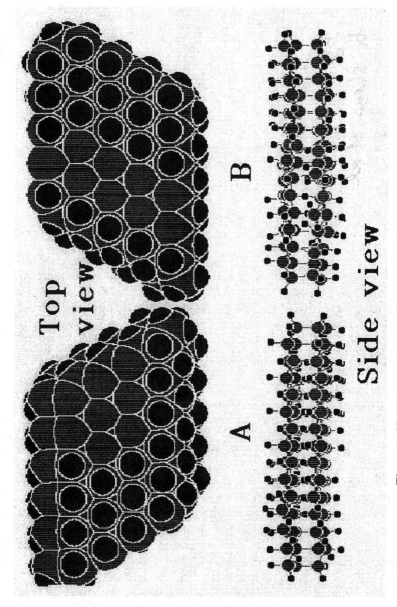

Figure 1. Model clusters for an alpha-C:H film. Fully optimised structure.

misation procedure accompanying the search of the potential energy minimum, both components were shown to retain their individuality to a great extent. This made allowance for splitting the computational problem into two related but separate sp3- and sp2-components. The corresponding subclusters, containing 56 and 94 atoms respectively, were used to study the interaction of the TCE molecule with diamond-like and graphite-like components of the film. The electric field was configured by a set of point charges whose arrangement simulated either a sharp or a blunt STM tip. The charges were placed in the position of the centres of tungsten atoms in the (111) plane. The sharp tip was modulated by a single point charge. The blunt tip configuration corresponds to the situation in which the area occupied by the point charges covers the cross section of the cluster under study. Electric voltage was varied by changing the charge values.

RESULTS

Substrates. The films, sp2- and sp3-components behave quite differently when applying an electric field. In the first case, no changes have been observed at either sharp or blunt configurations of the tip in the electric field strength (EFS) ranging up to 20 V/nm. The sp3-component reacts to the field at a sharp configuration only and results in a noticeable corrugation of the layered structure as well as in the breaking of surface C—H bonds at EFS of 10V/nm, which is accompanied by out-diffusion of the hydrogen atom from the surface [12]. This result may explain the poorly reproducible phenomena of some irreversible transformations occurring at a pure surface film when applying a pulsed electric field [8,12].

TCE + substrate complexes. No chemical reactions have been observed when applying an electric field of different configurations and polarity with EFS up to 25 V/nm to the adsorbed complex (ad-complex) involving the film sp2-component. The same result is obtained from field application to the ad-complex based on the sp3-component in the configuration of the above extreme limit of a sharp tip. A completely different situation occurs in the case of a blunt tip. Figure 2 shows a successive iteration series for the sp3-component cluster interacting with the TCE molecule when the tip is positively biased with EFS of 5 V/nm. As seen in the figure, a fully equilibrated starting configuration of the ad-complex is successively transformed in the following way: the TCE molecule approaches the surface, loses one chlorine atom and transforms into a $C_2HCl_2^{\cdot}$ radical. Simultaneously, one hydrogen atom leaves the surface and frees a dangling bond of the surface carbon atom. In the final stage, the radical saturates this dangling bond, forming a stable graft radical connected with the surface via C—C sigma-bond, 0.153 nm in length, while releasing chlorine and hydrogen atoms forming hydrochloric acid freely moving away from the reaction zone. The reaction is terminated when EFS exceeds 10 V/nm. Nothing happens if the bias polarity is inverted (see Figure 3). On the other hand, the adsorbate-surface complex formed in the course of the reaction shown in Figure 2 is not subjected to any changes when inverting and largely increasing the tip bias afterwards (see Figure 4).

Therefore, these quantum-chemical calculations have revealed an electric-field-stimulated chemical reaction of the TCE graft radical for-

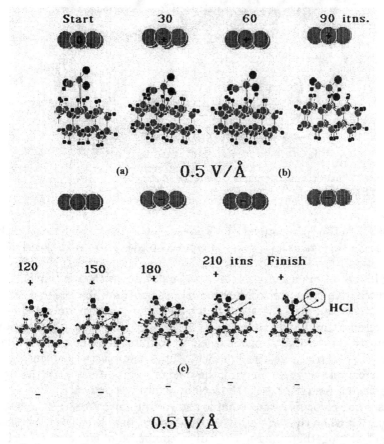

Figure 2. A consecutive series of structure transformations for ad-complex based on the sp3-component. The starting optimised configuration in zero electric field, is shown as are the intermediate and final optimised configurations at ESF 5V/nm. Positively biased STM tip.

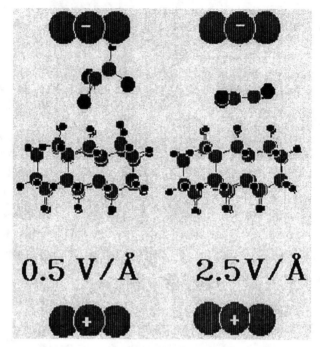

Figure 3. Final optimised configurations of the ad-complex in Figure 2, but for negatively biased STM tip.

mation on the sp3-component of the alpha-C:H film which is a threshold case and occurrs at EFS ranging from 5 up to 9 V/nm with positively biased tips.

This result makes it possible to understand how the binding of the adsorbate with the surface occurs but it does not show how the formed complex mass can grow. The solution was found when modeling chemical processes stimulated by the same electric field in the adsorbate medium. Figure 5 shows what happens when an equilibrated clustered configuration characteristic of the liquid medium is subjected to considerable transformation when applying the field of 5V/nm. As seen in the figure, the electric field stimulates the formation of a free radical $C_2HCl_2\cdot$ and releases one chlorine atom. What is shown in Figure 5 corresponds to the situation when the molecules are rather far from each other. When coming closer together, the radical will be composed of a two-member chain with a free molecule. One hydrogen atom is freed in the process. Obviously, this atom will interact with previously freed chlorine atoms forming a hydrochloric acid molecule. The next free radical will continue the

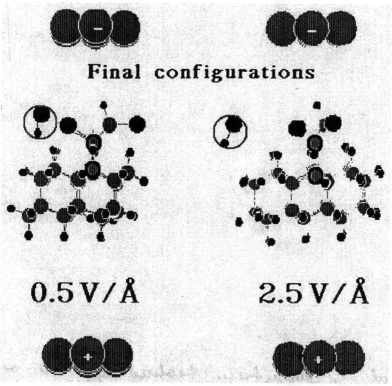

Figure 4. Effect of the electric field application to the final configuration shown in Figure 2. Negatively biased STM tip.

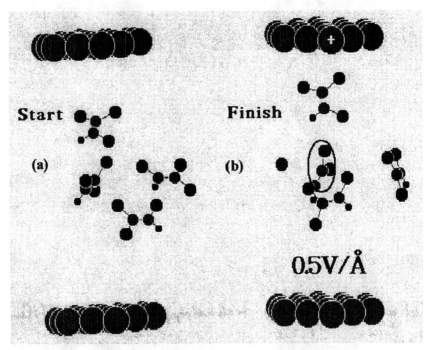

Figure 5. A cluster of four TCE molecules in an electric field. Fully optimised configuration in the absence (a) and with the application (b) of an electric field.

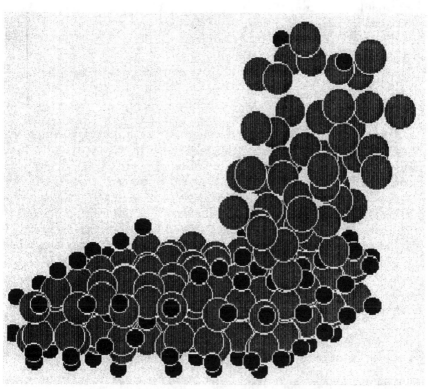

Figure 6. Graft TCE oligomer on the alpha-C:H film.

formation of the oligomer chain so that the final mass construction, or a "hill," under the tip should consist of free and cross-linked oligomeric chains attached to the diamond-like component of the film. One of the possible configurations is shown in Figure 6.

The exhibited field-stimulated and field-maintained radical grafted oligomerization laid the foundation for the definite recommendations of a technological process which has been successfully accomplished by fully controllable formation of nanosize building blocks, both individual and connected into desired "drawn lines," on the film surface [13]. The characteristics of the experimental findings are fully consistent with those given above.

CONCLUSION

As shown in the study, the nanostructures on the alpha-C:H film in the presence of the TCE medium can be formed in the course of a local electric-field-stimulated chemical reaction of graft oligomerization occurring via the formation of a graft and free radicals of the adsorbate. This leads to an understanding of what substances might be the most appropriate to particular experimental needs. When the "hills" need to be rather small, the related substance should be a poor polymerizer and should form an oligomer chain with a small number of members. When large hills are needed, substances with higher polymerizability should be chosen. A similar analysis of the electronic properties of the substances can be made.

The results allow one to look at the film surface two-component fragmentation, seemingly unfavourable at the first glance, from a new angle. It became obvious the components play different and equally important roles for the future application of the studied ad-complexes as nanoelectronic devices. Thus, the "aromatic" sp2-component controls the band gap width [9,10] providing the necessary regulated electroconductivity, while the sp3-component is mainly responsible for the formation of nanosize growths on the surface via strong chemical bonding of a certain surrounding with the surface. When the STM tip linear size exceeds that of the sp2-component, which is not larger than 1 nm, the heterogeneity of the film surface caused by embedded sp2-component fragments will not be noticeable due to flexibility of the oligomer chains attached to the sp3-components of the surface. These chains, whose length may be up to a few tens of a nm, can be easily cross-linked above the sp2-fragments forming a quite homogeneous nanosize structure strictly limited transversely in size by the cross section of the STM tip used.

The results highlight as well one more feature which is important in the context of further applications. The existence of the local chemical reaction on the alpha-C:H film surface is undoubtedly caused by a local character of interatomic interaction that, in turn, is a consequence of a strongly localised distribution of the electron density over surface atoms of the sp3-component. This may explain why the reaction does not occur on the sp2-component of the substrate and throws doubt on the applicability of the Si(111) 7 × 7 surface, which is widely studied [2,3], as a working surface for quantum devices due to highly delocalized character of electron density distribution over the surface atoms [14].

REFERENCES

1. R. S. Becker, J. A. Golovchenko, B. S. Schwartzentruber, *Nature (London)* 325, 419 (1987).

2. A. Kobayashi, F. Grey, R. S. Williams, M. Aono, *Science,* 259, 1724 (1993).

3. H. Kuramochi, H. Uchida, M. Aono, *Phys. Rev. Lett.,* 72, 932 (1994).

4. J. S. Fosler, J. E. Frommer, P. C. Arnelt, *Nature,* 331, No. 6154, 324 (1988).

5. G. Dujardin, R. E. Walkup, P. Avouris, *Science,* 255, 1232 (1992).

6. R. M. Penner, M. J. Heben, N. S. Lewis, C. F. Quate, *Appl. Phys. Lett.,* 58, 1389 (1991).

7. S.-T. Yau, D. Saltz, A. Wrickat, M. H. Nayich, *J. Appl. Phys.,* 69 2970 (1991).

8. I. Gorelkin, P. N. Luskinovich, V. I. Nikishin, *Elektron. Tekhn.,* Ser. 3, No. 2–3, 17 (1992).

9. J. Robertson, *Adv. Phys.* 35, 317 (1986).

10. J. Robertson, *Phil. Mag. Lett.* 57, 143 (1988).

11. E. F. Sheka, V. D. Khavryutchenko, V. A. Zayetz, *Intern Journ. Quant. Chem.,* 57, 741–755 (1996).

12. E. F. Sheka, V. D. Khavryutchenko, V. A. Zayetz, *Phys. Low-Dim Struct.,* 2/3, 59–78 (1995).

13. P. N. Luskinovich, V. D. Frolov, A. E. Shavykin, V. D. Khavryutchenko, E. F. Sheka, E. A. Nikitina, *JETP Lett.* 62, 881 (1995).

14. V. Khavryutchenko, E. Sheka, M. Aono, D. H. Huang, *Phys. Low-Dim. Struct.,* No. 12, 349 (1995).

DIAMOND FILM PROPERTIES AND CHARACTERIZATIONS

Mechanical Properties of Diamond Films and Related Materials

PETER J. GIELISSE
FAMU/FSU College of Engineering
Tallahassee, Florida 32310, USA

ABSTRACT: A brief review of the significant mechanical properties of diamond films is given. Current values for strength, hardness, Young's modulus and where possible for fracture toughness of diamond and DLC films have been tabulated.

1. GENERAL

Diamond is probably the only material in which size, morphology, shape and structure, influence both type and efficiency of its applications as critically as they do. In other words, the intrinsic physical (mechanical) properties are not sufficient to unambiguously analyze or explain the reasons for its successful use. A case in point is the shape, strength, structure and size differentiated (grit size) industrial diamond, which define different application areas. Diamond film products are likely to develop along similar lines, particularly as it concerns the diamond-like carbon films (DLC) which in their properties critically depend on synthesis parameters, allowing customizing or tailoring of the product to the application. A chart indicating the various diamond application categories based on size, covering a range from Angstrom to submeter, is given in Figure 1. A further problem in defining diamond's mechanical properties is that in most cases diamond is not available in a form required for conventional property testing, i.e., as ribbon, plate, rod or fiber. In the thin film area, the new shapes are starting to introduce al-

---

*Author to whom correspondence should be addressed.

281

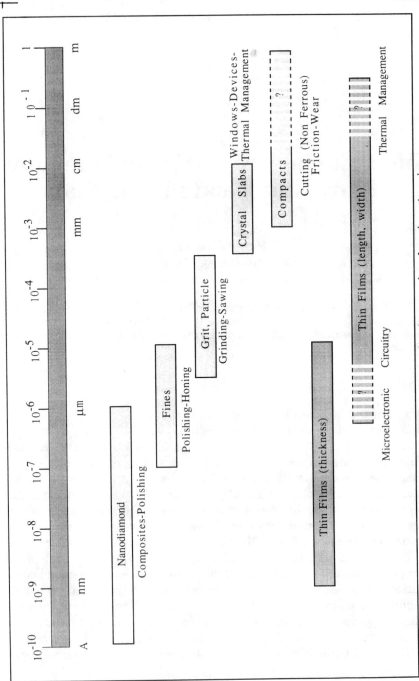

Figure 1. Particle and thin film application areas based on size categories.

ternative types of strength such as bursting or bulge strength (see further).

2. DEFINING DIAMOND AND DIAMOND-LIKE STRUCTURES

The microlevel arrangements of atoms in the two most important carbon structures—graphite and diamond, are very different. In graphite, six membered aromatic rings of sp^2-hybridized carbon atoms are arranged in layers, with the bond strength between the layers one to two orders of magnitude less than between the atoms within the layer. This imparts the structure with (solid) lubricant characteristics, due to the ease with which the interlayer bonds can be severed in shear. The carbon atoms in diamond, are covalently bonded via a network of tetrahedrally arranged aliphatic sp^3-hybridized carbon atoms. The essential difference lies, therefore, in both the type of hybrization, sp^2 or sp^3 (or as in mixed phases in the sp^3 to sp^2 ratio) and in the structure type.

Diamond-like carbon films contain a good portion of sp^2 bonding, often up to 60%, the exact extent of which is determined primarily by the type of deposition technique and the specific process parameters. The higher the aliphatic sp^3 content the more enhanced the mechanical and physical properties tend to be, although factors such as surface characteristics, foreign phases, impurities, structural inhomogeneities (voids, cracks, filling factors) and grain boundary effects, can make a significant impact on the actual values. DLC properties are also influenced by and differentiated on the basis of their (bonded) hydrogen content in hydrogenated films and their hydrogen and nitrogen content in amorphous/C:H:N films.

Diamond films currently are polycrystalline thin coatings consisting of a dense network of high density carbon (diamond) crystallites, generally of cubooctahedral morphology. Depending on purity and particularly on the non-diamond carbon content, free standing films range from essentially opaque to more commonly translucent and are in their best rendition, transparent. Synthesis of single crystal (epitaxial) films on substrates other than diamond, has not yet been accomplished. The films are often characterized with reference to their most common application as being of a tribological, thermal management, or device grade, with transparency (and thermal conductivity values) increasing in the same order.

3. THE STRENGTH OF DIAMOND AND DIAMOND FILMS

There exist no true tensile strength values for diamond. In the case of

Table 1. Fracture strength for diamond single crystals and polycrystalline films.

Material Type*	Strength (MPa)	Strength Type and Method	Reference Year
Single Crystal Natural Diamond	2800	Fracture Strength	[2] 1979
Single Crystal Natural Diamond	3500	—	[3] 1992
Polycrystalline Diamond Films CVD	500–1400	Tension	[4] 1993
Polycrystalline Diamond Films HFCVD	500–2200	Fracture Strength Bursting Pressure, Nitrogen on Diamond Membrane	[1] 1993
Polycrystalline Diamond Films BPCVD	746–1138	Disc Bursting Pressure. Simple Plate Theory	[5] 1994
Polycrystalline Diamond Films FACVD	340–1380	Fracture Strength Bursting Pressure, Air on Diamond Membrane	[2] 1979

*CVD, Chemical Vapor Deposition; HFCVD, Hot Filament CVD; BPCVD, Bulk Polycrystalline CVD; FACVD, Film Assisted CVD.

films, reliable strength data have normally been derived indirectly from hardness values, from compression tests, or from diaphragm bulge tests. In bulge tests the diamond disks, diaphragms or films, are held in a fixture along their periphery and are exposed to increasing air or gas pressure, till fracture occurs. The actual fracture strength values are then derived by combining strength of materials formulations, such as flat plate theory and bending moment approaches, with the experimental results. In evaluating such diamond film strength data, it should be kept in mind that they are influenced by geometric factors (thickness, grain size, morphology and disk diameter) by growth parameters (substrate treatment, gas type and concentration) as well as by the type and level of the surface roughness, surface crack size (1–10 micron) impurity concentration and residual stress levels.

The more important (fracture) strength data for diamond single crystals and polycrystalline films, as generated over the last decade or so, are compared in Table 1. In the HFCVD approach it was found that the crystal quality rather than the grain size had the greater effect on fracture strength. The use of a bias current between substrate and filament(s) produced fine agglomerated grains and a higher strength as compared to films with a columnar structure. The latter were more brittle and resulted when no bias was applied. The maximum fracture strength of 2200 MPa was obtained with the highest bias current, 700 mA. Fracture strength tends to be high for films of small grain size and with a low non-diamond content.

The BPCVD film products were evaluated polished and unpolished, with either the nucleation side or the growth side exposed to the pressurization direction. The latter condition did not produce significant differences. The highest bursting stress, 1138 MPa, was obtained for relatively thin (179 mm) small diameter (10 mm) discs, while the lowest bursting pressure (746 MPa) was reported for a thicker (296 μm) larger diameter (20 mm) disc.

The FACVD films with a fracture strength ranging from 340 to 1380 MPa, could be differentiated into two groups. Growth on the fine grained substrate (nucleation side) interfaces yielded a high(er) average strength of 880 MPa, while film hardness on the coarser grained diamond growth surfaces averaged 610 MPa.

It will be noted that the maximum fracture strength value for a diamond film of 2200 MPa, Table 1, comes close to that reported for type II natural diamond, (2800–3500) MPa. A more conservative analysis would indicate that the strength value, averaged over all the polycrystalline film data in Table 1, yields approximately 1000 MPa. This is about one third the value for single crystal diamond, in line with similar ratios for other non-metallic materials when comparing single crystal and polycrystalline strength values. The (1000 MPa) average diamond film strength favorably compares to values for other high technology industrial non-metallic materials such as sapphire (430 MPa) silicon nitride (830 MPa) and the other group IV elements, silicon (130 MPa) and germanium (90 MPa).

4. THE HARDNESS OF DIAMOND

4.1 Hardness Defined

The hardness of a material and particularly that of diamond, is not a rigorously defined physical quantity. The numerical quantity by which hardness is expressed, normally derives from a measurement technique that tests or incorporates more than what can truly be considered physical hardness. The results produced by the test techniques are thus sensitive to variables other than those intrinsic to "physical hardness." In practice, hardness relates to the resistance against a deformation process, which in its measurement methodology and the resultant unit expressions, can differ considerably from one approach to another. Hardness values are not likely to be comparable, even when the same type of units are used. Attempts at rigorous hardness formulations have over the years led to four fundamentally different approaches: deformation (scratch)-based hardness; wear rate-based hardness; energy-based hardness and the most commonly used, stress-based hardness.

4.1.1 SCRATCH HARDNESS

The first attempts at a relative hardness expression led to the well known ten point Mohs' scale. It ranked mineral specimen from a value of one for the softest, talc, to a value of ten for to the hardest known, diamond. The scale is currently only of historical significance.

4.1.2 WEAR RATE HARDNESS

The wear rate hardness expresses hardness in terms of the extent to which a material is worn away under standardized abrasion conditions. The abrasion during lapping experiments, resulting in the Wooddell hardness numbers and scale, appear to have been the first attempts [7].

4.1.3 THE ENERGY APPROACH

The energy approach was first introduced in the 1960's by Plendl and Gielisse [8]. They proposed the use of the quantity—lattice energy per unit volume—U/V, i.e., the energy to break down the molecular structure. It was an attempt to provide a fundamental underpinning to hardness, in other words to obtain a strictly physical expression of hardness. Data of this type was used to predict the hardness of c-BN (borazon) shortly after its initial synthesis [9]. A plot of physical hardness versus abrasion hardness for important industrial solids, is presented in Figure 2. The values for c-BN were slighty upward corrected [10] from the original values [8].

4.1.4 INDENTATION HARDNESS

Indentation hardness is defined as the resistance of a material against the static (low speed) indentation by an ultra hard indentor of known geometry, into the unrestrained volume of a less compressible body. The hardness determined in this way is related to the plastic work expended in creating the remnant impression on removal of the indentor.

The case of thin (micron-sized) films demands that, in determining their hardness, the contributions of both the substrate and the film are taken into account. Their micron size thickness generally implies that very low load-low penetration depth tests be made. This becomes highly problematic on surfaces with any degree of roughness (relative to film thickness) and for those that are polished but of marginal thickness. In any case the indentation depth cannot be large compared to the film

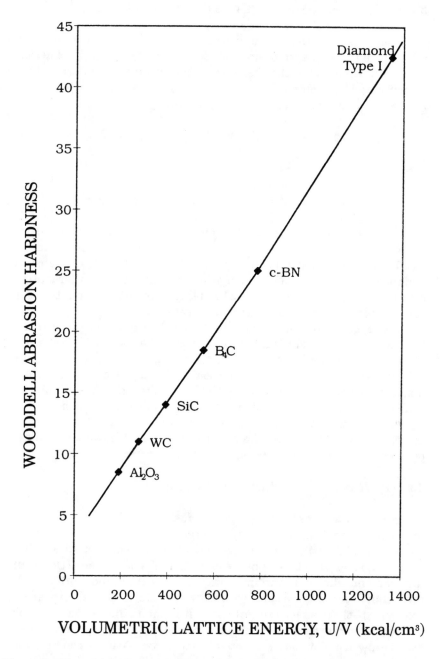

Figure 2. Volumetric lattice energy versus abrasion hardness for hard to superhard materials.

287

thickness. A rule-of-thumb value to prevent a substrate contribution, allows a penetration depth of no more than 10% of the film thickness.

It has been proposed that hardness of a coating on a substrate can be determined by a volume law of mixtures hardness model [11–14] for which the composite hardness H_c is given by,

$$H_c = (V_f/V)H_f + (V_s/V)H_s\alpha^3, \quad (H_s < H_f) \qquad (1)$$

or

$$H_c = (V_s/V)H_s + (V_f/V)H_f\alpha^3, \quad (H_s > H_f) \qquad (2)$$

where V_s and V_f are the deforming volumes in the substrate and coating (film) respectively, V is the total deforming volume, α is an interfacial constraint parameter, and HV_s and HV_s are the hardnesses of the substrate and the film. The (fractional) change in the volume of the softer material resulting from the presence of the harder component, is accounted for by α, an empirical interfacial constraint parameter.

The composite hardness expressions incorporate the dependence of hardness on load at small indentation sizes, the so-called "indentation size effect" (ISE) and a "plastic zone size term," to account for (indentor imparted) deviations from the spherical cavity approach. The load bearing profile under an indentor on a stiff film is likely to be non-uniform as well, with further adjustments necessary for substrate materials with a low yield stress to Young's modulus ratio (σ_y/E). All these effects need be incorporated in the value of the interface parameter, α, which is to be experimentally determined for each material.

4.2 The Hardness of Diamond Thin Films

Diamond thin films have been grown metastably from the vapor phase in low pressure environments by different deposition techniques. They all more or less involve a plasma growth environment. Depositions have been made on a variety of substrates each of which influence the resulting diamond film structure in a different way. As in most other multilayer thin film work, the choice of substrate material proves critical with reference to the desired results. So far, all diamond films have been polycrystalline and are normally columnar in structure. Single crystal epitaxial growth of diamond has not yet been accomplished except on itself or on very small grains of c-BN as substrates [33].

It is often stated or possibly assumed, that diamond films have the same properties as single crystal diamond. This is not true. The macrostructure of the films, principally consisting of intergrown crystallites, stressed, with a high defect concentration, a possible void content and the presence of other phases (graphite, carbon) make for a different product. As with other materials, a wide range of diamond film types and qualities are available, resulting in products from opaque to nearly glass clear with thermal conductivities from 500 W/mK to 2000 W/mK and strength variations that cover several orders of magnitude.

Beetz et al. [18] measured the hardness of diamond films generated with a hot filament CVD method, using an ultra low indentation technique. They obtained hardness values between 31 GPa for low partial pressures of CH_4 and 65 GPa at higher pressures. The latter figure would put their films at a hardness between that of c-BN and diamond. Natural diamond hardness has been reported over a wide range of values from around 50 to over 100 GPa.

Sussman et al. [5] in work primarily involved with the determination of the bursting strength of very thick (179-300) μm diamond films, felt confident enough to use 10,000 kg/mm^2 for the hardness, in calculating the fracture toughness values of their films.

Partridge et al. [21] in reviewing the status of diamond fibers tabulated the hardness of CVD films, apparently obtained from the open literature, as (31–90) GPa.

A tabulation of the mechanical properties of diamond films can be found in Table 2.

4.3 The Hardness of Diamond-Like Carbon (DLC) Films

From a physical property point of view, diamond is a rather well defined substance. This stands in sharp contrast to DLC type films and coatings which are not well defined substances. Their hardness characteristics depend on sp^2 to sp^3 bonding content, the various processing parameters in as many different growth techniques, the presence or absence of bonded hydrogen and nitrogen, the specific structure of the films and the surface character (roughness) to name a few. It is therefore relatively dangerous to evaluate the applicability of DLC films on the basis of one parameter, e.g., their hardness, only. It is, furthermore, well known that not only hardness, but also low wear resistance, low coefficients of friction, a relatively low stiffness and a high elasticity have created unique application areas for DLC films.

The currently available hardness values for a wide variety of DLC type films have been brought together in Table 3. In order to quickly compare DLC film hardness to those of diamond and diamond films, Table 4 is provided. The film descriptions attempt to describe the type of film and

Table 2. Mechanical properties of diamond films.

Material	Hardness (GPa)	Young's Modulus (GPa)	Poisson's Ratio	Bulk Modulus (GPa)	Density (g/cm^3)	K_{IC} (MPa·m$^{1/2}$)	[Reference] Year
Diamond film	80–100	500–533	—	—	—	—	[19] 1993
Diamond film (MPECVD)	40–75	370–430	—	—	—	—	[17] 1992
Diamond film (HFCVD)	31–65	—	—	—	—	—	[18] 1990
Diamond film (HFCVD)	75–111	—	—	—	—	—	[20] 1992
Diamond film (CVD)	31–90	700–1079	0.07	—	3.5	~6	[21] 1994
Diamond film (CVD) (0–5% CH$_4$)	—	—	—	380–500	—	—	[22] 1989
Diamond film (CVD)	—	891 (~820 at 800°C)	—	—	3.5	—	[23] 1993
Diamond film Bulk, Polycrystalline (BPCVD)	—	986–1079	0.1	—	—	6	[5] 1994
Diamond film (HFCVD)	—	250–1050	0.29	—	—	—	[1] 1993

Table 3. Mechanical properties of diamond-like carbon (DLC) films.

Material	Hardness (GPa)	Young's Modulus (GPa)	Poisson's Ratio	Internal Stress (GPa)	Bulk Modulus (GPa)	Density (g/cm³)	K_{1c} (MPa·m$^{1/2}$)	[Reference] Year
a-C:H (hydrogenated)	10.5–14	95–130	0.3	1.5–3.5	—	—	—	[28] 1992
a-C:N:H	~10 (average)	—	0.2–0.4	1.5–5 (3 average)	—	—	—	[31] 1994
a-C:H (diamond like)	—	127 (maximum)	0.2–0.4	0.4–4 (1.5 average)	196	2.0	—	[24] 1989
a-C:H (amorphous carbon film), (CH₄)	11–33	90–220	0.22–0.39	—	~135	>12.0	—	[26] 1990
a-C:H (amorphous carbon film), (CH₄)	13 (maximum)	140 (maximum)	—	—	—	>2.0	—	[32] 1989
a-C:H (PECVD-Sputtered)	7–14	35–136	—	—	—	1.7–1.8	—	[17] 1992

(continued)

291

Table 3. (Continued).

Material	Hardness (GPa)	Young's Modulus (GPa)	Poisson's Ratio	Internal Stress (GPa)	Bulk Modulus (GPa)	Density (g/cm^3)	K_{1c} (MPa·m$^{1/2}$)	[Reference] Year
a-C:H (PECVD, r.f.)	33–35	~200	—	—	—	1.6–1.8	—	[17] 1992
a-C:H (PECVD, C$_2$H$_2$)	5–40	—	—	—	—	—	—	[27] 1991
a-C:H (CVD-C$_2$H$_4$, C$_2$H$_2$)	7–20	—	—	1–6	—	up to 2	—	[25] 1990
DLC	12–30	62–213	—	—	—	—	—	[16] 1992
DLC (Hydrogenated amorphous carbon)	200–480 (?)	480–850 (?)	—	—	—	—	—	[15] 1989
DLC (Pulsed vacuum arc discharge)	30	300	—	—	—	—	—	[29] 1994
DLC, a-(C) (Amorphic Diamond)	37	369	—	—	—	—	—	[30] 1992

Table 4. Hardness and moduli for diamond, diamond films and DLC films.

Type	Hardness	Young's Modulus E (GPa)	Bulk Modulus B_o (GPa)
Diamond-like films DLC	5–49 (21)	35–369 (180)	100–196 (145)
Diamond films	30–110 (70)	250–1050 (730)	380–500 (440)
Natural diamond	55–113 (80)	900–1250 (1050)	433–540 (485)

its origin, using as much as possible the nomenclature, expressions and descriptive terminology used by the original authors. The reader is referred to the references if more in-depth knowledge is required. Specific information considered essential in the interpretation of the data in Tables 3 and 4 is given below.

Jiang et al. [24] studied the deformation of a-C:H films for a wide variety of deposition parameters and methods, producing polymer-like films with tensile strain and diamond-like films with compressive strain. Their elastic and bulk moduli were calculated from measured shear moduli, G, and micro indentation hardness measurements.

The work of Dworschak et al. [25] showed that low power (20 W) r.f. coupled plasmas with ion energies amounting to only a few electron volts, also led to polymer-like films. Higher power (50–500 W) and biased substrates produced their "hard," a-C:H, films with ion energies up to a maximum of 200 eV.

Further researches by the Jiang et al. team [26] could show that a high sp^3/sp^2 ratio was not sufficient to produce hard diamond-like films but that a low void content, which could be correlated with the filling factor, was even more important. The filling factor is defined as,

$$f = V_{OCC} / V_{film} \qquad (3)$$

where f is the filling factor, V_{film} is the total volume of an a-C:H film, and V_{OCC} is the volume occupied by carbon and hydrogen atoms in the respective films. Their hardness data decreased monotonically from 33 GPa to 11 GPa and the E-modulus from 220 to 90 GPa, as the C_2H_2 pressure was decreased. The bulk modulus value, 135 GPa, remained nearly constant.

The films produced by Kleber et al. [27] in an r.f. (13.56 MHz) discharge at 100 W, yielded the best Knoop hardness values, 40 GPa maximum, at the higher mean ion energies (250 eV) and the lowest hydrogen content (28 at.%). Overall hardness values ranged from 5 GPa to 40 GPa, with values increasing with a decreasing hydrogen content and an increasing mean energy.

The dc magnetron sputtered films and the r.f. PECVD a-C:H films produced by Bhushan et al. [17], differed considerably in hardness. The plain sputtered films showed a hardness of (7–15) GPa with an average of about 12.5 GPa. The PECVD films were considerably harder at (33–35) GPa. Their diamond films yielded (40–75) GPa. As might be expected the sputtered and PECVD films contain sp^2 bonding contributions and a considerable amount of hydrogen. The diamond films contained only a small amount of hydrogen.

A cooperative effort by a dutch and a belgian team of Dekempeneer et al. [28] produced a-C:H films with a capacitively coupled r.f. plasma assisted chemical vapor deposition (PACVD) set-up. Nanoindentation measurements, maximum depth of 0.18 μm on a 1.7 μm thick film, produced a maximum hardness of 13.5 GPa. The primary objective of the work was finding a correlation between hardness and film stress.

Savvides and Bell [16], have also stated that DLC can have extremely diverse properties. They determined that increasing hardness and modulus values correlate well with increasing energy per condensing carbon atom, to yield a hardness of a-C:H films of (12–30) GPa. This may be compared to the hardness of their diamond films of between 80 GPa and 100 GPa and the "accepted" values of H = 56–102 GPa for type IIa diamond. Low energy ion-assisted magnetron sputtering of a graphite target in Ar-H_2 mixtures was used.

Depositions of DLC films at room temperature in a dc plasma of methane and hydrogen gas by Hoshino et al. [15] showed different properties depending on substrate location. With the anode directly above the substrate cathode, Young's modulus of the film was found to be 480 GPa and with a "sideward" anode this became 850 GPa. The ultra high hardnesses quoted in the article appear out of line with all other values for DLC materials (Table 3).

Hydrogen free DLC coatings were deposited by a pulsed arc discharge method by Koskinen et al. [29] yielding a layered columnar growth structure. Their hardness analysis resulted in a value of 30 GPa and a modulus of 300 GPa.

Laser plasma discharge by Devanloo et al. [30] also produced hydrogen free diamond-like material called "amorphic" films. Averaging over twenty separate measurements, the hardness of the 5 mm thick films was found to be 37 GPa with a Young's modulus of 369 GPa.

The addition of nitrogen to the CH_4 growth chamber by Schwan et al. [31] produced amorphous hydrogenated carbon-nitrogen films, a-C:H:N. The presence of nitride groups was found to reduce hardness as well as stress in the film with increasing nitrogen content. Hardness drops from a value of approximately 17 GPa to about 7 GPa at a 1:1 ratio of N_2 to C_2H_2. Their nitrogen free DLC's show a hardness of (6–17) GPa, depending on gas pressure.

Specific comments with reference to Young's modulus values for DLC films do not appear necessary since most of them were derived from hardness data, more specifically from the slope value of the unloading curve in the load-displacement plot, dP/dh in which P = load (MN) and h the displacement (nm).

REFERENCES

1. Aikawa,Y. and Baba, K. (1993) *Jpn. J. Appl. Phys.* 32 (10), 4680.

2. Field, J. E. ed. (1979) *The Properties of Diamond*, J. E. Field, Academic Press London, 648.

3. Field, J. E. (1992) *Properties of Natural and Synthetic Diamond* (ed. J. E. Field), London, Academic Press, 474.

4. Field, J. E., Nicholson, E., Steward, C. R. and Feng, Z. (1993) *Philos. Trans. Roy. Soc.*, A, 342, 261–275.

5. Sussman, R. S., Brandon, J. R., Scarsbrook, G. A., Sweeney, C. G., Valentine, T. J., Whitehead, A. J. and Wort, C. J. H. (1994) *Diamond and Related Materials*, 3, 303.

6. Cardinale, G. F. and Robinson, C. J. (1992) *J. Mater. Res.*, 7 (6), 1432.

7. Wooddell, C. I. (1935) *J. Electrochem. Soc.*, 68, 111.

8. Plendl, J. N. and Gielisse, P. J. (1962) *Phys. Rev.*, 124, 828–832.

9. Gielisse, P. J. (1962). See also reference [8].

10. DeVries, R. C. (1972) Cubic Boron Nitride: Handbook of Properties, General Electric Internal Report No., 72 CRD 178.

11. Burnett, P. J. and Rickerby, D. S. (1987) *Thin Solid Films*, 148, 41–50.

12. Burnett, P. J. and Rickerby, D. S. (1987) *Thin Solid Films*, 148, 51.

13. Bull, S. J. and Rickerby, D. S. (1989) *British Ceram. Soc.Trans.*, 88, 177.

14. Bull, S. J. and Rickerby, D. S. (1990) *Surf. Coat. Technol.*, 42, 149.

15. Hoshino, S., Fujii, K., Shohata, N., Yamaguchi, H., Tsukamoto, Y. and Yanagisawa, M. (1989) *J. Appl. Phys.*, 65 (5), 1918.

16. Savvides, N. and Bell, T. J. (1992) *J. Appl. Phys.*, 72 (7), 2791.

17. Bhushan, B., Kellock, A. J., Cho, N. and Ager, J. (1992) *J. Mater. Res.*,7 (2), 404.

18. Beetz Jr, C. P., Cooper, C. V. and Perry, T. A. (1990) *J. Mater. Res.*, 5 (11), 2555.

19. Savvides, N. and Bell, T. J. (1993) *Thin Solid Films*, 228, 289.

20. Bull, S. J., Chalker, P. R. and Johnson, C. (1992) *Mater. Sci. and Techn.*, 8, 679.

21. Partridge, G. P., May, P. W. and Ashfold, M. N. (1994) *Mater. Sci. and Techn.*, 10, 177.

22. Sato, Y. and Kamo, M. (1989) *Surf. Coat. Technol.*, 39–40, 183.

23. Seino, Y. and Nagai, S. (1993) *J. Mater. Sci. Lett.*, 12, 324.

24. Jiang, X., Zou, J. W., Reichelt, K. and Grunberg, P. (1989) *J. Appl. Phys.*, 66 (10), 4729.

25. Dworschak, W., Kleber, R., Fuchs, A., Scheppat, B., Keller, G., Jung, K. and Ehrhardt, H. (1990) *Thin Solid Films*, 189, 257.

26. Jiang, X., Reichelt, K. and Stritzker, B. (1990) *J. Appl. Phys.*, 68 (3), 1018.

27. Kleber, R., Jung, K., Ehrhard, H., Muhling, I., Metz, K. and Engelke, F. (1991) *Thin Solid Films*, 205, 274.

28. Dekempeneer, E. H. A., Jacobs, R., Smeets, J., Meneve, J., Eersels, L., Blanpain, B., Ross, B. J. and Oostra, D. J. (1992) *ibid.*, 217, 56.

29. Koskinen, J., Hirvonen, J. P., Hannula, S. P., Pischow, K., Kattelus, H. and Suni, I. (1994) *Diamond and Related Materials*, 3, 1107.

30. Davanloo, F., Lee, T. J., Jander, D. R., Park, H., You, J. H. and Collins, C. B. (1992) *J. Appl. Phys.*, 71 (3), 1446.

31. Schwan, J., Dworschak, W., Jung, K. and Ehrhardt, H. (1994) *Diamond and Related Materials*, 3, 1034.

32. Jiang, X., Reichelt, K. and Stritzker, B. (1989) *J. Appl. Phys.*, 66, 5805.

33. Yoshikava, M., Ishida, H., Ishitanis, A., Murakami, T., Kaizumi, S. and Inuzuka, T. (1990) *Appl. Phys. Letters*, 57, 428.

Physio-Chemical Properties of Shock-Wave Synthesized Nanometric Diamond

G. P. BOGATYREVA AND V. L. GVYAZDOVSKAYA
V. N. Bakul Institute for Superhard Materials of the
National Academy of Sciences of Ukraine
2, Avtozavodskaya St.
Kiev, 254074
Ukraine

V. V. DANILENKO*
I. N. Frantsevich Institute of Materials Science Problems
of the National Academy of Sciences of Ukraine
3, Krzhizhanovsky St.
Kiev, 252180
Ukraine

ABSTRACT: The physico-chemical properties of nanocrystalline shock wave-synthesized diamond—ultradisperse diamond (UDD)—with a surface area of 178 m^2/g, have been investigated. The specific magnetic susceptibility, electrophysical adsorption and thermodynamic properties were compared to similar properties of statically-synthesized diamond.

KEY WORDS: ultradisperse diamond, electrophysical adsorption, thermodynamic characteristics, specific magnetic susceptibility.

---

*Author to whom correspondence should be addressed.

INTRODUCTION

The production of ultradisperse diamond which could be used for making high-quality polycrystalline nanocrystalline material structures is of great importance. Owing to an unusual combination of micro-properties with a specific structure and superstructural organization typical of cluster materials, ultradispersed powder (UDP) has unique properties and can affect structure and properties of UDD-base composite materials. In the present work, physico-chemical properties of ultradispersed diamond (UDD) made of TNT RDX-based explosive materials have been studied.

MATERIALS AND METHODS OF INVESTIGATION

Subjects of investigations were UDD samples after deep chemical cleaning using mineral acids to remove metal impurities and liquid-phase oxidation by hexavalent chromium to remove various graphite phases. Auger and IR spectroscopic examinations revealed oxygen and silicon in small proportion as well as hydrogen-containing functional groups. Picnometric density (d) of such UDD was determined by gas picnometry using a Coultronics Company (France) Autopycnometer 1320. A specific magnetic susceptibility $(\chi, m^3/kg)$ was determined by the Faraday method in a magnetic field with the intensity of $H = 3 \cdot 10^7\,A/m$. The method sensitivity is $0.3 \cdot 10^{-3}\,m^3/kg$, the measurement error being less than 3%. The specific electrical resistance (ρ) was measured using a method of measuring a current passing through a diamond powder layer of a known area and thickness with a fixed voltage. To avoid the effect of adsorbed moisture, prior to measurements diamond powders were dried at 393–423 K. Adsorption studies were conducted by the method of low-temperature nitrogen adsorption in the relative pressure range from 0.05 to 1.0 according to the procedure of measuring specific surfaces of powder materials [1] and plotting adsorption isotherms on a Coultronics Acusorb 2100 device. Prior to measurements, the powder was degasified at 473 K and $1 \cdot 10^{-4}$Hg. Specific surface (S_{BET}) was found by the expression $S_{BET} = 0.2687\,S(s + i)$, where S is the area of the cross section of an adsorbed nitrogen molecule $(1.62 \cdot 10^{-13}\,mm^2)$, s and i the coefficient found from the graphic representation of the BET equation. The relative error in our measurements was no more than 10%. Water adsorption was assessed from the free energy of diamond surface saturation (ΔC_s) with water vapors using the expression:

$$\Delta C_s = \int_0^n \ln P/P_s \cdot dn, \, mJ/mole \cdot g$$

Table 1. Basic physico-chemical characteristics of disperse diamonds.

Diamond Type	S_{BET}, m^2/g	χ, 10^{-3}, m^3/kg	ρ, Ohn·m	ΔC_s, mJ/mole·g	d, kg/m^3
UDD	178.1	1.0	$7.7 \cdot 10^9$	3100	3.40
ACM 1/0	13.5	0.5	10^{11}	1480	3.52

where n is the number of moles of water adsorbed at the surface unit at the water vapor pressure P; and P_s is the pressure of saturated water vapor. The water vapor saturation was performed in a dessicator, at the bottom of which was distilled water at a constant temperature. The adsorbed quantity of water was found from the increase of diamond mass good to ± 0.00001 g. Before adsorption, diamonds were dried to a constant mass at 423 K. For comparison, similar studies were carried out on statically synthesized ACM 1/0 micron diamond powders. Before studies ACM 1/0 powders were also subjected to the deep chemical cleaning used on UDD.

EXPERIMENTAL RESULTS

Basic physico-chemical characteristics of UDD and ACM 1/0 type disperse diamond powders are listed in Table 1, while Figure 1 shows experimental isotherms of nitrogen adsorption on their surface. Basic thermodynamic characteristics of the powders under study were defined from the experimental isotherms. These characteristics are: ultimately saturated volume of pores (V_{por}), which is found from isotherms at $P/P_s = 1$; adsorption potential (A), that includes both the activity of the surface adsorption centers and their quantity; interphase reaction energy at the diamond-nitrogen interface ($\Delta\sigma_{d-n}$), which is energetic characteristic of the centers, and the distribution of the adsorbed potential, that illustrates quantitatively the surface energetic nonuniformity. All these characteristics were found based on approaches and methods described in

Table 2. Basic thermodynamic characteristics of disperse diamonds.

Diamond Type	V_{por}, ml/g	$\Delta\sigma_{d-n} = RT \cdot \dfrac{V}{P \, P_s \cdot S_{BET}}$, J/m^2 at		$A = RT \displaystyle\int_0^1 \dfrac{V}{P \, P_s}$	
		$P/P_s = 0.05$	$P/P_s = 0.7$	$\Delta\sigma$, J/m^2	$d(P/P_s)$, J/G
UDD	0.830	3.4	0.64	6.0	384
ACM 1/0	0.045	3.6	0.60	6.0	14.2

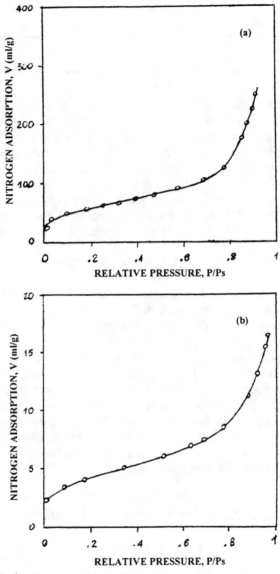

Figure 1. Experimental isotherms of nitrogen adsorption on the UDD (a) and ACM 1/0 (b) disperse diamond powders, both at T = 77K (ml/g).

300

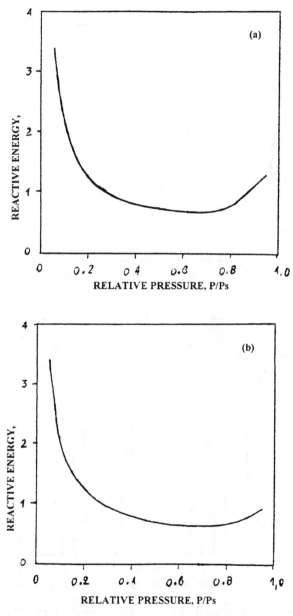

Figure 2. Dependance of $\Delta\sigma_{d-n}$ on P/Ps for UDD (a) and ACM 1/0 (b) diamond powders. J/m^2.

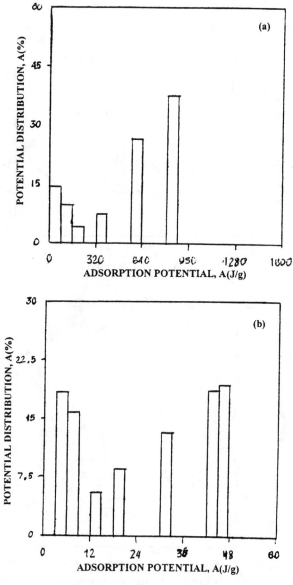

Figure 3. The adsorption potential (A) vs. distribution over the surface of UDD (a) and the same for ACM 1/0 diamond powders (b). Distribution % A J/g.

Reference [2]. Calculated results are given in Table 2 and Figures 2 and 3. Table 2 shows the $\Delta\sigma$ values at the lowest P/P_s value of 0.05, and at $P/P_s =$ 0.7, that correspond to one-layer coating. Extrapolation of the curve in Figure 2 to the $\Delta\sigma_{d-n}$ axis, obtains the value of interphase surface energy of diamond at the diamond-vacuum interface, i.e., the real value of the interphase surface energy of disperse diamond powders ($\Delta\sigma$).

CONCLUSIONS

The examination of physico-chemical properties of UDD show that they differ greatly from statically synthesized disperse diamond, though the chemical nature of the surface of these diamond crystals is similar as demonstrated by close values of $\Delta\sigma_{d-n}$ and $\Delta\sigma$. UDD crystals show a higher adsorption potential and a higher scattering of its values from less than 10 to 900 J/g, while ACM 1/0 crystals exhibit absorption potential values between 4 and 48 J/g. Just these differences are responsible for a higher adsorption activity towards water, which is supported by virtually complete coverage of the UDD surface by the various kinds of functional groups, predominately oxygen- and hydrogen-containing ones, that is observed by different methods. These facts should be taken into consideration when UDD is used to produce various materials.

REFERENCES

1. Gregg, S. J., K. S. W. Sing. 1982. *Adsorption, Surface Area and Porosity,* Academic Press.
2. Aleshin, V. G., A. A. Smekhnov, G. P. Bogatyreva. 1990. V. B. Kruk, *Chemistry of Diamond Surface* [in Russian], Naukova Dumka, Kiev.

On the Calculation of Dielectric and Optical Properties of Wide Band Gap Semiconductors

S. YU. DAVYDOV* AND S. K. TIKHONOV

A. F. Ioffe Physico-Technical Institute
Russian Academy of Sciences
194021, Polytechnicheskaya str. 26
St. Petersburg, Russia

ABSTRACT: The bond orbital model has been used to calculate linear and quadric dielectric susceptibility, electro-optical coefficient and static permittivity for diamond and cubic modification of SiC, BN, AlN, and GaN.

ANALYSIS AND SUMMARY

The well known bond orbital model (BOM) of Harrison [1–3] is a simple but a direct first principle method to calculate a wide range of semiconductor properties. Here we use the extended version of BOM [4,5] to describe linear and quadric dielectric susceptibility, electro-optical coefficient and static permittivity for diamond and cubic modification of SiC, BN, AlN, and GaN.

For the calculation of the terms $(X_1, X_2,$ etc.) in the expansion of polarization $\vec{P}$ in power of electric field $\vec{E}$ one can use the BOM expression for the bond energy (per one bond)

$$E = E_{ic} + V_0 + E' \tag{1}$$

where

---

*Author to whom correspondence should be addressed.

$$E_{ic} = -2(V_2^2 + V_3^2)^{1/2} \tag{2}$$

is the ionic-covalent component of the energy, V_0 is the repulsion energy, and V_2 and V_3 are the covalent and ionic energies. The term E' includes all contributions that do not depend on bond length d of the external electric field $\vec{E}$.

It is easy to show now [4,5] that the electronic contribution to the linear and quadric susceptibilities X_1, X_2, i.e., X_1^{el}, X_2^{el}, are

$$X_1^{el} = -\frac{N(q\gamma d_0)^2}{12 V_2} \alpha_c^3 \tag{3}$$

$$X_{14}^{el} = \frac{\sqrt{3} N(q\gamma d_0)^3}{48 V_2^2} \alpha_c^4 \alpha_p \tag{4}$$

where d_0 is the equilibrium bond length, q, positronic charge, γ, scaling factor, accounting the overlapping and field effects [1], N, valence-electron density, α_c, covalency, defined as

$$\alpha_c = -V_2/(V_2^2 + V_3^2)^{1/2} \tag{5}$$

and α_p is the ionicity ($\alpha_p^2 = 1 - \alpha_c^2$). The electronic component of linear susceptibility determines the high-frequency dielectric permittivity $\epsilon_\infty = n^2$ (n is the refractive index), since

$$\epsilon_\infty = 1 + 4\pi X_1^{el} \tag{6}$$

The dependence of X_1^{el} on pressure can be conveniently characterized by the parameter $\eta = \partial(\ln X_1^{el})/\partial(\ln d)$ [6]. Since $\gamma^2 \propto d$ [1,4], we obtain

$$\eta = 2 - 6\alpha_p^2 \tag{7}$$

Note also that the parameter η is related to the elasto-optical coefficients p_{ij} through

$$\eta = -\frac{\epsilon_\infty^2}{\epsilon_\infty - 1}(p_{11} + 2p_{12}) \tag{8}$$

The contribution of the ionic subsystem to the susceptibilities X_1 and X_{14}, i.e., the lattice response to the external electric field $\vec{E}$, can be calculated, if we take into account the dependence of bond vector $\vec{d}$ ($\vec{d}$

points from the cation to anion) on $\vec{E}$. The corresponding contributions read [5]

$$X_1^{ion} = -\frac{N(q\gamma d_0)^2}{24V_2} \cdot \frac{\alpha_p(1 + 2\alpha_c^2)}{\alpha_c} \tag{9}$$

$$X_{14}^{ion} = -\frac{\sqrt{3}N(q\gamma d_0)^3}{48 V_2^2}\alpha_p\alpha_c^2(1 - 2\alpha_c^2) \tag{10}$$

The total susceptibilities, which describe the response of a crystal to a low frequency field, are the sum of the electronic and ionic contributions

$$X_1 = X_1^{el}(1 + \vartheta) \tag{11}$$

where $\vartheta = \alpha_p^2(1 + 2\alpha_c^2)/2\alpha_c^4$, and

$$X_{14} = \frac{\sqrt{3} N(q\gamma d_0)^3}{48 V_2^2} \alpha_p^3\alpha_c^2 \tag{12}$$

Thus the static dielectric permittivity is $\epsilon_0 = 1 + 4\pi X_1$.

The linear electro-optical coefficient r_{41}, referring to the variation of the refractive index of a non-centrosymmetric crystal in a low-frequency electric field, is related to X_{14} through [5,7]

$$r_{41} = -4\pi X_{14}/n^4 \tag{13}$$

The results of our calculation of the dielectric characteristics are presented in Table 1. The geometrical parameters (d_0 and N) were taken from Reference [1], the energy parameters, from Reference [3], and the experimental values, from References [1] and [8–10]. We chose for γ the value 1.27, which is the mean of all the compounds considered here. Note that when used within the nearest-neighbor approach, the BOM can be applied to description not only of sphalerite structures but of wurtzite structures in the "cubic approximation" as well [1]. A comparison of the calculated ϵ_∞ and ϵ_0 with experimental data shows that while for covalent crystals (C, SiC) the calculated values of ϵ_∞ lie below the experimental figures, for the nitrides, which have a higher ionicity, they lie above. The theoretical values of ϵ_0 are underestimated relative to the experimental values, but the agreement for the ionic compounds appear to be better than for the covalent ones. The published value of

Table 1. Coupling parameters and values of the dielectric and optical characteristics of wide band gap semiconductors, $X_1^{el}, X_1, X_{14}^{el}, X_{14},$ and r_{41} given in units 10^{-8} cgse.

		C	SiC	BN	AlN	GaN
	α_p	0	0.26	0.34	0.59	0.60
	α_c	1	0.97	0.94	0.81	0.80
Theory	X_1^{el}	0.31	0.38	0.32	0.38	0.40
	ϵ_∞	4.90	5.83	5.02	5.72	5.03
	X_1	0.31	0.43	0.38	0.73	0.80
	ϵ_0	4.90	6.36	5.83	10.12	11.05
	η	2	1.59	1.31	−0.09	−0.16
	X_{14}^{el}	—	3.87	2.18	4.36	4.68
	X_{14}	0	0.27	0.29	2.31	2.63
	r_{41}	—	0.10	0.23	1.30	1.31
Experiment	ϵ_∞	5.7	6.5	4.5	4.8	5.8
	ϵ_0	5.7	9.7	—	—	12.2

η for diamond [6] is 2.2, which is in good agreement with our value of 2. As follows from Reference [4], the theory shows much better consistency with experimental data on η for low-ionicity crystals. This warrants the conclusion that $\eta = 1.59$ for SiC is a reliable value. As for the high-ionicity compounds, here the disagreement between theory and experiment may be quite large. The value $X_{zzz}^{(2)} = 2.88 \times 10^{-8}$ cgse electrostatic units found for GaN [11] is in satisfactory agreement with our result. Note that the theory [12] based on the model of Phillips-Van Vachten gives a value 22 times smaller than experiment. We have not been successful in finding any experimental information on r_{41}.

The above results have been obtained within the BOM approximation. It can be refined by taking into account the bonding between two $|sp^3\rangle$ orbitals on the same atom by adding to the bond energy E [Equation (1)] an additional term E_m called the metallization energy [3]

$$E_m = \frac{3}{8} \alpha_m^2 \alpha_c^2 V_2 \tag{14}$$

where

$$\alpha_m = -\sqrt{2}[(V_1^a)^2 + (V_1^c)^2]^{1/2} \tag{15}$$

is the metallicity, and V_1^a and V_1^c are the metallic energies of the anion and the cation, respectively [3]. Using this extended BOM, one can show [4] that Equations (3) and (4) for the linear and quadric electronic susceptibilities should be supplemented, respectively, by

$$g_1 = 1 + \frac{9}{16} \alpha_m^2 \alpha_c^2 (1 - 5\alpha_p^2) \tag{16}$$

and

$$g_{14} = 1 + \frac{21}{16} \alpha_m^2 \alpha_c^2 \left(1 - \frac{27}{7} \alpha_p^2\right) \tag{17}$$

and the term

$$\frac{9}{4} \frac{\alpha_m^2 \alpha_c^4 (1 - 10\alpha_p^2)}{1 + \frac{9}{16} \alpha_m^2 \alpha_c^2 (1 - 5\alpha_p^2)} \tag{18}$$

should be added on the right-hand side of Equation (7) for η.

The results of the extended BOM calculations are given in Table 2. The parameter γ was determined as this was done by Harrison in Chapter 4 of Reference [1]. The correction factors g_1 and g_{14} increase the values of susceptibility for the covalent crystals and reduce them for ionic crystals. The inclusion of metallicity affects particularly significantly the parameter η for crystals with a high bond ionicity. It should be pointed out that taking into account that metallicity is identical to the inclusion of band effects, since it is the matrix elements V_1^a and V_1^c that control the width of the band making electron delocalization possible [1]. The inadequacy of experimental information does not,

Table 2. *Results of calculations of the dielectric and optical properties of wide band gap semiconductors,* X_{14}^{el} *given in units* 10^{-7} *cgse.*

	C	SiC	BN	AlN	GaN
α_m	0.41	0.56	0.46	0.65	0.72
g_1	1.09	1.11	1.04	0.88	0.85
g_{14}	1.22	1.28	1.14	0.88	0.83
γ	1.33	1.51	1.33	1.51	1.62
ϵ_∞	5.70	7.87	5.03	4.15	4.61
η	2.35	1.78	1.25	−1.24	−1.63
X_{14}^{el}	0	0.85	0.29	0.66	0.83

however, permit a comprehensive analysis of the significance of inclusion of the metallicity.

In summary, we have calculated the dielectric and optical characteristics of wide band gap materials by the BOM, which was successfully employed in describing many properties of semiconductors. This method permits circumventing complicated computations that are usually difficult to avoid in calculations of dielectric and optical properties [13–15], and this is particularly useful in trying to predict and estimate the properties of new materials. The satisfactory agreement of our results with the available experimental data, regrettably very sparse, gives us grounds to hope that the values of physical characteristics predicted by us will likewise prove to be reasonable.

This work was supported by Arizona University (USA) and INTAS grant 93-543.

REFERENCES

1. Harrison, W. A. 1980. "Electronic Structure and the Properties of Solids," Freeman, San Francisco.
2. Harrison, W. A. 1981. *Phys. Rev.,* B24:5835.
3. Harrison, W. A. 1983. *Phys. Rev.,* B27:3592.
4. Davydov, S. Y. and E. I. Leonov. 1987. *Sov. Phys. Solid State,* 29:1662.
5. Davydov, S. Y. and E. I. Leonov. 1988. *Sov. Phys. Solid State,* 30:768
6. Kastner, M. 1972. *Phys. Rev.,* B6:2273.
7. Sirotin, Yu. I. and M. P. Shaskol'skaya. 1979. "Fundamentals of the Physics of Crystals," In Russian, Nauka, Moscow.
8. 1976. "Tables of Physical Quantities: A Handbook," I. K. Kikoin, ed., Atomizdat, Moscow.
9. Gavrilenko, V. I., A. M. Grekhov, D. V. Korbutyak and V. G. Litovchenko. 1987. "Optical Properties of Semiconductors: A Handbook," In Russian, Naukova Dumka, Kiev.
10. 1991. "Physical Properties: A Handbook," I. S. Grigor'ev and E. Z. Meilokhov, eds., Energoizdat, Moscow.
11. Miragliotta, J., W. A. Bryden, T. J. Kistenmacher and D. K. Wickenden. 1993. In. *Proc. Fifty Conf. Silicone Carbide and Related Materials,* M. G. Spencer et al., eds. Institute of Physics Conference Series, No. 137. Bristol and Philadelphia. p. 537.
12. Levine, B. F. 1973. *Phys. Rev.,* B7:2600.
13. Moss, D. J., J. E. Sipe and H. M. van Driel. 1987. *Phys. Rev.,* B36:9708.
14. Dovgii, Ya. O. and I. V. Kityk. 1991. *Sov. Phys. Solid State,* 33:238.
15. Kityk, I. V. 1991. *Sov. Phys. Solid State,* 33:1026.

Properties of Carbyne Obtained by Various Methods

K. VILCEV*
SPRG "Powerchnostj" Metalophysical Institute 9/23
2nd Baumanskaja str.
107005, Moscow, Russia

Y. YAMADA, M. SHIRAISHI AND N. KIKUKAWA
National Institute for Resources and Environment
16-03 Onogawa, Tsukuba
Ibaraki 305, Japan

M. B. GUSEVA AND V. G. BABAEV
M. V. Lomonosov Moscow State University
Department of Physics
Division of Electronics
Moscow, Russia

ABSTRACT: Carbyne films were obtained by graphite evaporation in a vacuum chamber by two methods: simultaneous irradiation of the growing material with fluxes of Ar^+ and C atoms; and by "capping" with CF_3 radicals. Some differences in the electron diffraction patterns and the Raman spectra were found.

INTRODUCTION

S olid carbon with triple or adjacent double bonds, carbynes, possess very interesting properties, with applications in superconductivity, electronics, catalysis and other areas. Already these compounds are applied in the development of medical prostheses [1].

---

*Author to whom correspondence should be addressed.

There exist at least two methods for obtaining carbyne through vaporization of carbon.

1. Deposition with simultaneous irradiation of Ar^+ [3]
2. Stabilization of chains using "capped species" [4]

EXPERIMENTAL

We used the second method. Graphite rods (99.999%), a He (99.99%) high energy pulsed CO_2 laser (Lumonics TEA-103) with a pulse of 10 Joules, at a frequency of 1 Hz, were used. The energy was focused on the graphite rod with a ZnSe lens (F = 25.4 cm). Samples were deposited on glass substrates at –20°C (253 K) and on the surface of a copper tube at 70 K.

Deposition took place in a vacuum chamber (Figure 1) that was previously evacuated to 0.05 Torr. Experiments were started after backfilling the chamber with He (0.2 atm). C_2F_6, 1%, was introduced for "capping" of the carbon chains.

Laser evaporation took 1.5 hours and was followed by a 100 W microwave discharge for 30 min. Films were deposited on the glass substrate at a temperature of 100 K. Speed of evaporation was in the range of 0.003

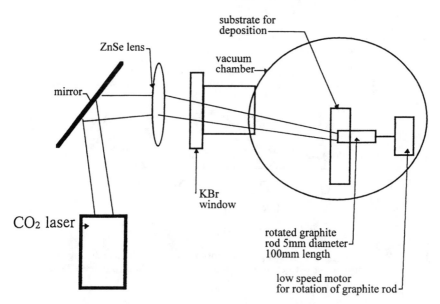

Figure 1. Setup for preparation of carbon chains by laser irradiation with capped species.

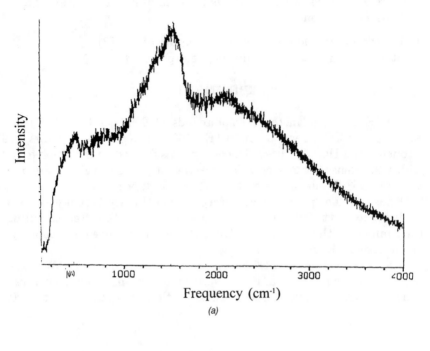

Frequency (cm⁻¹)

(a)

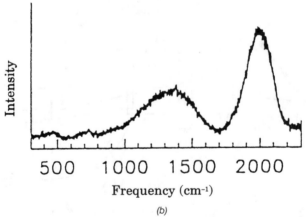

Frequency (cm⁻¹)

(b)

Figure 2. Raman spectra. (a) Copper substrate with coating formed during laser evaporation of graphite target with capped species. (b) Carbyne obtained from PTFE [4].

312

g/hour. Significant amounts of material were dispersed in the volume of the chamber. For this report only material from the surface of the glass substrate and from the copper were used. Raman spectra were obtained with a Reinishaw Raman imaging microscope.

RESULTS AND DISCUSSION

The electron diffraction patterns were typical for this material, but differed in the lattice constant which was 0.481 ± 0.4 nm. This is slightly higher than usually observed [5]. It could be due to a greater amount of material with end groups. The microcrystals that were observed in other cases were absent in this experiment.

The Raman spectra show a broad peak with a maximum at about 1900 cm^{-1}. This peak is due to the triple bond of carbon. The exact position of this band can significantly vary depending on the number of chains and the type of end group in the molecule. In the case of carbyne from a fluoropolymer [6] a very sharp peak was observed (Figure 2). This can be explained by the uniformity of the chain sizes. The frequency of this peak changed only little with substitution of groups $-F$, $-CF_3$, $-(CF_2)_n-CF_3$ on the end of the carbon molecules. This dependence was already established theoretically for polyacetylenes and carbynes.

In the case of carbyne from polytetrafluoroethylene (PTFE), it is the opinion of some authors [6] that the number of chains in one molecule is not more than four. In our case we can expect formation of a mixture of chains with various lengths [7]. When growing with a simultaneous flux of Ar^+ we cannot obtain long lengths because of insufficient stabilization without "capped species."

CONCLUSION

The carbyne materials obtained by three methods differ in lattice constant, level of crystallinity, and of Raman spectra details. This is likely to be due to different mechanisms of formation and stability of these materials.

ACKNOWLEDGEMENTS

One of the authors (V.C.) would like to thank the Agency of International Science and Technology (Japan) for awarding him a Research Fellowship. He also thanks the staff of Division of Carbon for support.

The authors gratefully thank Prof. Inagaki M. for useful discussion, and the staff of Gunma Prefectural Industrial Research Technical Laboratory for the Raman investigations.

REFERENCES

1. Guseva, M. B. and V. G. Babaev, et al. 1995. *Diamond and Related Materials,* 4:1442.

2. Heimann, R. B. 1994. *Diamond and Related Materials,* 3:1151–1157.

3. Guseva, M. B., N. F. Savchenko and V. G. Babaev. 1985. "New Data on a Structure of Carbynes," *Physika,* pp. 1336–1339.

4. Lagow, R. J. 1995. *Science,* January, vol. 267.

5. Digilov, M. and K. Vilcev, et al. 1993. *Carbon,* 31(N2):365–371.

6. Kavan, L. and J. Hlavaty. *Carbon,* 33(N9):1321–1329.

7. Wang, C.-R. and R.-B. Huang, et al. 1994. "Chemical Physics Letters," N227:103–108.

Structural Modification of Diamond Subjected to High Energy Ion Irradiation

V. A. MARTINOVICH, D. P. ERCHAK, V. S. VARICHENKO,* E. G. EFIMOV,
N. M. PENINA AND V. F. STELMAKH
Belarussian State University
220050 Minsk, Belarus

A. F. ALEXANDROV AND M. B. GUSEVA
Moscow State University
Moscow, Russia

INTRODUCTION

High energy ion implantation in diamond leads to formation of specific extended defects—tracks along the ion paths [1,2]. The ion track is believed to be a rod-like macrodefect with a lower atom density along the ion path and an envelope of which has an enhanced atom density. It is proposed that modified regions with nontetrahedral interatomic bondings are created inside individual track due to atomic reconstruction processes during high energy ion implantation. Transport phenomena in these structures can be realized by means of mobile quasiparticles like solitons.

The aim of this work is a study of track-like quasi-one-dimensional nontetrahedral structures in diamond.

---

*Author to whom correspondence should be addressed. Dr. V. Varichenko, HEII, Belarussian State University, Minsk 220080, Belarus. Fax: 375/0172/265940. E-mail: VALERY@HEII.BSU.MINSK.BY.

EXPERIMENTAL

Ia and IIa natural diamond samples irradiated by ions of copper (63 MeV, 5×10^{14} cm^{-2}) and nickel (335 MeV, $5 \times 10^{12} - 5 \times 10^{14}$ cm^{-2}) have been investigated by EPR technique at the room temperature. The signal of Cr^{3+} ions of a ruby standard sample permanently located in the cavity was used for the quality factor control and for calibration of amplitude of magnetic component H_1 of microwave field.

RESULTS AND DISCUSSION

High energy ion irradiation of diamond leads to the appearance of weak anisotropic paramagnetic centres (PC) having intensive singlet lines with g-value near to 2.0027 and a set of strong anisotropic narrow lines with relatively low intensity. The strong anisotropic narrow lines result from radiation point defects of diamond lattice. The strong intensive lines characterized by anisotropic intensities, linewidths and asymmetry of A/B (A and B-amplitudes of high- and low-field parts of EPR line) are attributed to resonance absorption on mobile quasiparticles, namely, solitons in track-like quasi-one-dimensional structures produced during high energy ion implantation in diamond [1,2]. These structures are stable enough and some part of them was detected after annealing at a temperature of 1400°C.

Annealing investigation of diamond implanted by 335 MeV nickel ions with a dose of 5×10^{14} cm^{-2} has demonstrated a nonmonotonic dependence of the singlet line width of dominating PC versus increase of temperature (Figure 1). These changes are high enough. At temperatures within a range of 650–700°C the linewidth increases from 5.3 G up to 7.7 G and then decreases within a range of 750–850°C up to 3.3 G. Such behaviour of linewidth speaks in favor of quasiparticle origin of the dominating PC. Indeed, for deep level PC having dangling bonds (for example, Si-G8 centre in Si [3]) the linewidth changes during an annealing are usually negligible and cannot be nonmonotonic in any case. Paramagnetic quasiparticles have also deep level in gap but they possess a more wide distribution of wave function than above mentioned PC with dangling bonds. So, for example, for topological solitons in polyacetylene wave function spreads all over the 45 carbon atoms [4]. Besides the quasiparticles are mobile. It determines their high sensitivity to structural rearrangements of track-like regions caused more probably by activation of vacancy movement during annealing [5].

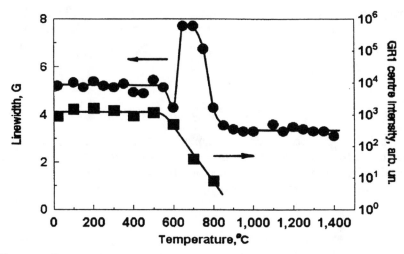

Figure 1. Change of singlet line width in diamond irradiated with 335 MeV Ni ions (5×10^{14} cm^{-2}) versus annealing temperature ($\bullet$) in comparison with the annealing curve of intensity of cathodoluminescence vacancy centre GR1 [5] ($\blacksquare$).

An anisotropy of EPR spectra, relative to the H_1 microwave field resulting in appearance of new additional lines, has been found for the first time. So, Figure 2 demonstrates the spectra of diamond irradiated by 63 MeV Cu ions. In the first case [Figure 2(a)] the spectrum is registered when H_1 is perpendicular to the implanted plane of the sample, i.e., H_1 coincides with irradiation direction. In the second case [Figure 2(b)] the spectrum is registered when H_1 is parallel to the implanted plane of the sample. One can see that additional broad lines L_2 and L_3 are appeared as left sidebands of the main resonance line L_1 in the first case only.

The same tendency of appearance of additional sidebands L_2, L_3 of the main resonance line has been found in diamond subjected to 335 MeV Ni implantation in the case when H_1 was parallel to irradiation direction. A thermal treatment of the sample leads to the increase of the tendency. It should be noted that after 20 min. isochronal annealing at 350°C and higher temperatures the lines L_2, L_3 are also detected when H_1 is perpendicular to irradiation direction.

According to the results obtained during investigations of implanted diamond films [6] the lines, L_1, L_2 and L_3 are ascribed to resonance modes of anisotropic spin waves excited at modified structures along the irradiation direction. A source of excitation of the spin waves is proposed to be a spin system of mobile quasiparticles. The registered spectrum has the similar spectral distribution of the amplitudes and linewidths of the modes which is a characteristic of spin wave resonance [6]. The distance between the modes is a quadratic dispersion relation like Kittel relation for thin ferromagnetic films [7]:

$$\omega = A\kappa^2 + w_0 \tag{1}$$

$$k = \frac{n\pi}{d} \tag{2}$$

where d is an effective thickness of measured structure; n is an integer and equals to 3 and 5 for modes L_3 and L_5, respectively; A is a coefficient characterizing a magnitude of electron-electron correlations.

As it has been mentioned above spin wave resonance in as-implanted

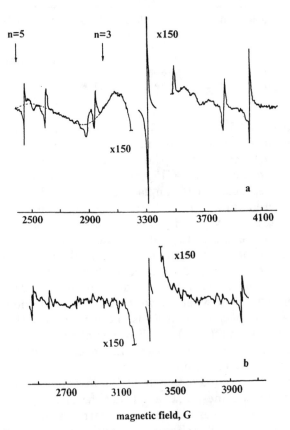

Figure 2. EPR spectrum of diamond irradiated with 63 MeV Cu ions (5×10^{14} cm^{-2}): (a) sample implantation plane is perpendicular to magnetic component H$_1$, $\nu = 9316.18$ MHz; (b) sample implantation plane is parallel to H$_1$, $\nu = 9318.47$ MHz.

diamond is observed only in the case when magnetic component H_1 coincides with ion track direction. A calculated ratios A/d^2 obtained accordingly with Equations (1) and (2) are equal to 6.6×10^7 and 6.4×10^7 c^{-1} for 335 MeV Ni and 63 MeV Cu, respectively. The anisotropy of EPR spectra relative magnetic microwave field as a consequence of the anisotropy of the excitation of the spin waves can be explained by taking into account quasi-one-dimensional nature of the track-like modified structures. This can be considered as additional argument for the formation of such structures during high energy implantation in diamond.

Registration of spin waves in annealed Ni-implanted sample, in case when H_1 is perpendicular to irradiation direction, speaks about increase of electron-electron correlations connected with carbon atoms in track-like modified structures and can be a consequence of the rearrangement of defect structure during annealing. In this case, A/d^2 value varies from 6.6×10^7 (for as-implanted diamond sample) to 1.2×10^8 c^{-1} (for implanted and annealed at 700°C).

CONCLUSION

EPR studies have revealed the formation of track-like quasi-one-dimensional nontetrahedral structures along ion path after high energy ion implantation in diamond. The intensive singlet EPR lines corresponding to PC of these ion-modified structures are the result of EPR absorption on mobile quasiparticles like solitons.

Considerable nonmonotonic changes of EPR singlet line width during annealing are explained in terms of mobile quasiparticles model of corresponding PC.

Spin waves associated with mobile quasiparticles have been observed in track-like ion-modified structures of diamond single crystals at room temperature. For the first time an anisotropy of EPR spectra relatively magnetic component of microwave field has been observed. This effect is determined by the anisotropy of the spin waves excitation by the microwave field.

This work was partially supported by INTAS project No. 94-1982.

REFERENCES

1. Varichenko, V. S., A. Yu. Didyk, V. A. Martinovich, D. P. Erchak, et al. 1995. *Coommun. JINR,* P14-95-145, Dubna, p. 11.

2. Erchak, D. P., V. G. Efimov, A. M. Zaitsev, V. F. Stelmakh, N. M. Penina, V. S.

Varichenko and V. P. Tolstykh. 1992. *Nucl. Instr. and Meth. in Phys. Res.* B69:443.

3. Watkins, G. D. and J. A. Corbett. 1964. *Phys. Rev.,* 134A:1359.

4. Mehring, M., P. Hofer, H. Kass and A. Grupp. 1988. *Europhys. Lett.,* 6:463.

5. Varichenko, V. S., A. Yu. Didyk, A. M. Zaitsev, V. I. Kuznetsov, et al. 1988. *Coommum. JINR,* P14-88-44, Dubna, p. 12.

6. Erchak, D. P., M. B. Guseva, A. F. Alexandrov, H. Alexander and A. Pilar von Pilchau. 1993. *JETR Letters,* 58:275.

7. Kittel, C. 1958. *Phys. Rev.,* 110:836.

Photothermal Measurements of Thermal Conductivity of Thin Amorphous Films

J. BODZENTA,* J. MAZUR AND Z. KLESZCZEWSKI

Institute of Physics
Silesian Technical University
ul. Krzywoustego 2
44-100 Gliwice, Poland

ABSTRACT: A nondestructive method for the determination of thermal conductivity of transparent, thermally thin coating on opaque, thermally thick substrates is proposed. The sample is heated by intensity modulated, unfocused laser beam. The photodeflection detection is used for signal registration and the "mirage" signal dependence on modulation frequency is measured. The method is based on 1-D model of heat propagation in layered structures. The light interference in coating layer is taken into account. Two geometries of photothermal experiments are considered. The modulated light may illuminate the coated surface of sample or the opposite one. It is shown that the ratio of signals measured in these geometries does not depend on parameters of the probe beam. The proposed method is verified experimentally for silicon wafers coated with DLC. These results are very promising.

KEY WORDS: thermal properties, thin films, photothermal measurement.

1. INTRODUCTION

Thermal properties of diamonds and diamond films were investigated by many authors using different experimental techniques, e.g., thermocouple method [1], optically induced transient grating experi-

---

*Author to whom correspondence should be addressed.

ment [2]. In the last few years photothermal measurements of thermal properties of diamond layers has become popular [3–7].

Potential uses of photothermal measurement techniques strongly depend on the method used for signal detection. The mirage effect or the photothermal beam deflection (PBD) method is used for signal detection in our laboratory. Intensity modulated laser beam illuminates the sample surface and causes local changes of temperature distribution in the sample and the air above it. The probe beam which passes parallel to the sample surface is deflected by refractive index gradient caused by temperature gradient in the gas. Two different dependencies of PBD signal can be measured—dependence on distance between the probe and the power beam (spatial dependence) and dependence on modulation frequency.

Boudina and co-workers [6], and Bachmann and co-workers [7] had measured spatial dependence of the PBD signal and then used rather complicated multiparameter fitting methods for determination of thermal parameters of polycrystalline diamond films (PCD) layers. Moreover, the measurements were performed for free standing PCD membranes, which required special preparation of samples. It may be difficult to determine thermal properties of thin diamond-like carbon (DLC) coatings using this method.

In this paper a new measuring technique is proposed for thermal conductivity determination of thin DLC layer on thermally thick substrate. The proposed method is based on PBD signal frequency dependence and simple one dimension (1-D) theoretical model. The method is nondestructive and noncontact.

2. THEORY

Let us consider a layered structure which consists of gas over the sample, coating layer of thickness w, substrate with thickness d, and gas below the sample. Thermal properties of each layer are characterized by its thermal conductivity κ_i and thermal diffusivity β_i (where $i = s, c, g$ denotes the substrate, the coating layer and the gas respectively). Optical properties of coating and substrate are described by their complex refractive indexes. The sample is illuminated by intensity modulated light (the angular frequency of modulation is equal to ω). Two orientations of the sample are considered (Figure 1). The light beam illuminates either the coated surface of the sample ("top illumination") or the opposite ("bottom illumination"). The probe beam is passing parallel to the sample surface at height h above it.

In PDB detection the measured signal is proportional to the temperature gradient in the gas over the surface of sample [8]

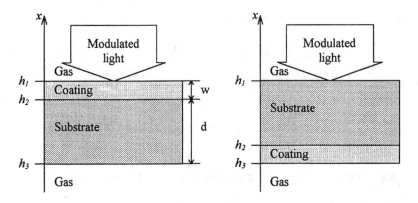

Figure 1. "Top" and "bottom" illumination of the sample.

$$S(\omega) = K \left.\frac{\partial T_g(x,\omega)}{\partial x}\right|_{x=h} \tag{1}$$

where K is a constant which depends on thermooptical properties of gas and sensitivity of detector.

As one can see the signal is completely determined by distribution of temperature in gas over the sample. The following section is devoted to description of the temperature field in considered system.

2.1 Description of Temperature Field in Gas and Sample

2.1.1 Fourier-Kirchhoff Equation and Boundary Conditions

The distribution of temperature in each layer of the sample and gas is described by the Fourier-Kirchhoff heat transfer equation. After Fourier time-frequency transformation this equation can be written in the form

$$\frac{\partial^2 T_j(x,\omega)}{\partial x^2} - \frac{i\omega}{\beta_j} T_j(x,\omega) = \Lambda_j(x,\omega) \tag{2}$$

where $j = g$ (gas over the sample), s (substrate), c (coating) or gb (gas below the sample), $\Lambda_j(x,\omega)$ is the volume density of heat source in jth medium and $T_j(x,\omega)$ is the Fourier component of temperature field on frequency ω.

These differential equations must be completed with the following boundary conditions

- temperature field continuity

$$T_j(h_j, \omega) = T_{j+1}(h_j, \omega) \qquad (3a)$$

- conservation of heat flux

$$\kappa_j \left. \frac{\partial T_j(x, \omega)}{\partial x} \right|_{x=h_j} = \kappa_{j+1} \left. \frac{\partial T_{j+1}(x, \omega)}{\partial x} \right|_{x=h_j} - Q_j(x, \omega) \qquad (3b)$$

where $Q_j(x, \omega)$ is a heat flux source on $j - j + 1$ interface.
It is also assumed that

$$T_g(+\infty, \omega) = 0, \qquad T_{gb}(-\infty, \omega) = 0 \qquad (3c)$$

The solutions of Equation (2) can be written as:

$$T_j(x, \omega) = \lambda_j(x, \omega) + A_{j,1}(\omega) \ \exp\left(-\sqrt{\frac{i\omega}{\beta_j}} x\right) + A_{j,2}(\omega) \ \exp\left(\sqrt{\frac{i\omega}{\beta_j}} x\right) \qquad (4)$$

where $\lambda_j(x, \omega)$ is the inhomogeneous solution caused by volume heat source, and $A_{j,n}$ are functions of ω. In particular for gas above the sample we have

$$T_g(x, \omega) = A_{g,1}(\omega) \ \exp\left(-\sqrt{\frac{i\omega}{\beta_g}} x\right) \qquad (5)$$

and then we obtain from Equation (1) an expression for PDB signal

$$S(\omega) = -K \sqrt{\frac{i\omega}{\beta_g}} A_{g,1}(\omega) \ \exp\left(-\sqrt{\frac{i\omega}{\beta_g}} h\right) \qquad (6)$$

where h is the height of the probe beam over the surface of the sample. The function $A_{g,1}(\omega)$ depends on thermal and optical properties of substrate and coating.

2.1.2 Heat Sources in the Sample

In photothermal experiments heat sources arise as a result of absorption of energy from a light beam. In the case of opaque media, energy is absorbed near illuminated surface. Resulting heat source can be incorporated in the theoretical model as additional heat flux in flux-continuity condition [Q_j in Equation (3b)]. For transparent media, absorption of energy leads to volume heat sources [Λ_j in Equation (2)].

As noted above, the sample is illuminated in two ways. For the "top illumination" one should consider two heat sources. The first one is a volume heat source in coating. Proper description of this source requires analysis of light interference in the layer. It can be shown that a general form of such a source is,

$$\Lambda_l(x,\omega) = \frac{\alpha I_1}{\kappa_c}\left(B_1\,\exp(\alpha x) + B_2\,\exp(-2\alpha w + \alpha x) + B_3\,\cos\left(\frac{4\pi n_c}{\lambda}x - \phi\right)\right) \quad (7)$$

where I_1 = intensity of light entering the coating, B_1, B_2, B_3, ϕ = constants depending on optical properties of gas, the coating and the substrate, α = absorption coefficient of coating layer, n_c = refractive index, λ = light wavelength. The second source is placed on the coating-substrate interface and is proportional to the energy absorbed in the substrate

$$Q_2 = \frac{I_2}{\kappa_s} \quad (8)$$

where I_2 = energy flux entering the substrate.

For the "bottom illumination" only one plain source arises on the bottom surface of sample. This source depends on the intensity, I_3, of light illuminating the sample

$$Q_1 = \frac{I_3}{\kappa_s} \quad (9)$$

2.2 Analysis of Experimental Data

It follows from Equation (6) that the ratio of signals measured for "top" and "bottom" illuminations is of the form

$$R(\omega) = \frac{S^{top}(\omega)}{S^{bottom}(\omega)} = \frac{A_{g,1}^{top}(\omega)}{A_{g,1}^{bottom}(\omega)} \quad (10)$$

where $A_{g,1}^{top}$ and $A_{g,1}^{bottom}$ can be derived from Equations (3a), (3b), (3c) and

Equation (4) with volume heat source Equation (7) and heat flux sources Equations (8) and (9).

The analytical formula for $R(\omega)$ obtained from Symbolic Computation System is very complicated, but based on some assumptions one can obtain simpler form.

The following properties of this system can be used:

- The thermal conductivity of gas is much smaller than the thermal conductivity of coating and substrate.
- For this range of modulation frequency, the coating layer is thermally thin.
- The coating is transparent.

These simplifications lead to a more convenient formula for signal ratio

$$R(\omega) = \frac{\alpha I_1 (C_4 - 2\pi C_1 w) + I_2 \dfrac{\kappa_c}{\kappa_s} - \pi \alpha I_1 C_1 w^2 \dfrac{\kappa_s}{\kappa_c} \Gamma_s(\omega) \tanh(\Gamma_s(\omega)d)}{I_3 \left(\kappa_s + \dfrac{\rho_c c_c}{\beta_s} w \Gamma_s(\omega) \tanh(\Gamma_s(\omega)d) \right)} \tag{11}$$

where $\Gamma_j(\omega) = \sqrt{i\omega / \beta_j}$ and $C_1, C_6 = $ constants.

3. EXPERIMENTAL VERIFICATION

The measurements were carried out for two silicon wafers coated with thin DLC layers. Optical parameters of the coatings and their thickness have been determined from independent optical measurements. Results are shown in Table 1. Thermal parameters of silicon substrates have been measured for a non-coated silicon wafer (reference sample) using method described in Reference [9]. The thermal diffusivity of silicon obtained is 0.95 cm²/s.

Table 1. Parameters of investigated samples.

Sample	Substrate Thickness d [μm]	Coating Layer Thickness w [μm]	Coating Layer Refractive Index n
1	480 ± 5	1.32 ± 0.06	2.52 ± 0.02
2	465 ± 5	0.65 ± 0.01	2.46 ± 0.02
Reference	480 ± 5		

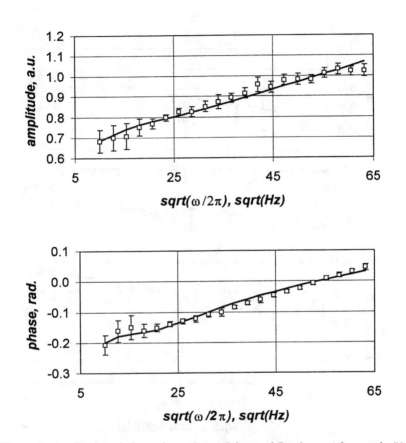

Figure 2. Amplitude and phase of experimental data and fitted curves for sample #1.

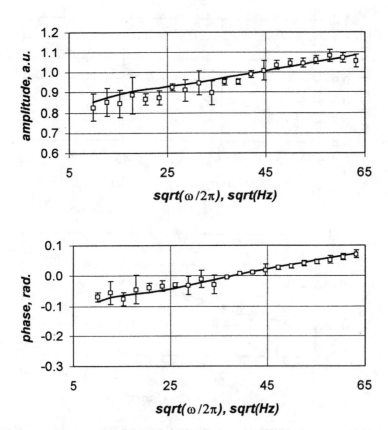

Figure 3. Amplitude and phase of experimental data and fitted curves for sample #2.

The PBD signal dependence on square root of frequency has been measured for top and bottom illumination in frequency range 100 Hz–40 kHz for example sample. Then, the ratio of signals obtained for "top" and "bottom" illumination has been calculated. Results are shown in Figure 2.

Theoretical curves Equation (11) have been fitted to the experimental data using least square method. Fitted curves are shown in Figures 2 and 3 together with experimental data. Preliminary fitting results show that the thermal conductivity of the DLC layer is about 0.2–0.3 $Wm^{-1}K^{-1}$.

4. CONCLUSIONS

A new method of nondestructive determination of thermal conductivity of thermally thin layers on thermally thick substrates is proposed. An analytical formula for description of measured signal ratio has been derived. Experimental results support the possibility of using the method for investigation of DLC carbon layers.

Values of thermal conductivity of investigated DLC layers on silicon substrates are much smaller than the thermal conductivity of pure diamond. This small value of thermal conductivity of thin diamond-like carbon layers is probably caused by small grain size, which restricts phonon-free path. Since the phonon-free path at room temperature is much larger than the grain dimensions in DLC films it would be expected to scatter phonons. Moreover, estimated thermal conductivity values should be treated as effective values for diamond and interface amorphous material from grain boundaries.

One should point out that the results presented previously [10] are not correct. This was caused by an over simplified model which neglected the problem of heat source structure in the coating layer.

ACKNOWLEDGEMENTS

We would like to thank professors S. Mitura (Technical University of Łódź), A. Sokołowska and J. Szmidt (Warsaw University of Technology) for helpful discussion of this work. We gratefully acknowledge the Polish State Committee for Scientific Research (KBN) for support under grant no. 2 P302 084 06.

REFERENCES

1. Graebner, J. E., J. A. Mucha, L. Seibles and G. W. Kamlot. 1992. "The Thermal Conductivity of Chemical-Vapor-Deposited Diamond Films on Silicon," *J. Appl. Phys.*, 71(7):3143–3146.

2. Tokmakoff, A., W. F. Banholzer and M. D. Fayer, 1993. "Thermal Diffusivity Measurements of Natural and Isotopically Enriched Diamond by Picosecond Infrared Transient Grating Experiments," *Appl. Phys.*, A56:87–90.

3. Visser, E. P., E. H. Versteegen and W. J. P. van Enckevort. 1992. "Measurement of Thermal Diffusion in Thin Films Using a Modulated Laser Technique: Application to Chemical-Vapor-Deposited Diamond Films," *J. Appl. Phys.*, 71(7):3238–3248.

4. Chen, Z. and A. Mandelis. 1992. "Thermal-Diffusivity Measurements of Ultrahigh Thermal Conductors with Use of Scanning Photothermal Rate-Window Spectrometry: Chemical-Vapor-Deposition Diamonds," *Phys. Rev.*, B46(20):13526.

5. Petrovsky, A. N., A. O. Salnick, D. O. Mukhin and B. V. Spitsyn. 1992. "Thermal Conductivity Measurements of Synthetic Diamond Films Using the Photothermal Beam Deflection Technique," *Mater. Sci. Eng.*, B11:353–354.

6. Boudina, A., E. Fitzer, U. Netzelmann and H. Reiss. 1993. "Thermal Diffusivity of Diamond Films Synthesized from Methane by Arc Discharge Plasma Jet CVD," *Diamond Relat. Mater.*, 2:852–858.

7. Bachmann, P. K., H. J. Hagemann, H. Lade, D. Leers, D. U. Wiechert, H. Wilson, D. Fournier and K. Plamann. 1995. "Thermal Properties of C/H-, C/H/O-, C/H/N- and C/H/X-Grown Polycrystalline CVD Diamond," *Diamond Relat. Mater.*, 4:820–826.

8. Murphy, J. C. and L. C. Aamodt. 1980. "Photothermal Spectroscopy Using Optical Probing: Mirage Effect," *J. Appl. Phys.*, 51(9):4580–4588.

9. Bodzenta, J., J. Mazur, R. Bukowski and Z. Kleszczewski. 1995. "Photothermal Measurement for Plates," *Proc. SPIE*, 2643:286–292.

10. Bodzenta, J., J. Mazur, S. Mitura and P. Niedzielski. 1995. "Investigation of DLC Coated Wafers," *J. Chem. Vap. Deposit.*, 3(4):311–323.

Optical Absorptivity of Diamond Powders and Polycrystalline Films

A. N. OBRAZTSOV* AND I. YU. PAVLOVSKY
Physics Department of Moscow State University
119899 Moscow, Russia

H. OKUSHI AND H. WATANABE
Electrotechnical Laboratory
Tsukuba, Ibaraki 305, Japan

ABSTRACT: Optical absorptivity of diamond powders and polycrystalline CVD films have been obtained by using photoacoustical spectroscopy in the spectral diapason from 200 to 1500 nm with different light modulation frequencies (from 20 to 1500 Hz). The both powders and films demonstrate spectral features corresponded to fundamental absorption edge near the diamond band gap (5.5 eV) and optical absorption in visual light region which was similar to one in neutron or electron irradiated single diamond. Specific "downfall" in optical absorption have been detected in UV region for powders and polycrystalline films with smallest diamond crystallites. The spectra dependence from light modulation frequency allow to conclude that UV "downfall" is due to mirror light reflection from randomly oriented diamond grains.

1. INTRODUCTION

Diamond possesses a combination of excellent properties: optical transparency from the ultraviolet to the far infrared, greatest hardness, highest room temperature thermal conductivity and essential chemical inertness. These properties make it a desirable material for many optical applications. The high cost and limited dimension of optical quality diamond have limited its use to a few specialized optical

---

*Author to whom correspondence should be addressed.

applications. The best optical quality diamond still comes from the natural sources. However, chemical vapor deposition (CVD) method offers the promise to synthesize the diamond films or plates of large sizes, making possible the wide use of diamond optical coatings and bulk optics [1].

The investigation of the optical properties of CVD diamond films is of interest both for furthering our knowledge of the physics of diamonds and for the application of these materials. One of the fundamental parameters of materials which determines their optical properties is the electronic structure. Thus, the optical properties of CVD diamond films are of great interest to study the possibility for their electronic applications also.

The optical properties of diamond are strongly affected by defects. The first group includes the points defects, the impurities, and the defect complexes. Single-atom impurities occur in the diamond including nitrogen, boron, silicon, and hydrogen. Defects and impurities may cause the optical absorption in several ways. They may break the lattice symmetry, causing one-phonon infrared absorption, which is forbidden in pure diamond. And transition between the localized electronic or vibration states of a defect may cause the absorption lines or bands that are specific to the defect. The second group includes the lattice defects as dislocations, stacking faults, twin boundaries, and grain boundaries. It is not well known how these defects affect the optical transmission, although the dislocations have been associated with luminescence. But these defects may be the reason for diffuse light scattering making polycrystalline films nontransparent even in case of the relatively low concentration of impurities and points defects [2].

From the point of view of the light scattering, the optical properties of CVD polycrystalline films should be very similar to ones of the diamond powders. Usually the optical absorptivity of powders and polycrystalline films is difficult to obtain by common methods due to the strong diffuse light scattering.

In this paper we present the results of the diamond powders and polycrystalline CVD films study of Photoacoustical (PA) spectroscopy. The basic peculiarity of the photoacoustical method is the sample illumination by the periodically modulated light. As a result the sample is heated due to the nonradiative transitions associated with the optical absorption. The periodic heating generates a pressure variation in the ambient gas due to the periodic heat flow to the gas. The pressure variation (acoustic wave) can be detected with a sensitive microphone by keeping the sample in an enclosed cell. The PA spectroscopy is known as a very effective method especially for the powders. This spectroscopy has the important advantage of allowing the study of the optical absorption of the thin films without their separating from the substrates [3].

2. EXPERIMENTAL DETAILS

Here we used the CVD diamond films fabricated in General Physics Institute (Moscow). The first sample set included the diamond films deposited by microwave (MW) plasma enhanced CVD using the ASTEX MW system. The diamond films were deposited onto Si(100) substrate and then have been separated. This procedure provided the free-standing films with thickness of 50, 100 and 200 μm grown by CVD process with durations 4, 12 and 29 hours correspondingly. The second set included the diamond films deposited by arc discharge techniques onto the silicon substrates. The thickness of the samples was of 50 μm (6 hours growth) and 10 μm (1 hour growth). Also, the thin nanograined diamond films (thickness of 1 μm) onto Si substrates have been investigated.

The third material was the HPHT synthetic diamond powders with crystallite dimension ranging from 0.25 μm to 125 μm and the ultrafine explosive diamond (UFD) powders with grain size of 4 nm. To provide the comparative data for the diamond powders and films the 3 × 3 × 0.3 mm³ size HPHT synthetic diamond single crystals (Sumitomo Electric Co.) with boron and nitrogen contamination were measured in the same conditions.

The amplitude and phase of PA signals were measured as a function of the wavelength with the modernized PA spectrometer manufactured by the Princeton Applied Research Corporation (Model 6001). A 1 kW Xenon arc lamp was used as a light source. To minimize the influence of output variation the compensation with the pyroelectric detector was utilized. All PA amplitude spectra were normalized to a carbon black standard. The spectra were recorded at various frequencies ranging from 20 to 1500 Hz in the spectral diapason from 200 to 1500 nm. The measurements were carried out at room temperature in the air filled chamber.

3. RESULTS AND DISCUSSION

Figure 1(a) and (b) shows PA spectra of synthetic diamond single crystals containing nitrogen and boron correspondingly. The spectra are similar to the well known optical spectra for type Ib [Figure 1(a)] and type IIa [Figure 1(b)] diamonds. They exhibit a fundamental near

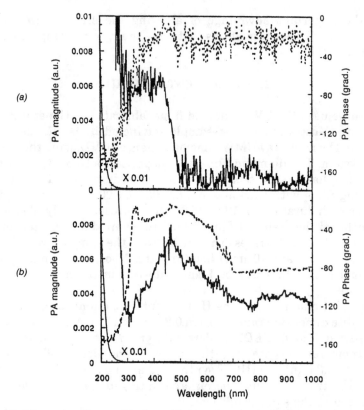

Figure 1. PA spectra (for amplitude–solid line, for phase–dashed line) of synthetic diamond single crystals with nitrogen (a) and boron (b) contaminations.

band gap absorption edge at 5.5 eV (~220 nm) and a spectral feature corresponding to the nitrogen associated defect centers (in the spectral range from 300 to 500 nm in Figure 1(a)). The spectral feature centered at 500 nm in Figure 1(b) may be attributed to the small regions of amorphous carbon since their optical absorption is very similar to that introduced into diamond by neutron irradiation [2].

The PA spectra of CVD diamond films typically have the features corresponded to the fundamental absorption edge near the diamond band gap (5.5 eV) and the absorption on the diamond lattice defects (having the same origin like defects in diamond single crystals appearing as a result of neutron and electron irradiation). Figure 2 shows the spectra for free-standing CVD diamond films prepared by using MW discharge plasma (ASTEX) with process duration 4, 12 and 29 hours. The general feature of the spectra is significant decreasing of the PA signal amplitude in UV region (of 250 to 350 nm). The analogous decreasing has been detected for the films produced by other methods.

The example of such spectrum is presented in Figure 3. It has been recorded for the very thin (near 1 μm) diamond film fabricated on Si substrate by arc discharge CVD method. It was so called "nanocrystalline" diamond film containing the diamond grains of nanometer size. The Raman spectrum of the film displayed the band near 1140 cm^{-1} in addition to the usual 1332 cm^{-1} band of well-crystallized diamond and 1580 cm^{-1} band of nondiamond carbon phase. It should be noted that the PA spectrum in Figure 3 has a spectral feature at 1000 nm which apparently corresponds to the fundamental absorption edge of Si and shows that in our case the light partly passes through the diamond film, is absorbed in Si substrate and excites the PA signal. Special measurements has shown that the spectral curve in the range of 800 to 1200 nm coincides well with the PA spectrum of Si wafer.

To understand the nature of PA decreasing in UV region we have studied the PA spectra of diamond powders with grain size ranging from 125 μm down to 4 nm. Possessing the large amplitude of PA signal and allowing to measure the spectra at the various chopper frequencies [3] the powder samples are very convenient for the Photoacoustical Spectroscopy. In our research we used the light modulation frequency ranging from 20 to 1500 Hz. A few spectra are presented in Figure 4.

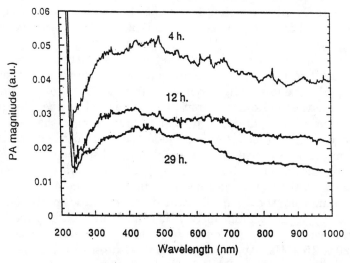

Figure 2. PA spectra of free-standing CVD diamond films fabricated by MW discharge plasma (ASTEX) process with the different process durations: 4, 12, and 29 hours correspondingly.

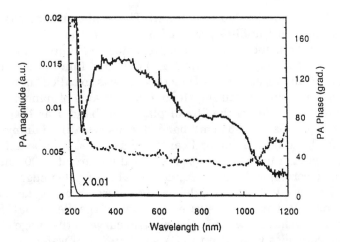

Figure 3. PA spectrum (for amplitude–solid line, for phase–dashed line) of the thin CVD diamond film on Si substrate fabricated by arc discharge plasma and characterized by Raman scattering as a "nanocrystalline" film.

As well as CVD polycrystalline films the diamond powders demonstrated the PA spectra with the specific shape in UV region for the low modulation frequency (spectrum in Figure 4 corresponds to 90 Hz frequency) even for the crystallites larger than 100 μm. This value is much higher than the light wavelength. But for the higher chopper frequency (spectrum in Figure 4 corresponds to 1350 Hz) the above mentioned specific spectral feature in PA spectrum was absent.

The similar frequency dependence of PA spectrum shape in UV region was detected for the powders up to 14 μm grain size and could be explained by the mirror light reflection from the faces of the diamond grains. The thermal diffuse length (characterized by the effective PA signal excitation) in diamond powder is much longer than the grain size for the low light modulation frequency, and after the several mirror reflecting surfaces only a small part of light is absorbed by the material producing the PA signal. The diamond reflectance has a tendency to increase with the wavelength decrease in UV region [4].

Beginning from the diamond grain size of 7 μm the spectral shape of PA spectra changes significantly. It is evident in Figure 4: the PA signal decreasing in UV region is detected for all applied frequencies (from 20 to 1500 Hz). Apparently the crystallite size of the fine powders is smaller than the thermal diffusion length even at the highest chopper frequency (1500 Hz). The second feature is the shape of PA spectrum in the visual region which corresponds to the increase of light

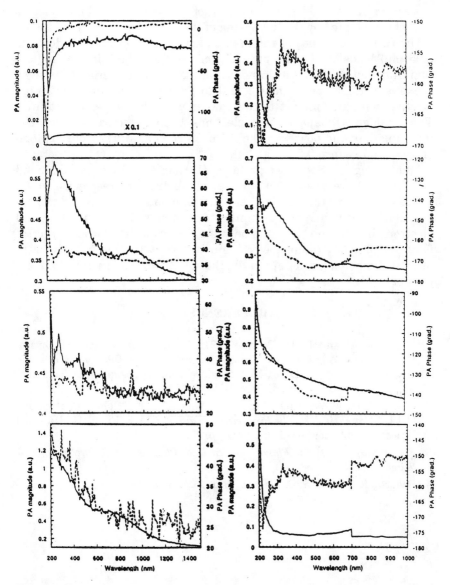

Figure 4. PA spectra (for amplitude–solid line, for phase–dashed line) of diamond powders with grain size 100–125 μm; 3–5 μm, <0.25 μm, and 4 nm registered with 90 Hz (left side) and 1350 (right side) chopper frequency.

absorption in the subband gap region of the fine diamond particles. It could be explained by the increasing of the point defect specific density. The further decrease of the diamond crystallite grain size (down to 0.25 μm) leads to the disappearing of the "downfall" in UV region. Simultaneously, the optical and scanning electron microscopies show the crystallite shape change. The 0.25 μm size crystallites have non-faceted shape in contrast to the powder with the bigger crystallites. The analogous shape of particles has been detected also for the ultrafine diamond powders. Their 4 nm size crystallites were agglomerated usually into 0.2 μm globules.

Thus using PA Spectroscopy for CVD diamond films and diamond powders we have observed a few spectral features corresponded to the fundamental absorption edge near the diamond band gap and the absorption by the point defects (similar to ones caused by the neutron and electron irradiation in diamond single crystals). Other spectral features could be explained by the mirror light reflection from the well-faceted diamond grains either in the CVD films or in the diamond powders.

REFERENCES

1. Feldman, A. and L. H. Robins. 1991. "Critical Assessment of Optical Properties of CVD Diamond Films," in *Application of Diamond Films and Related Materials,* Y. Tzeng, M. Yoshikawa, M. Murakawa and A. Feldman, eds., Elsevier Science Publishers, B. V., p. 181.

2. Clark, C. D., E. W. Mitchell and B. J. Parson. 1990. "Colour Centres and Optical Properties," in *The Properties of Diamond,* J. E. Field, ed., Harcourt Brace J., Publisher: Academic Press, p. 23.

3. Rosencwaig, A. 1980. *Photoacoustics and Photoacoustic Spectroscopy.* J. Wiley & Sons.

4. Roberts, R. A. and W. C. Walker. 1961. *Phys. Rev.,* 161:730.

Optical Characterization of Thick Diamond Films Produced by Microwave Plasma Chemical Vapor Deposition

A. V. KHOMICH

Institute of Radio Engineering & Electronics of RAS
Academician Vvedensky sq. 1
Fryazino, Moscow Region 141120
Russia

V. G. RALCHENKO, A. A. SMOLIN, V. V. MIGULIN,
S. M. PIMENOV, I. I. VLASOV
AND V. I. KONOV
General Physics Institute
Russian Academy of Sciences
ul. Vavilova 38
Moscow 117942
Russia

ABSTRACT: High quality diamond films of 2.25 inch diameter with thickness from 60 to 600 μm were grown on Si substrates by Microwave Plasma Chemical Vapor Deposition. The influence of growth regimes on optical properties of the diamond films has been examined with IR, UV-visible and Raman spectroscopies. The spectral shapes of the defect-induced one-phonon and CH_x-stretching bands are discussed. A mapping of SiC layer that was observed in some samples on substrate side was performed over the whole area of diamond wafer with IR spectroscopy.

KEY WORDS: diamond films, optical properties, polishing, silicon carbide.

339

INTRODUCTION

There is increasing interest in optical applications of diamond films as their properties may approach those of natural diamond. The growth of thick (>100 microns) CVD diamond wafers over large areas has been realized at a number of laboratories, so the use of CVD diamond for optical coatings, windows, domes and lenses can become a reality. There is a need in careful optical characterization of diamond films to evaluate absorption and scattering losses, impurities and defects concentration which determine diamond quality. In the present paper thick free-standing diamond films produced from microwave plasma at different deposition conditions were characterized with transmission/absorption optical spectroscopy.

EXPERIMENTAL DETAILS

An 8 kW power MW plasma deposition system ASTeX PDS-19 has been installed at GPI-Diagascrown Ltd. joint research center in 1995. This CVD reactor allows production of high quality diamond films of 2.25 inch diameter using $CH_4/H_2/O_2$ feed gas mixture at typical parameters: pressure 90–100 Torr, substrate temperature 700–800°C, methane content 0.5–2.5%. It takes about 150 hours to grow 0.5 mm thick diamond wafer of optical grade. Substrate temperature and gas composition were systematically varied to produce diamond material of different grades and establish optimum deposition conditions.

Optical spectroscopy measurements were carried out using an IR and UV-vis "Specord" spectrometers with ranges 0.185–0.9 μm and 2.5–50 μm (200–4000 cm^{-1}). The films were examined with Raman spectroscopy (488 nm excitation wavelength, 2 μm laser beam spot size).

Diamond films grown by CVD are composed of diamond grains of sizes in the range of 10–80 microns, and the surface roughness of grown films is on the order of several tens of microns. This roughness leads to a significant optical scattering loss resulting in a strong decrease in transparency. The optical transmission can be improved by laser polishing of as-grown surface of CVD diamond, as this technique allows to reduce roughness down to 100 nm [1].

Translucent free-standing diamond films were obtained by removing chemically the silicon substrate, then cut by Nd:YAG laser, and finally were polished either mechanically or by KrF excimer laser. Friction and wear properties of such polished films are found to be comparable to those of natural diamond [1,2]. To remove graphitized layer after the laser polishing the samples were annealed in air at 530°C during 2–3 hours.

RESULTS

The IR transmission spectra of a 0.23 mm thick sample before and after laser polishing are shown in Figure 1. The transmission at 10 μm was increased by 45 times after laser treatment. The rms roughness of this film changed as a result of laser polishing from 3000 nm to 70–50 nm, as determined from the UV-vis spectra (not shown here) taking into account exponential decay of transmittance with roughness.

One-Phonon Band

The intrinsic two- and three-phonon absorption bands, which have been well-studied in natural diamonds, are seen clearly between 2670 and 1500 cm^{-1} and just below 4000 cm^{-1}, respectively (Figure 1). Diamond absorbance near 10 μm is of particular interest because many infrared sensors operate within this atmospheric window. Usually, in de-

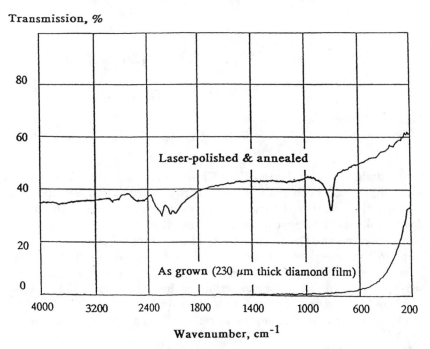

Figure 1. Infrared transmission spectra of a 230 μm thick CVD diamond film, deposited at T = 800°C (2.5% CH$_4$; 0.5% O$_2$; 97% H$_2$) before and after laser polishing. The band at 800 cm^{-1} is due to a silicon carbide layer which formed at the diamond/substrate interface.

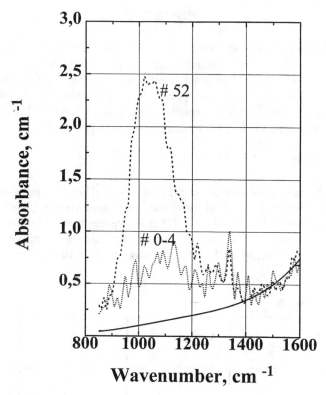

Wavenumber, cm $^{-1}$

Figure 2. The one-phonon spectra for two samples: #0-4 ("white," T_{gr} = 725°C) and #52 ("gray," T_{gr} = 900°C). The intrinsic two-phonon background is shown by the the bold line.

fected diamond crystals and CVD films some absorption can occur at wavelengths of less than 1350 cm^{-1} due to impurities, imperfections, or disorder that destroy the transitional symmetry and cause the breakdown of the momentum conservation law allowing lattice modes to couple with electromagnetic radiation. The absorption coefficient of our films in the one-phonon band was calculated from the difference between measured transmission and transmission background due to the surface scattering with the correction on the tail of the two-phonon absorption [3]. The one-phonon bands for two films of "white" and "gray" grade, obtained at different growth temperatures, T_{gr}, are shown in Figure 2.

The spectral shapes of one-phonon band for gray CVD films (#52, Figure 2) differ significantly from the characteristic, recognizable one-phonon infrared absorption bands connected with nitrogen-related defects in diamond [4] and is close to that observed in brown Type IIa diamonds [5]. The brown Type IIa diamonds showed birefringence patterns of the "tatami" type between crossed polars that are indicative of plastic deformation. Other workers [5] suggest, therefore, that the one-phonon IR absorption by the brown Type IIa diamond specimens are likely to be due to dislocations.

The Raman spectra of the same white and gray films are shown in Figure 3. According to Raman analysis the film quality improves with lowering deposition temperature and/or oxygen addition to the feed gas, the width of diamond peak at 1332 cm^{-1} becoming as narrow as 2.5 cm^{-1}.

Hydrogen and Nitrogen

The rather broad absorption bands due to sp^3-bonded CH$_x$ stretching

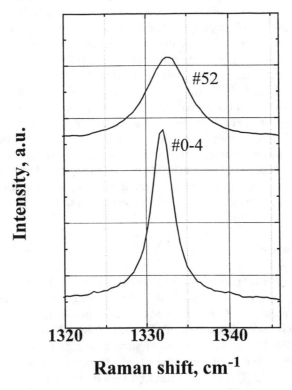

Figure 3. Raman diamond lines of the samples shown in Figure 2. Linewidth of #0-4 is 2.9 cm^{-1} and #52 is 6.5 cm^{-1}.

modes are observed between 2800 and 3000 cm^{-1} (Figure 1). The concentration of carbon-hydrogen groups in our films was in the range of $(1.5–9.0) * 10^{19}$ cm^{-3} as estimated from integral intensity of CH$_x$ stretching mode vibrations in the IR spectra [6]. The sp^3-hybridized CH$_2$ groups (antisymmetric and symmetric stretches at 2920 cm^{-1} and 2850 cm^{-1}, correspondingly) and the absorption band with the peak near 2836 cm^{-1} give the main contribution to the band intensity, while the absorption due to CH$_3$ and CH sp^3-groups and due to sp^2-hybridized CH$_x$ groups were lower in our films. The interpretation of the nature of the absorption band with the peak at 2836 cm^{-1} is less obvious. This band was never observed in IR spectra of diamond-like carbon films, but is usually observed in the CVD diamond films. Some authors [7] assigned this band to hydrogen located at (111) diamond surfaces, while others [8] assigned this strong C-H stretch band to N-CH$_3$ groups in CVD diamond films. We analyzed UV-vis spectra and IR one-phonon defect-induced band in films under study and calculated the maximum amount of nitrogen content. For example, we estimate the concentration of nitrogen impurity in the form of single substitutional N atoms from the amplitude of 4.6 eV (270 nm) absorption band. This absorption is mainly due to vibronic transitions associated with zero-phonon line at 4.059 eV occurring at the single-substitutional nitrogen atom [9]. It was found that an absorption coefficient of 1 cm^{-1} at 270 nm is produced by $0.98 * 10^{17}$ cm^{-3} of nitrogen in this form [9,10]. For our films the N$_C$ concentration varied between $5 * 10^{17}$ to $2 * 10^{18}$ cm^{-3}. For some films the overall nitrogen content occurred to be several times lower than the concentration of the bonds, responsible for the 2836 cm^{-1} band, which we determined from the area of the corresponding CH stretching mode. We assigned the 2836 cm^{-1} band to the hydrogen located at (111) diamond grain surface, because the incorporation of so small nitrogen alone cannot be responsible for the whole 2836 cm^{-1} band in our films.

SiC Interlayer

In the transmittance (Figure 1) and reflectance (Figure 4) spectra of the majority of free-standing films we observed a band near 800 cm^{-1}, originated from an SiC layer, located at diamond/silicon interface. We were able to remove this thin layer either by laser ablation, or RF plasma etching in SF$_6$ during 10 minutes of substrate side of the free-standing films. Reflectance measurements in the infrared range were carried out at near-normal (20°) incidence using 4 mm aperture.

In order to analyze the reflection spectra due to SiC layer at the diamond film's back surface we used the expressions given in Reference

[11] for buffer layers (less than 50 nm thick) of SiC on silicon. We ignore, in our calculations, multiple reflections within the diamond film due to light scattering effects. The damping constant usually varied from 35 to 100 cm^{-1}. The average thickness of SiC layer was of the order of 20–150 nm for different samples. Such values seem to be reasonable taking into account previous investigations of SiC interlayer formation upon diamond deposition in MW plasma [12].

The intensity, shape, position and width of the SiC band varies across the sample surface (Figures 4 and 5), the intensity having a maximum somewhere between the center and edge of the wafer. A similar SiC distribution was observed in Reference [13] being ascribed to nonuniformity of plasma ball. There are two main possible reasons for such changes: the structural changes across the sample surface (the presence of β-SiC and amorphous SiC phases and/or discontinuous silicon carbide interlayer. This is currently the subject of further investigations.

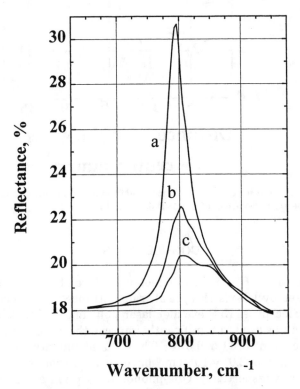

Figure 4. The reflectance spectra from different locations (a, b and c) on the diamond wafer (substrate side), indicating the presence of SiC.

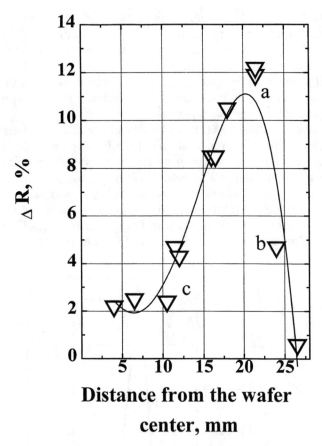

Distance from the wafer center, mm

Figure 5. The amplitude of SiC reflectance band from different locations (a, b and c correspond to spectra in Figure 4) of the CVD diamond film.

CONCLUSIONS

Diamond films of 2.25 inch diameter produced by MW plasma were characterized by UV-vis-IR optical transmittance/reflectance, and Raman spectroscopy. A KrF excimer laser polishing of free-standing as-grown films was used to significantly improve the optical transmission. Hydrogen content of $(1.5-9.0) * 10^{19}$ cm^{-3} and substitutional nitrogen concentration $(0.5-2.0) * 10^{18}$ cm^{-3} have been determined from IR and UV-vis spectra respectively. One-phonon absorption at 1000 cm^{-1} was about 0.2–0.4 cm^{-1}, and Raman peak width as small as 2.5 cm^{-1} was obtained for most translucent films.

REFERENCES

1. Pimenov, S. M., A. A. Smolin, V. G. Ralchenko and V. I. Konov. 1993. "Excimer Laser Polishing of Diamond Films," *Diamond Films and Technology,* 2(4):201–214.

2. Bhusman, B., V. V. Subramaniam and B. K. Gupta. 1994. "Polishing of Diamond Films," *Diamond Films and Technology,* 4(2):71–97.

3. Thomas, M. E., W. J. Tropf and A. Szpak. 1995. "Optical Properties of Diamond," *Diamond Films and Technology,* 5(3):159–180.

4. Clark, C. D. and S. T. Davey. 1984. "One-Phonon Infrared Absorption in Diamond," *J. Phys. C: Solid State Phys.,* 17(6):1127–1140.

5. Kiflawi, I., C. M. Welbourn and G. S. Woods. 1993. "One-Phonon Absorption by Type IIa Diamonds," *Solid State Communications,* 85(6):551–552.

6. Coudberg, P. and Y. Catherine. 1987. "Structure and Physical Properties of Plasma-Grown Amorphous Hydrogenated Carbon Films," *Thin Solid Films,* 146(1):93–107.

7. Chin, R. P., J. Y. Huang, Y. R. Chen, T. J. Chuang and H. Seki. 1995. "Interaction of Atomic Hydrogen with the Diamond C(111) Surface Studied by Infrared-Visible Sum-Frequency- Generation Spectroscopy," *Physical Review B,*52(8):5985– 5994.

8. McNamara, K. M., B. E. Williams, K. K. Gleason and B. E. Scruggs. 1994. "Identification of Defects and Impurities in Chemical-Vapor-Deposited Diamond through Infrared Spectroscopy," *J. Applied Physics,* 76(4):2466–2472.

9. Vaz de C. Nazare, M. H. and A. J. T. das Neves. 1987. "Paramagnetic Nitrogen in Diamond: Ultraviolet Absorption," *J. of Physics C: Solid State Physics,* 20(18):2713–2722.

10. Kiflawi, I., A. E. Mayer, P. M. Spear, J. A. van Wyk and G. S. Woods. 1994. "Infrared Absorption by the Single Nitrogen and A-Defect Centers in Diamond," *Philosophical Magazine B,* 69(6):1141–1147.

11. Holm, R. T., P. H. Klein and P. E. R. Nordquist, Jr. 1986. "Infrared Reflectance Evaluation of Chemically Vapor Deposited β-SiC Films Grown on Si Substrates," *J. Applied Physics,* 60(4):1479–1485.

12. Meilunas, R., M. S. Wong, K. C. Sheng and R. P. H. Chang. 1989. "Early Stages of Plasma Synthesis of Diamond Films," *Applied Physics Letters,* 54(22):2204–2206.

13. Friedrich, M., S. Morley, B. Mainz, H.-J. Hinneberg and D. R. T. Zahn. 1995. "The Local Distribution of SiC Formed by Diamond Heteroepitaxy on Silicon Studied by Infrared Spectroscopy," *Diamond and Related Materials,* 4(7):944–947.

Optical Studies of Graphitized Layers in Ion-Implanted Diamond

R. A. KHMELNITSKIY,* V. A. DRAVIN AND A. A. GIPPIUS

P.N. Lebedev Physical Institute of the Academy of Sciences of Russia
Leninsky prospect 53
Moscow 117924, Russia

ABSTRACT: Optical interference studies of graphitized layers formed in diamond by ion bombardment and annealing provided the data on the depth and the thickness of the layers as well as their optical parameters. The latter were found to be close to those of dispersed graphite.

1. INTRODUCTION

Due to metastability of diamond it tends to transform to graphite if lattice damage density exceeds some critical value [1]. In this case, a sufficient number of diamond sp^3 bonds are broken which is a prerequisite for transformation (enhanced by heating) into a graphite phase with sp^2 bonds. This puts a severe limitation on ion-implantation doping. On the other hand, ion beam induced graphitization opens new prospects of defect engineering. The possibility to graphitize selected areas of diamond subjected to radiation damage provides an additional degree of freedom in the development of diamond based electronic devices. One can think of built-in graphite resistors, capacitors, connections between elements of circuits etc. Certainly, extensive material research is needed before these suggestions can be realized. In this paper we present the first

---

*Author to whom correspondence should be addressed.

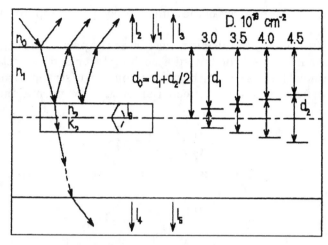

Figure 1. Light beams and layer boundaries in ion-implanted samples.

data of optical studies of graphitized layers formed in diamond by ion implantation and annealing.

2. EXPERIMENT

Polished samples of natural diamond were bombarded at room temperature by He$^+$ ions with the energy of 350 keV and fluence of the order of 10^{16} cm^{-2} and subsequently annealed in vacuum at 1400°C. The implanted area was 0.5 × 1.5 mm. Starting from the critical fluence of 2.8 × 10^{16} cm^{-2}, the absorbing and reflecting buried graphitized layers were formed in ion-implanted diamond. We performed optical measurements taking into account the absorption in the buried damaged layer and interference in the part of the crystal between the surface and the buried layer to obtain the data on the depth and the thickness of the layers as well as their conductivity at optical frequencies. The scheme of measurements is illustrated in Figure 1. I_1 is the intensity of incident monochromatic light, I_2 and I_4 are the intensities of light reflected and transmitted within the implanted area, I_3 and I_5 are those outside of this area, I_6 represents the absorption in the buried layer. Also shown are the boundaries of graphitized layer at various implantation fluences. Values $a_1 = I_2/I_3$, $a_2 = I_4/I_5$ and $T_0 = I_5/I_1$ were measured and used in subsequent analysis.

3. RESULTS AND DISCUSSION

In the range 400–800 nm we observed four interference minima and three maxima. The analysis of the data is based upon the approach developed in [2].

Taking into account multiple reflection in the part of the crystal between the surface and the buried absorbing layer and neglecting the contribution of beams coming back from this layer (strong absorption) one can write the complex reflected amplitude for normal incidence [2]:

$$\hat{r} = \frac{r_1 + \hat{r}_2 \exp(-2i\delta)}{1 + r_1 \hat{r}_2 \exp(-2i\delta)} \tag{1}$$

where $r_1 = (1 - n_1)/(1 + n_1)$ is the Fresnel reflection coefficient for the boundary "air-diamond,"

$$\hat{r}_2 = \frac{n_1 - (n_2 - ik_2)}{n_1 + (n_2 - ik_2)} = r_2 \exp(i\varphi) \tag{2}$$

is the complex Fresnel reflection coefficient for boundary "diamond-graphitized layer," n_1—refractive index of diamond, $n_2 - ik_2$—complex refractive index of the absorbing graphitized layer, $\delta = 2\pi n_1 d_1/\lambda$—change of the phase of the beam traversing the distance d_1 between the surface and the boundary of the buried layer, φ—that on the reflection from this boundary, λ—wavelength in the vacuum.

From Equations (1) and (2) we obtain

$$\hat{r} = \frac{r_1 + r_2 \exp[i(\varphi - 2\delta)]}{1 + r_1 r_2 \exp[i(\varphi - 2\delta)]} \tag{3}$$

Interference extrema are observed for integer m

$$2\delta - \varphi = \pi m \tag{4}$$

Taking Equation (4) for mth and $(m + 1)$th extrema at λ_m and λ_{m+1} and assuming that $\varphi(\lambda_m) = \varphi(\lambda_{m+1}) = const$ we obtain:

$$d_1 = \frac{1}{4[n_1(\lambda_{m+1})/\lambda_{m+1} - n_1(\lambda_m)/\lambda_m]} \tag{5}$$

In Table 1 the values of d_1 found from Equation (5) for various fluences are given.

The thickness d_2 of graphitized layer can be found from the optical losses $I_6 = (I_1 - I_2)[1 - \exp(-\alpha d_2)]$ and absorption coefficient $\alpha = 2\omega k_2/c = 4\pi k_2/\lambda$:

$$k_2 d_2 = -\frac{\lambda}{4\pi} \ln[1 - I_6/(I_1 - I_2)] \tag{6}$$

The value $I_6/(I_1 - I_2)$ can be determined from conservation of energy ($I_1 = I_3 + I_5$, $I_1 = I_2 + I_4 + I_6$) and experimentally measured ratios $a_1 = I_2/I_3$, $a_2 = I_4/I_5$, $T_0 = I_5/I_1$:

$$\frac{I_6}{I_1 - I_2} = \left[1 + \frac{a_2 T_0}{(a_1 - a_2)T_0 - a_1 + 1}\right]^{-1} \tag{7}$$

Using Equations (6) and (7) the values $k_2 d_2$ were determined for various fluences (see Table 1).

The next step is based upon the approximation that the position d_0 of the middle of the absorbing layer and the value of k_2 are the same for all fluences. Then it is possible to choose the value d_0 to fit the proportionality between $k_2 d_2$ and $(d_0 - d_1)$ for all fluences and to determine $d^2 = 2(d_0 - d_1)$. The value of d_0 and values of d_2 for various fluences are given in the table. Additional data on optical parameters of the absorbing layer can be obtained from the analysis of ratios of reflection coefficients at neighbouring maxima and minima which provides the value of $r_2 \approx 0.15$. Using Equations (2), (4) and (5) we can find φ and then n_2 and k_2 to compare the latter with the value k_2 already determined from the optical losses. Based upon the statistics for various samples and fluences, the optical parameters of the buried absorbing layer at the wavelength 700 nm were found to be $n_2 = 2.21 \pm 0.06$, $k_2 = 0.70 \pm 0.03$, which correspond to the absorption coefficient $\alpha = 4\pi k_2/\lambda = 1.2 \times 10^5$ cm^{-1} and conductivity $\sigma = 7.4 \times 10^4$ Ω^{-1} cm^{-1}. These parameters are close to those of dispersed graphite ($n = 2.05$, $k = 0.66$).

In Figure 2 the distribution of radiation damage calculated using TRIM program for 350 keV He$^+$ ions implanted into diamond is compared with the position of absorbing graphite layer. With all the limitations of the calculation (which does not take into account channeling, diffusion and interaction between defects) and approximations used in our analysis it is evident that the graphite layer starts to form at the maximum of

Table 1. Sample DR9.

Fluence, 10^{16} cm^{-2}:	3.0	3.5	4.0	4.5	Comment
d_1, nm	726	700	685	677	Interference
($k_2 d_2$), nm	50.6	89.3	108	125	Absorption
$d_2 = 2(d_0 - d_1)$, nm	70	122	152	168	$d_0 = 761$ nm, selected to fit proportion $(d_0 - d_1)/$ ($k_2 d_2$) for all fluences
$d_1 + d_2$, nm	796	822	837	845	
k_2	0.73	0.73	0.71	0.74	$k_2 = 0.73 \pm 0.05$

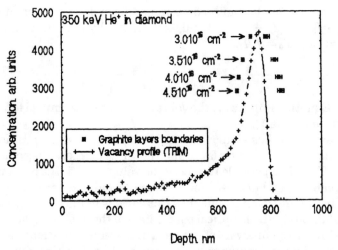

Figure 2. Damage calculated by TRIM simulation and boundaries of graphitized layers.

radiation damage distribution, when defect density exceeds some critical value. The critical fluence was estimated to be 2.8×10^{16} cm^{-2}.

4. CONCLUSIONS

In ion bombarded and annealed diamond buried graphitized layers are formed starting from critical fluence of 2.8×10^{16} cm^{-2} of 350 keV He$^+$. The thickness of the layers increases with the increase of fluence. Optical parameters of the buried layers are close to those of dispersed graphite.

ACKNOWLEDGEMENTS

This work was supported by the Russian Foundation for Basic Research (project N 95-02-04278-a).

REFERENCES

1. Kalish, R. 1993. *Diamond and Related Materials*, 2:621.
2. Heavens, O. S. 1955. *Optical Properties of Thin Solid Films*. London. Butterworths Scientific Publications.

ESR Study of Neutron Irradiated Doped Diamond Films

V. S. Varichenko,* A. A. Melnikov and N. M. Penina
Belarussian State University
Minsk 220050, Belarus

M. A. Prelas, S. Khasawinah, T. Sung and G. Popovici
University of Missouri
Columbia, Missouri 65211

INTRODUCTION

The doping of diamond and diamond films with electrically active impurities can be accomplished during synthesis [1], or by ion implantation [2]. An alternate method of doping with impurities in diamond is transmutation doping. It is especially important for realization of a n-type in diamond, the creation of which is complicated. Recently an opportunity for Li doping of diamond films by means of the reaction ^{10}B (n,α) ^7Li [3,4] was shown. Transmutational doping is connected to formation of defects, rendering significant influence on the electrical and optical characteristics of the diamond films.

In this work paramagnetic defects in the doped ^{10}B initial diamond films, irradiated by neutrons and subjected to subsequent thermal annealing have been studied.

---

*Author to whom correspondence should be addressed. Dr. V. Varichenko, HEII, Belarussian State University, Minsk 220080, Belarus. Fax: 375/0172/265940. E-mail: VALERY@HEII.BSU.MINSK.BY.

EXPERIMENTAL

Diamond films, grown in a hot filament reactor [5], contained 95% isotopically enriched ^{10}B with a total concentration of about 10^{20} cm^{-3}. Irradiation was carried out during 4 weeks by a flux of thermal neutrons 1.1 $\times$ 10^{14} n/cm^{2}sec and fast neutrons (E > 0.1 MeV) 1.3 $\times$ 10^{14} n/cm^{2}sec. The concentration of the Li atoms after irradiation by means of the reaction ^{10}B (n,α) ^{7}Li was 6.3 $\times$ 10^{19} cm^{-3} and the total concentration of B was 4.2 $\times$ 10^{18} cm^{-3}. Annealing of the irradiated diamond films was carried out in argon at temperature 1000°C for 8 hours and then in vacuum at temperature 1300°C–1600°C for 20 minutes.

The measurements of ESR spectra were carried out at room temperature on "Radiopan" and "Varian" spectrometers 3 cm range with rectangular TE102 by the resonator and 100 kHz by modulation of a magnetic field. The signal of Cr^{3+} ions of a ruby standard sample permanently located in the cavity was used for the quality factor control and for calibration of amplitude of magnetic component H1 of microwave field. The measurements of absolute concentration spin were carried out with the use of the coal standard.

RESULTS AND DISCUSSION

The ESR spectrum from the as grown films is due to three centers. First, a single isotropic line with g = 2.0026 ± 0.0002 and width 3.42 G [Figure 1(a)] is characteristic of a defect structure with C—C dangling bonds [6]. Second, a single line with g value close to that of the first center. The width of the second center is 11.4 G and it's intensity is significantly lower than of the first center. This center is observed particularly clear on the side wings of the total spectrum. A possible origin of the second center may be nitrogen combined with vacancy or, similar to the first center, dangling C—C bonds [6]. Areal concentration of the paramagnetic defects comprising the main contribution in the ESR spectrum has been measured as 1.6 $\times$ 10^{15} cm^{-2}. The third type of the observed paramagnetic defects is a characteristic of the centers with long relaxation times (>10^{-5} s), g-factor of which is equal to 2.0023 ± 0.0002.

Spectrum ESR represents a triplet of superfine structure with rather narrow symmetric central peak and asymmetric lateral components [Figure 1(b)]. The asymmetry of the lateral component may indicate anisotropy of superfine structure. These centers can be detected only when the modulation phase differs from the signal phase by 90°. The corresponding spectrum of the third center is the triplet which is often seen in polycrystalline diamond. An origin of this center is dispersed substitutional nitro-

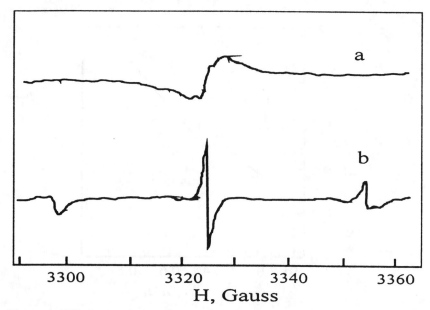

Figure 1. ESR spectra of unirradiated diamond film: (a) signal is detected in-phase with 100 kHz modulation of magnetic field; (b) signal is detected in 90° out of phase.

gen atoms [7]. Areal concentration of these centers is very low, not exceeding a value of 5×10^{12} cm^{-2}.

The dependence of the ESR signal on the intensity of microwaves was investigated for unirradiated diamond film (Figure 2). At small power of microwave paramagnetic centers, saturated with growth of power, give input in ESR signal. With increased power nonsaturated centers give a substantial affect. It is clear that the curve is influenced by centers with various relaxation times.

The irradiation by neutrons gives rise to a strong single line with g value 2.0015 ± 0.0002 (Figure 3). The width of this line is 22.7 ± 0.1 G. The spin concentration measured for this center has been measured as 4.4×10^{17} cm^{-2}. Such paramagnetic center is a characteristic of mechanically damaged diamond powder [8]. The broadening of the spectrum may be caused by different processes. In samples containing nitrogen, the broadening may occur due to the interaction with ^{15}N ($I = 1/2$) and the nearest ^{13}C nuclei [9]. Since the spin density in C-N complexes is higher for carbon atoms as compared with that of nitrogen's [10], the quadrupole, dipole [11] and exchange [12] interaction must be considered as origin of the broadening. Besides, the broadening may be additionally

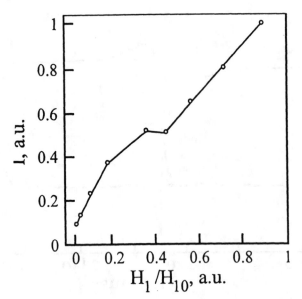

Figure 2. Change of ESR signal amplitude of unirradiated diamond film versus microwave power level.

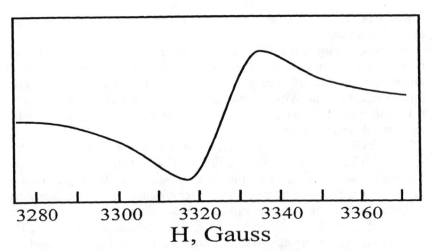

Figure 3. ESR spectrum of neutron irradiated diamond film.

caused by local nonhomogeneous magnetic fields due to ferromagnetic inclusions [13]. In case of the neutron irradiated diamond films the line broadening can be connected with relaxation processes which do not interfere with the fine and hyperfine structure of the point defects normally observed in neutron irradiated single crystals.

After the subsequent annealing at 1000°C–1600°C ESR spectrum of neutron irradiated diamond film becomes similar to the spectrum of as grown sample (Figure 4). The line width of dominating isotropic center with g value 2.0023 ± 0.0002 have become 2.1 G and is not changing with annealing temperature. The decrease of the line width may be caused by an ordering of the internal structure of the irradiated area during annealing. The paramagnetic centers, which give input in a wider spectrum line, do not change their parameters during annealing. A real spin concentration of both types of paramagnetic centers remains equal to 1.6×10^{16} cm^{-2} up to the annealing temperature 1500°C. At the annealing temperature 1600°C their concentration decreases and is equal to 6×10^{15} cm^{-2}. The concentration of the paramagnetic centers, due to nitrogen, with a large relaxation time decreases to the level lower than the registration (2×10^{12} cm^{-2}).

CONCLUSION

Radiation effects of neutrons on ^{10}B doped diamond films grown by hot

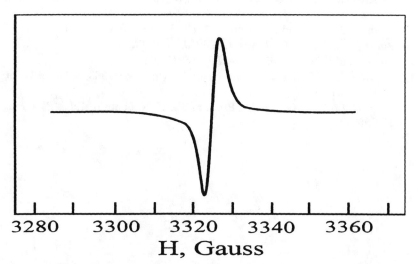

3280   3300   3320   3340   3360

H, Gauss

Figure 4. ESR spectrum of neutron irradiated diamond film after annealing at 1000°C.

filament CVD have been investigated. Irradiation of the diamond films by neutrons causes changes to the ESR-line parameters (increase line width and g-factor change). After annealing at 1000°C–1600°C parameters of ESR-line approaches the parameters of ESR-line of the initial diamond film.

REFERENCES

1. Davies, G., ed. 1994. *Properties and Growth of Diamond.* London, UK: King's College.

2. Prins, J. F. 1992. *Mat. Sci. Rep.,* 7(7,8):271.

3. Spytsyn, B., G. Popovici and M. A. Prelas. 1993. *II Int. Conf. On the Appl. Of Diamond Films and Rel. Mat.,* T. Yoshikawa, M. Murakawa, Y. Tzeng and W. A. Yarorough, eds., Aug. 25–27, Tokyo, Japan, pp. 57–64.

4. Sung, T., M. A. Prelas, A. V. Denisenko, N. M. Penina, V. A. Martinovich, E. N. Drozdova, A. M. Zaitsev and W. R. Fahrner. 1995. *Conf. DF-95,* September 10–15, Barcelona, Spain.

5. Khasawinah, S., T. Sung, B. Spytsyn, W. H. Miller, G. Popovici, M. A. Prelas, E. J. Charlson, E. M. Charlson, J. Meese and T. Stacy. 1993. *Diam. Mat.,* J. P. Dismukes and K. V. Ravi, eds., *Electrochem. Soc. Proc.,* 93-17:1032.

6. Brosious, P. R., J. W. Corbett and J. C. Bourgoin. 1974. *Phys. Stat. Solidi,* A21:677.

7. Erchak, D. P., V. G. Efimov, I. I. Azarko, A. V. Denisenko, N. M. Penina, V. F. Stelmakh, V. S. Varichenko, A. M. Zaitsev and A. A. Melnikov. 1993. *Diam. and Rel. Mat.,* 2:1164.

8. Vlasova, M. V. and N. G. Kakazei. 1979. *Electron Spin Resonance in Mechanically Damaged Solid States,* Kiev: Naukova dumka, 200 pp.

9. Loubser, J. H. N. and L. Du Preez. 1965. *J. Appl. Phys.,* 16:457.

10. Smith, M. J. A. and B. R. Angel. 1982. *J. Appl. Phys.,* p. 53.

11. Samsonenko, N. D. 1965. *Soviet Phys. Solid State,* 6:2460.

12. Loubser, J. H. N., W. P. Van Runeveld and L. Du Preez. 1965. *Solid State Commun.,* 3:307.

13. Dyer, H. B., F. A. Raal, L. Du Preez and J. H. N. Loubser. 1965. *Phil. Mag.,* 11:763.

ESR of Defects in Surface Layer of As-Grown and Ion Implanted CVD Diamond Films

T. IZUMI* AND Y. SHOW
Dept. of Electronics
Tokai University
1117 Kitakaname, Hiratsuka
Kanagawa, Japan

M. DEGUCHI AND M. KITABATAKE
Central Research Laboratories
Matsushita Electric Industrial Co., Ltd.
Hikaridai, Seika
Soraku, Japan

T. HIRAO
Matsushita Technoresearch, Inc.
3-15 Yagumo-Nakamachi, Moriguchi
Osaka, Japan

Y. MORI, A. HATTA, T. ITO AND A. HIRAKI
Dept. of Electrical Engineering
Osaka University
2-1 Yamada-oka, Suita
Osaka, Japan

ABSTRACT: Defects in diamond films prepared by chemical vapor deposition (CVD) method have been studied using electron spin resonance (ESR) measure-

---

*Author to whom correspondence should be addressed.

ment. Two kinds of ESR centers were observed in CVD diamond films. One is a narrow line (P_{dia}-center: $g = 2.003$, $\Delta H_{pp} = 3$ Oe originating from carbon dangling bonds in diamond phase carbon. The other is a broad Lorentzian line (P_{ac}-center: $g = 2.003$, $\Delta H_{pp} = 8$ Oe) originating from carbon dangling bonds in non-diamond phase carbon region. The P_{ac}-center exists in the surface region of CVD diamond films with spin density of 10^{20} spins/cm^3. Diamond surface is converted to an amorphous carbon layer by N$^+$ ion implantation.

1. INTRODUCTION

CVD diamond films have superior properties that would be useful for a number of electronic and optical applications as high temperature, high power and high frequency devices. Considerable recent interest has been centered particularly on the growth of doped diamonds films. However, it is difficult to locate the dopant impurities into electrically active sites (n- and/or p-type dopants) in CVD diamond films, because of the extremely low solubility of most elements except nitrogen and boron [1]. Ion implantation seems to be one effective doping technique for diamond films. However, damaged regions such as graphite and amorphous carbon are formed in the diamond surface region by ion implantation [2]. The graphitization and amorphorization formed in the diamond surface region leads to dramatic changes in certain physical properties of diamond [3]. Therefore, it is important to understand the structure of implanted diamond layers, as well as as-grown diamond surface layers. In this paper, we discuss the structural defects in diamond films and the effect of thermal treatment on the surface layer.

2. EXPERIMENTAL

Diamond films were prepared by the microwave plasma CVD method on p-type silicon substrate that had undergone ultrasonic treatment with diamond powders. The film deposition was performed at 2.45 GHz, 300 W microwave source, 800°C substrate temperature, diluted 5% CO with H$_2$ 110 sccm gas flow and at a pressure of 26 torr. Ion implantation was carried out at room temperature with nitrogen ions at an energy of 100 keV with a dose of $1 \times 10^{14} \sim 1 \times 10^{16}$ ions/cm^2.

ESR measurements were performed at room temperature using an X-band spectrometer. The g-value, the linewidth (ΔH_{pp}) and the signal intensities were determined using the signal of Mn^{2+} and DPPH as the calibration reference.

3. RESULTS AND DISCUSSION

3.1 Defects in As-Grown Diamond Films

Figure 1 shows a typical ESR signal with g-value of 2.003 for as-grown diamond films, having a thickness of 4.5 μm. The ESR signal of Figure 1(a) consists of two kinds of ESR lines as shown in Figure 1(b) and (c). One is a narrow Lorentzian line (P_{dia}-center) with g = 2.003 and ΔH_{pp} = 3 Oe.

The other is a broad Lorentzian line (P_{ac}-center) with g = 2.003 and ΔH_{pp} = 8 Oe. The P_{dia}-center originates from carbon dangling bonds in the diamond layer, and the P_{ac}-center originates from carbon dangling bonds in the non-diamond phase carbon region [4,5]. It has become apparent from the measurement of the depth distribution of the defect centers, that the P_{dia}-

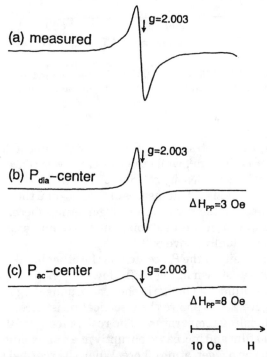

Figure 1. (a) The ESR signal for as-grown diamond film with film thickness of 4.5 μm. The calculated ESR lines for the ESR signal in Figure 1(a) using computer simulation (b) P_{dia}-center: carbon dangling bonds in the diamond layer. (c) P_{ac}-center: carbon dangling bonds in the non-diamond phase carbon region.

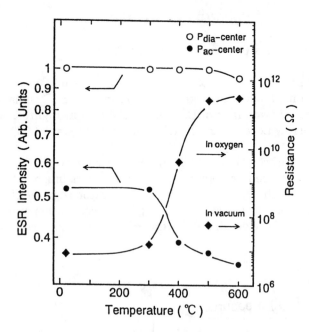

Figure 2. Intensity of the ESR centers (P_{dia}- and P_{ac}-centers) and resistance of CVD diamond films as a function of the thermal treatment temperature in oxygen and vacuum atmosphere.

center was uniformly distributed in the deposited diamond films with spin density of order of 10^{17}/cm^3, while the P_{ac}-center was concentrated at the surface region with spin density of order of 10^{20}/cm^3.

Figure 2 shows the signal intensities of the P_{dia}- and the P_{ac}-centers as a function of annealing temperature in oxygen atmosphere. The signal intensity of the P_{dia}-center was held constant at a temperature up to 500°C, and decreased gradually above 500°C.

The signal intensity of the P_{ac}-center, on the other hand, decreased rapidly above 300°C as shown in Figure 2. However, the signal intensities of these ESR centers did not change in the case of annealing in a vacuum atmosphere. Figure 2 also shows the electrical resistance of the film as a function of annealing temperature. The resistance was 10^7 Ω for the as-grown diamond film and increased abruptly at an annealing temperature above 300°C in the oxygen atmosphere. When the thermal treatment for the diamond films carried out at 600°C, the resistance of the film increased to 4×10^{11} Ω. In the case of the thermal treatment in vacuum at 500°C, the resistance of the film increased slightly as shown in Figure 2.

From these results mentioned above, the following becomes clear. Hydrogen termination of the surface cannot compensate completely the dangling bonds around the vacancies of the defects. Therefore, the P_{ac}-center with high density was distributed in this surface layer. This surface layer can be removed by the thermal treatment in an oxygen atmosphere at 600°C. At the same time, oxygen atoms can compensate the dangling bonds around the vacancies near the surface with oxygen atoms. Therefore, the resistance of the film surface increased from 10^7 to 4×10^{11} Ω.

3.2 Defects in Ion Implanted Diamond Films

Figure 3 shows the ESR signal for the diamond film, which was prepared by N ion implantation (10^{14} ions/cm² at 100 keV) into the sample of Figure 1. The ESR signal was an isotropic line with a dissemble shoulder, which had $g = 2.003$ and $\Delta H_{pp} = 3$ Oe. The ESR signal of Figure 1 was eliminated from the signal for ion implanted diamond films to investigate the defect structures in ion implanted layers. Then, the ESR signal of an isotropic Lorentzian line shape with $g = 2.003$ and $\Delta H_{pp} = 20$ Oe appeared as shown in Figure 4(a). The spin density of the signal was $\sim 1.3 \times 10^{19}$ spins/cm³. Furthermore, Raman spectrum had a broad peak at 1500 cm^{-1} in addition to the line at 1333 cm^{-1}. Amorphous carbon films usually give a broad ESR line with g-value of 2.0027 and ΔH_{pp} of ~ 16 Oe [6]. Therefore, the above results indicate that the diamond surface is converted to an amorphous carbon layer (graphitelike) by ion implantation.

When the ion dose was increased from 10^{14} to 10^{16} ions/cm², the spin density in the implanted layer also increased from 1.3×10^{19} to 1.2×10^{21} spins/cm³, while the ΔH_{pp} decreased from 20 to 2.6 Oe as shown in Figures 4(a–c). The abrupt decrease of ΔH_{pp} is due to the exchange narrowing for a high spin density of 1.2×10^{21} spins/cm³ [6,7].

Figure 5 shows the isochronal annealing curves of the spin density cre-

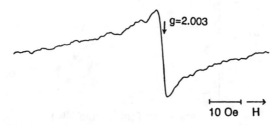

Figure 3. The ESR signal for diamond films prepared by N$^+$ ion implantation with a dose of 10^{14} ions/cm² at 100 keV into the sample of Figure 1.

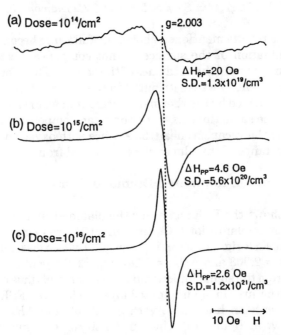

Figure 4. The P_{ac}-center observed in diamond films in the dose range 10^{14}–10^{16} ions/cm^2 (P_{dia}-center was eliminated from the signal).

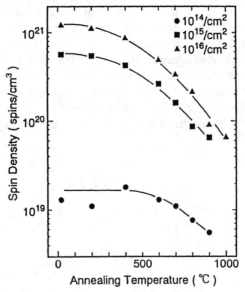

Figure 5. The isochronal annealing curves of spin density (annealing time = 10 min).

ated in the diamond surface layer by N^+ ion implantation. With increasing annealing temperature, the spin density in the ion implanted layer with a dose of 10^{14} ions/cm^2 disappeared at 1000°C, while those with doses of 10^{15} and 10^{16} ions/cm^2 did not disappear even at 1000°C. Furthermore, the resonant lines of these samples did not saturate at a microwave power of up to 200 mW. The amorphous carbon layer contacting many carbon dangling bonds produced by high dose implantation, therefore, remains stable even after annealing at 1000°C, although the a-center produced in silicon by implantation generally disappeared at temperatures above 600°C [8].

4. SUMMARY

The results are summarized as follows.

1. The ESR analyses revealed the presence of two kinds of defect centers in deposited diamond films. One is P_{dia}-center ($g = 2.003$, $\Delta H_{pp} = 3$ Oe), which originates from carbon dangling bonds in the diamond layer, and the other is P_{ac}-center ($g = 2.003$, $\Delta H_{pp} = 8$ Oe), which originates from carbon dangling bonds in the non-diamond region.

2. The P_{dia}-center is uniformly distributed in the deposited diamond films, with 10^{17} spins/cm^3. On the other hand, the P_{ac}-center is concentrated in the surface conductive layer, with a high density of 10^{20} spins/cm^3. In this case, the electrical resistance of the surface layer was 10^7 Ω.

3. When the thermal treatment was carried out at 600°C for 10 min in an oxygen atmosphere, the P_{ac}-center was removed, and then the electrical resistance increased to 4×10^{11} Ω.

4. Diamond surface is converted to an amorphous carbon layer by N^+ ion implantation. In the case of high dose implantation over 10^{15}/cm^2, the ESR center in the amorphous carbon layer remained stable even after annealing at 1000°C.

ACKNOWLEDGEMENTS

The authors gratefully thank Dr. A. Sandhu of Tokai University for helpful discussions. They also would like to thank Mr. F. Matsuoka for technical assistance on ESR measurements.

REFERENCES

1. Fujimori, N., T. Imai and A. Doi. 1986. *Vacuum,* 36:99.

2. Kalish, R., T. Bernstein, B. Shapiro and A. Talmi. 1980. *Rad. Eff.* 52:153.

3. Vavilov, V. S., V. V. Krasnopevtse, Y. V. Miljutin, A. E. Gorodtsky, A. P. Zakharov. 1974. *Rad. Eff.* 22:141.

4. Show, Y., Y. Nakamura, T. Izumi, M. Deguchi, M. Kitabatake, T. Hirao, Y. Mori, A. Hatta, T. Ito and A. Hiraki. 1996. *Thin Solid Films,* to be published.

5. Show, Y., M. Iwase and T. Izumi. 1994. *Advance in New Dia. Sci. and Tech.,* S. Sito, ed., (MYU, Tokyo, 1994), p. 407.

6. Miller, D. J. and D. R. Mckenzie. 1983. *Thin Solid Films,* 108:257.

7. Adel, M. E., R. Kalish and S. Prawer. 1987. *J. Appl. Phys.* 62:4096.

8. Izumi, T., T. Matsumori, T. Hirao, Y. Yaegashi, G. Fuse and K. Inoue. 1981. *Nucl. Instr. and Meth.,* 182/183:483.

Electron Paramagnetic Resonance of Natural Diamond Irradiated with 92 MeV Boron Ions

E. N. DROZDOVA, N. M. PENINA AND V. S. VARICHENKO*

Belarussian State University,
Minsk 220050, Belarus

ABSTRACT: EPR method has been used for investigation of defect production in Ia type natural diamond irradiated with 92 MeV B ions within a doze range of 1×10^{15} to 1×10^{16} cm^{-2}. It is shown that the ion irradiation leads to the appearance of a characteristic singlet line with g-value at 2.0027 assigned to the paramagnetic centres of amorphous regions (PC AR). The presented experimental results are compared with earlier data obtained on diamond irradiated with 150 KeV and 13.6 MeV B ions.

1. INTRODUCTION

The production of nanoscale elements in semiconductors is considered to be necessary for present and future electronics. Ion beam technology gives the possibility to produce quasi-one-dimensional track-like structures with specific electronic properties and can be used in superhard semiconductors. The best candidate for such a material is diamond. Diamond is well known as a crystal possessing excellent thermal conductivity, lattice stability, stability of any kind of defects as well as very high electrical.

The aim of the present work is to study the defect formation and defect

---

* Author to whom correspondence should be addressed. Dr. V. Varichenko, HEII, Belarussian State University, Minsk 220080, Belarus. Fax: 375/0172/265940. E-mail: VALERY@ HEII.BSU.MINSK.BY.

paramagnetic properties in diamond subjected to high energy implantation of 92 MeV B ions.

2. EXPERIMENTAL DETAILS

Samples of type Ia natural diamond irradiation with 92 MeV boron ions within a dose range of 1×10^{15} to 1×10^{16} cm^{-2} have been investigated. The temperature of the samples during the irradiation did not exceed 300 K.

EPR measurements were performed using Varian and Radiopan spectrometers with a TE_{102} cavity. The Cr^{3+} ion signal for a ruby standard sample permanently located in the cavity was used for modulation field phase fine tuning and for calibration of the relative changes in the magnetic component of the microwave field H_1.

3. RESULTS AND DISCUSSION

It is well known that ion irradiation of diamond leads to the appearance of an EPR signal assigned to the paramagnetic centres of amorphous regions (PC AR) [1].

The presented results are compared with data obtained on diamond irradiated with 150 keV and 13.6 MeV ions within a dose range of 1×10^{15} to 1×10^{16} cm^{-2}[2]. In all cases an anisotropic PC AR single line with g-value at 2.0025 ± 0.0002 is observed. In the samples irradiated with 150 keV and 13.6 MeV ions the line intensity is not saturated and its spectral width (4.9 and 16.8 G in the cases of 150 keV and 13.6 MeV respectively) does not change with increasing microwave power from 0.05 μW up to 180 mW.

In the case of 92 MeV irradiation the saturation dependence of the line as measured at room temperature reveals two types of paramagnetic centres giving rise to the PC AR line (Figure 1). The first type centres are strongly saturating and therefore they dominate at low microwave power. These centres result in a homogeneously broadened line of Lorentz shape. At increased power the paramagnetic centres of the second type become dominant causing nonhomogeneous broadening of the line. These centres may be characterised by an unresolved super fine structure and/or g-value anisotropy. The dominant role of the second type centres becomes more pronounced at liquid nitrogen temperature measurements. This is also the reason for a sub-linear dependence of the linewidth on the microwave power.

The EPR spectrum of these samples represents the single line having the linewidth $\Delta H \approx 5$ G and $g = 2.0027$, which is characteristic for amor-

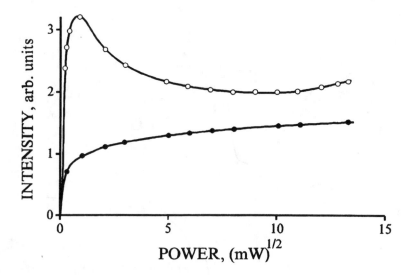

Figure 1. PC AR line intensity versus microwave power in diamond irradiated with 92 MeV boron ions. The open circles are the room temperature measurements; the full circles are the liquid nitrogen temperature measurements.

phous carbon (as well as for the case of low energy ion implantation). A peculiarity of the 92 MeV irradiation is the appearance, with decreasing microwave power, to a number of 200 μW isotropic narrow lines (width about 0.3 G) with g-values within a range of 2.0004 to 2.0036 (Figure 2). These lines are detected on phase of the 100 kHz magnetic field modulation.

Moreover, the signal in quadrature (phase of modulation is 90°) was observed. This fact points at very long relaxation times of the corresponding centres. The spin-spin relaxation time T_2 calculated from the linewidth in the case of saturation absence has complied about 10^{-7} s. The spin-lattice relaxation time $T_1 \cong 10^{-5}$ s was determined in the case of saturation. The angular dependence of EPR signal was studied at the maximal microwave power. The amplitude of EPR line was normalized on signal of a ruby standard sample. The angular dependencies of the linewidth (Figure 3) and PC concentration were constructed.

The angular dependence of the g-value has the following feature: at rotation of sample from 0° to 70° g-value almost does not change and equals 2.003 ± 0.0002. At the orientation 80° g-value is equal to 2.0052 (Figure 4).

The angular dependence of the line shape asymmetry A/B (Figure 5)

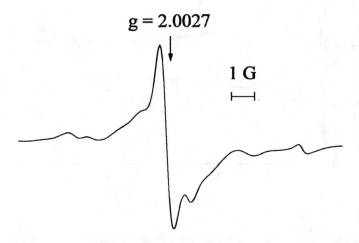

Figure 2. The EPR spectrum of 92 MeV boron ions implanted diamond; microwave power is 200 μW.

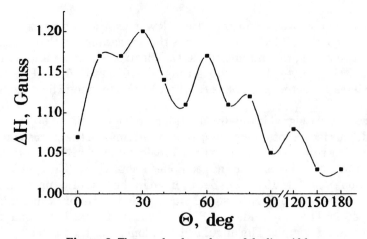

Figure 3. The angular dependence of the linewidth.

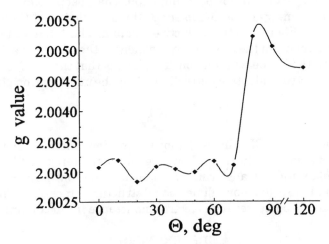

Figure 4. The angular dependence of the *g*-value.

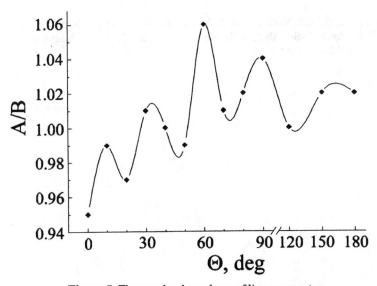

Figure 5. The angular dependence of line asymmetry.

371

speaks in favour of the model for the tracklike zones as regions extended preferentially along the ion beam direction. The most probable reason for such an asymmetry is the Dyson effect. It means that the PC regions are situated in a highly electroconductive medium. The A/B asymmetry has its maximum when the electrical component of the microwave field coincides with the ion beam direction. This means that the maximum of microwave electroconductivity is realized along the implantation direction.

4. SUMMARY

On the basis of EPR investigations of ion implanted diamond it has been proposed that the amorphous regions characteristic of low energy ion implantation transform by high energy ion irradiation (≥ 1 MeV/a.m.u.) to a quasi-one-dimensional structure (tracklike). This structure probably possesses high electroconductance in this direction.

REFERENCES

1. Brosiaus, P. R., J. W. Corbett and J. C. Bourgoin. 1974. *Phys. Stat. Solidi* A21:677.
2. Azarko, I. N., V. S. Varichenko, A. M. Zaitsev, N. M. Penina, V. F. Stelmakh and V. P. Tolstych. 1994. Abst. of X Inter. Conf. on Ion Impl. Technology. Catania-Italy, 13-17 June. 3.28

Auger-Spectroscopic Investigation of a-C:H Films on Micro-Roughness Substrates

A. V. VASIN, L. A. MATVEEVA
AND T. Y. GORBACH
Institute of Semiconductor Physics of the
National Academy of Sciences of Ukraine
45, av. Nauki, Kiev
252028, Ukraine

A. G. GONTAR AND A. M. KUTSAY
V. N. Bakul Institute for Superhard Materials of the
National Academy of Sciences of Ukraine
2, Avtozavodskay Kiev
254074, Ukraine

INTRODUCTION

A number of methods for the determination of carbon specific weight in different chemical states (sp^2, sp^3) are known [1–3]. Some papers describe a quantitative evaluation of the sp^2/sp^3 ratio in a-c:H films. A factor analysis of Auger spectra and a calibrating method were used, in which the reference spectra were those of diamond, graphite and a-C:H films with the sp^2/sp^3 ratio available from NMR studies. In Reference [4], the sp^2/sp^3 ratio as determined from the factor analysis of the shapes of Auger spectra were compared with the NMR measurements. The comparison has shown good agreement between the two techniques. The method of analysis of the shape of a high resolution Auger spectrum may thus be considered sufficiently reliable for the assessment of the carbon chemical state. In the present work, the impact of the silicon surface structure on the sp^2/sp^3 bond ratio has been studied using the above mentioned method.

EXPERIMENTAL

A specific microrelief at the surface of a (100)-oriented silicon plate has been produced by anisotropic chemical etching in HF + HNO_3 acid (A and B-type surfaces) and 10% KOH alkali (C-type surface) etchants. The

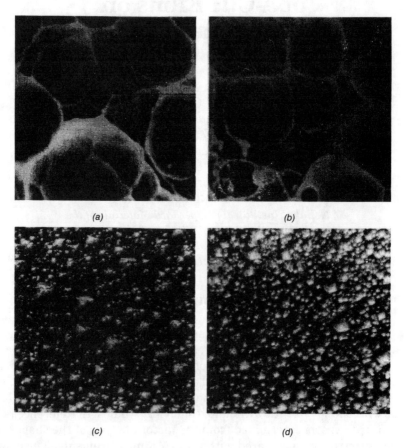

Figure 1. Different types of substrate surfaces: (a) A-type with film 80 nm thick; (b) B-type with film 80 nm thick; (c) C-type with film 80 nm thick; (d) C-type with film 240 nm thick.

microrelief morphology and scale were regulated by etching time and temperature. Figure 1 shows the different types of substrate surfaces that resulted. Surfaces of types A and B differ as to the presence of small "pits" and are formed at different etching times. Auger spectra were recorded using a JEOL JAMP-10s setup, with an electron energy of 10 keV, a modulating voltage of 3 V and an electron beam current (absorbed by the sample) of 1–7 mA. To remove the absorbed layer, an argon ion beam of 2.5 mm in diameter was used at 3 keV and a current from 1–7 mA. To avoid graphitization of the amorphous film caused by the ion bombardment, etching was done for no more than 10 s.

RESULTS AND DISCUSSION

High resolution Auger spectra were analyzed by: (1)—a visual comparison of the peculiarities of the low-energy region of the spectrum of carbon, taken from the film, with standard spectra of amorphous carbon, graphite and diamond, taking into account specimen charging; (2)—a measurement of the displacement of the peak plasma losses relative to the KLL peak; (3)—a measurement of the distances between spectrum bending points and their comparison with the standard spectra.

The analysis has allowed us to assess the sp^2/sp^3 ratio (see Table 1). Figure 2 shows Auger-spectra of standard bulk samples. Figure 3 shows films graphitized by argon ions bombardment. The initial spectrum represents that of amorphous carbon with a pronounced peak at the distance of 23 eV from the main one "M", and a small peak at a distance of 17 eV. With an increase in the dose (time) of etching, peak "a" flattens gradually and transforms into a flat, which is characteristic of graphite. Figure 4 shows Auger spectra of 80-nm thick films on flat and relief substrates. The shape of the spectrum from the film on a type A substrate [Figure 4(line 1)] is intermediate between those of amorphous carbon and graphite. This spectrum exhibits one strong peak at a distance of 23 eV from the main peak.

The spectrum eV of a film on a type C substrate is quite different. It is characterized by a well pronounced peak at 13 eV and a weak one at 23 eV. This is in good agreement with standard spectra of natural diamond

Table 1. Characteristic values for various sample types.

Sample number	A 80 nm	B 80 nm	C 80 nm	Flat 80 nm
Distance of peak of plasma losses from the main peak (eV)	35.4	33.4	36.3	37.4
sp^2/sp^3 ratio	5.3	0.1	0	1.0

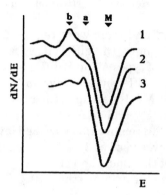

Figure 2. Auger-spectra from standard bulk samples: 1-graphite; 2-glassy carbon; 3-natural diamond.

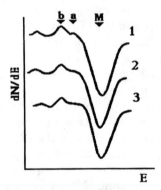

Figure 3. Variation of the Auger-spectra from the sample on a type B substrate under ion bombardment: 1–10 s; 2–20 s; 3–180 s.

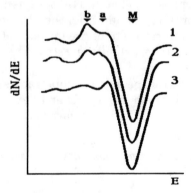

Figure 4. Auger-spectra from samples with films 80-nm thick: 1-a type A substrate; 2: a flat substrate; 3: a type C substrate.

and the data for diamond polycrystalline films [5]. Spectra of all the specimen under study revealed no signs of silicon. The effect of carbon bound in the form of silicon carbide is, therefore, excluded. The spectrum of the specimen on the flat substrate shows a double peak at 18 and 23 eV. The peak "b" of amorphous carbon is, however, more intense than the characteristic sp^3 peak "a". Spectra from surfaces of thicker films (240 nm) do not exhibit fine structure and have the shape typical of amorphous carbon. Reflection electron diffraction at 80 keV from films on flat substrates and films of 240 nm in thickness on relief substrates, exhibits highly diffused low-information haloes.

CONCLUSIONS

The character of the substrate structure might have an important effect on the type of carbon bonding at the early stages of film growth. Depending on the crystallographic orientation and the relief mode, both sp^2 (type A substrate) and sp^3 (type C substrate) hybridization might be stimulated. With an increase in film thickness up to 240 nm, Auger-spectra become typical of amorphous carbon in all specimen including the standard one.

REFERENCES

1. Derry, T. E. and J. P. F. Sellschop. 1981. *Nucl. Instr. and Meth.*, 291:23–26.
2. Pate, B. B. 1986. *Surf. Sci.*, 165:83–142.
3. Lurie, P. G. and J. M. Wilson. 1977. *Surf. Sci.*,65:476–498.
4. Fuchs, A., J. Scherer, K. Jung and H. Ehrhardt. 1993. *Thin. Sol. Films.*, 232:51–55.
5. Normand, F. Le., A. Ababou, N. Braul, B. Carriere, L. Faylette, B. Marcus, M. Mermoux, M. Romeo and C. Speisser. 1994. *Appl. Surf. Sci.*, 81:309–324.

Amorphous Carbon: Free and Bound Hydrogen

I. N. KAPITONOV,
O. I. KONKOV, I. N. TRAPEZNIKOVA
AND E. I. TERUKOV

A. F. Ioffe Physico-Technical Institute
Russian Academy of Sciences
Polytechnicheskaya 26
192401 St. Petersburg
Russia

(Received March 28, 1997)
(Accepted April 25, 1997)

ABSTRACT: We have studied the effusion of hydrogen and other gases from amorphous carbon films while annealing in the temperature range 100–1000°C. The amount of hydrogen in the material (28 at.%) has been determined by the direct method. The amount of free hydrogen in the investigated material was determined as about 4 at.%. The temperature dependence of free hydrogen effusion had a maximum at 400°C. A wide band of hydrogen effusion was observed at temperatures greater than 400°C. The amount of hydrogen evolved after weak C—H bond breaking was of the order 10 at.%. The hydrogen amount evolved after strong C—H bond breaking was 12 at.% in these a-C:H films.

KEY WORDS: a-C:H, hydrogen, effusion.

INTRODUCTION

In recent years amorphous hydrogenated carbon films have been studied extensively. Particular attention has been devoted to a study of a-C:H film's physical properties and to numerous problems of its syn-

378

thesis. However the mechanism of film structure formation and its correlation with the physical properties are not yet finally understood. It is known that, depending on growing methods, a-C:H or DLC films contain a significant amount of free and bonded hydrogen. The role of hydrogen in carbon films is not well understood but it is believed to stabilize the tetrahedral bonding. The hydrogen content critically determines the film structure, e.g., the ratio between carbon atoms in the different coordinations and thus film properties. The hydrogen content is the key factor to the film's structure and its physical properties. The results of hydrogen spectroscopy of a-C:H is extremely important.

IR-spectroscopy is a traditional research method. Hydrogen concentration, type of carbon and hydrogen atom bonding, hybridization and degree of carbon atoms, were estimated with its help [1]. However, it was shown recently that the extent of sp^2-coordinated carbon atoms as well as the complete amount of hydrogen in the film cannot be determined precisely by IR-spectroscopy [2]. Preliminary data indicate that from 1/3 up to 1/2 of hydrogen atoms are not bound to carbon atoms but occur in the atomic or molecular forms in the film.

Nuclear magnetic resonance [3] is also used for hydrogen content estimation and carbon atoms bonding determination. An sp^3/sp^2 coordinated carbon atom ratio can be readily determined, however, the technique requires rather thick films, which limits the method's application.

Effective methods of materials electronic structure research are the optical spectral-kinetic methods, such as temperature dependence of the fundamental absorption edge, luminescence spectra and luminescence excitation spectra. The information on material structural features can be obtained from the energy location of the photoluminescence (PL) band maximum, the half-width of the PL band, and the recombination kinetic and spectrum of luminescence excitation [4,5].

The present work is devoted to a-C:H film hydrogen bonding study by the direct method, hydrogen effusion from the film during the film heating.

SAMPLE PREPARATION AND TECHNIQUE

Samples of a-C:H films were grown on Si substrates by HF decomposition of a methane-hydrogen mixture in a capacitor type reactor. The substrate temperatures (T_s) ranged from 20–150°C. Preliminary film characterization has been discussed in Reference [6].

Step heating in vacuum was applied to study effusion from the films. A vacuum extraction system with external heating and calibrated volumes was used. The empty extraction volume was heated to 1200°C to obtain a hydrogen "free" system. The samples were located in a drop-device and were maintained in vacuum for 24 hours for removal of surface sorbed gases and moisture.

The samples were serially dropped into the reactor by the drop-device with an external magnetic drive. Temperature control was carried out with a Pt-Pt/Rh thermocouple. High temperature calibration was done with an optical pyrometer. The total effused gas amount was determined manometrically, using previously graduated thermocouple manometers and thermoresistance manometers. Hydrogen was removed from the extraction volume by diffusion through a palladium membrane heated to 500°C. The amount of hydrogen was measured by comparing the difference of the pressure in the system before and after hydrogen diffusion.

RESULTS

The experimental results of the hydrogen content determination in a-C:H films are shown in Figure 1 (curve 2). Various parts of the same sample were used in all measurements.

The hydrogen content in pure (without a-C:H film) silicon substrates is very small. The results were corrected for system background. Stepheating of a-C:H films on silicon substrates was carried out in 100°C steps. One temperature step heating duration was 30 minutes. The reactor temperature was decreased during the time required for hydrogen content determination. The residual gas fraction was removed from the system volume after the completion of the measurements.

The temperature effusion dependence of other gas products (except hydrogen) is also shown in Figure 1 (curve 1). These gas products are: (1) Ar, which was incorporated into carbon network during the film growth or was adsorbed by the film surface imperfections such as microvoids, column and grain boundaries and did not form chemical bonds with carbon atoms; (2) methane and more complex hydrocarbons which were incorporated during the methane decomposition process.

We shall call all gases introduced or adsorbed in carbon film but not chemically bound with carbon atoms as being in condition "A." It is obvious that some hydrogen may be in condition "A" also. Such atomic or molecular hydrogen is thus free.

We suppose that hydrogen effusion from the film during the heating may occur stepwise. First hydrogen in condition "A," then weakly bonded hydrogen and then strongly bonded hydrogen (causing the reorganization of the carbon structure), will effuse. The mechanism of such a process was discussed in Reference [7].

The hydrogen in condition "A," the most weakly bonded one, begins to effuse at low but higher than T_s temperatures. Film structure

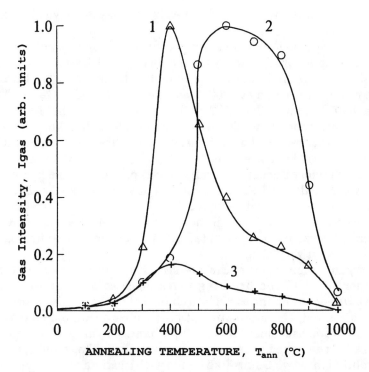

Figure 1. Temperature dependence of gas effusion from a-C:H films. Curve 1—effusion of all gas products except hydrogen. Curve 2—effusion of all hydrogen. Curve 3—effusion of free hydrogen.

transformation does not occur at this stage in accordance with IR-spectroscopy data. The intensities, maximum position and half-width of the absorption bands with maxima at 2920 cm⁻¹ (CH₂), 2860 and 2950 cm⁻¹ (CH₃), 1450 cm⁻¹ (CH₂) and 1370 cm⁻¹ (CH₃) remain constant in the 100–300°C range. Annealing up to 400°C leads to a change of the refractive index from 1.56 at 20°C to 1.62. It indicates an increase of film density, without a chemical bonding change. The initial, gently sloping part of the hydrogen effusion curve corresponds to such a process (Figure 1, curve 1).

The quantitative estimation of hydrogen effusion from condition "A" can be carried out by comparison of non-hydrogen effusion from condition "A" (curve 2) with the initial part of hydrogen one (curve 1). It is obvious, that at low temperatures (20–200°C) the argon and hydrogen effusion curves should be described by the same law. The effusion of hydrogen from condition "A" (i.e., atomic and free molecular hydrogen) obtained in this way is shown on Figure 1 as curve 3.

This curve peaks and nonmonotonity must correspond to carbon structure reorganization and to the increasing of hydrogen, Ar and other gases' effusion from the states corresponding to condition "A." Such features can be discovered at 400°C—as a peak, at 550–600°C—as a "tail" and at 800°C—as a weak peak. It will be shown further, that the two last features correspond to weak and strong C—H bond breaking, to carbon network structure reorganization and to significant hydrogen effusion during chemical bond breaking.

The area under the curve permits an estimate of the free hydrogen in the investigated a-C:H samples at the 4 at.% level and the amount of bonded (C—H bonds) hydrogen at the 24 at.% level. The IR-spectroscopy data give the amount of optically active hydrogen in the same samples at the 21 at.% level. It should be noted that the amount of free hydrogen in our films is not very high.

Reorganization of the carbon structure with weak and then stronger (C—H) bond breaking takes place at temperatures above 400°C. Under annealing to 800°C a change of the refractive index up to 1.8 is observed; the intensities of all IR absorption bands decrease down to the measurement limit. A wide band of hydrogen effusion (see Figure 1) is observed in this temperature range. This band can be deconvoluted into two Gauss-like bands with maxima at 500 and 750°C. The structural reorganization is also accompanied by hydrogen effusion from condition "A."

The maximum position of the first band changes from 300 up to 500°C according to some authors. This band is linked with hydrogen effusion from neighboring sites [8] or breaking of (C—H) weak bonds in polymer-like chains [9,10]. Amount of hydrogen evolved from the films is of the order of 10 at.%.

The second band observed at temperatures higher than 700°C is usually linked to breaking of strong (C—H) bonds [8,10,11] and with transformation into a graphite-like structure. The amount of hydrogen evolved during the breaking of strong (C—H) bonds is 12 at.%.

CONCLUSION

Hydrogen and other gases effusion from amorphous carbon films under annealing in the range 100–1000°C was studied. The amount of hydrogen in the material (28 at.%) was determined by the direct method. The value is close to that obtained from IR spectroscopy. The amount of free hydrogen in the investigated material is estimated at 4 at.%. The temperature curve of free hydrogen effusion has a maximum at 400°C. A wide band of hydrogen effusion is observed at temperatures higher than 400°C. The amount of hydrogen evolved during C—H weak bond breaking is of the order 10 at.%. The hydrogen amount effused under strong (C—H) bond breaking is 12 at.%.

ACKNOWLEDGEMENT

This work was supported in part by Arizona University and an RFBR grant N 96-02-16851-a.

REFERENCES

1. Discher, B., A. Burenzer and P. Koidl. 1983. *Sol. State Commun.*, 48:105.
2. Grill, A. and V. Patel. 1992. *Appl. Phys. Lett.*, 60(17):2089.
3. Grill, A., B. Meyerson and V. Patel. 1987. *J. Appl. Phys.*, 61:2874.
4. Vassilyev, V. A., A. S. Volkov, E. Musabekov, E. I. Terukov and V. E. Chelnokov. 1989. *Mater. Res. Soc. Symp.*, Pittsburgh, 149:347.
5. Vassilyev, V. A., A. S. Volkov, E. Musabekov, E. I. Terukov, S. V. Chernyshov and Y. M. Shernyakov. 1989. *J. Non-Cryst. Sol.*, 114:507.
6. Ataev, J., O. I. Konkov, E. I. Terukov, V. A. Vassilev and A. S. Volkov. 1990. *Phys. Status Solidi A*, 120:519.
7. Trapeznikova, I. N., O. I. Konkov, E. I. Terukov and S. G. Yastrebov. 1994. *Phys. Solid State*, 36(9):1519.
8. Robertson, J. 1992. *Prog. Solid State Chem.*, 21:199.
9. Beyer, W. and H. Wagner. 1982. *J. Appl. Phys.*, 53:8745.
10. Logothetidic, S., G. Kiriakidis and E. C. Palowra. 1991. *J. Appl. Phys.*, 70:2791.
11. Logothetidic S., G. Petalas and S. Ves. 1996. *J. Appl. Phys.*, 79:1040.

DIAMOND-LIKE CARBON AND OTHER MATERIAL

CVD of AlN Films and Some of Their Properties

V. P. STOYAN AND I. I. KULAKOVA
Chemical Department
Moscow State University
Vorobjevy Gory
119899 Moscow
Russia

E. I. KOCHETKOVA* AND B. V. SPITSYN
Institute of Physical Chemistry
Russian Academy of Sciences
31 Leninsky Prospekt
117915 Moscow
Russia

ABSTRACT: CVD of AlN by thermal decomposition of NH_4AlCl_4 and NH_4AlBr_4 in argon flows have been studied in the temperature range 973 to 1373 K. The thermodynamic yield of the reactions in the temperature range is of the order of 0.90 to 0.98 in conditions typical for our experiment (ammonia tetrahaloaluminate pressure was $2.1*10^{-3}$ atm). The activation energy of the process in 973 to 1173 K range are equal to 90 ± 4 and 92 ± 4 kJ/mole, respectively. Polycrystalline AlN films on silica glass, polycrystalline diamond films and glassy carbon and heteroepitaxial ones at 1223 to 1273 K on single crystal Al_2O_3 and 6H-SiC were grown. Measured optical and electrical properties of the film are close to those known from literature.

KEY WORDS: AlN film, thermodynamics and kinetics, stoichiometry.

---

*Author to whom correspondence should be addressed.

1. INTRODUCTION

Aluminum nitride is a material having a unique combination of physical, chemical and mechanical properties. Utilization of CVD as a flexible method of synthesis for growing aluminum nitride films on different substrates.

2. THERMODYNAMICS OF REACTIONS WITH AlN

The temperature dependence of the Gibbs free energy for ammonia tetrachloraluminate and ammonia tetrabromaluminate decomposition predicts aluminum nitride formation at a deposition temperature 1000 to 1400 K. Thermodynamic yield of the reactions is of the order of 0.90 to 0.98 for both compounds. Also of interest is aluminum nitride reactions with some compounds which may be in contact with the growing AlN films. The large negative Gibbs potential for reactions of AlN with oxygen containing compounds was calculated. On the other hand, AlN interaction with solid carbon (graphite) is not a thermodynamically favorable process and we have used it as a substrate material.

3. EXPERIMENTAL

CVD of AlN films proceeds according to the reaction:

$$NH_4AlCl_{4,g} \rightarrow AlN + 4HCl_g \qquad (1)$$

where CVD-CL or Br were performed in a quartz tube reactor. Initial substances evaporate in Ar flow 2.7 (slm) at ~573 K and then decompose in the temperature range 973 to 1273 K with AlN film formation on different substrates: quartz glass, glassy carbon, Al_2O_3, Si, SiC, Si_3N_4 and diamond coated Si substrates.

It is remarkable that the deposition rate depends on the nature of the substrate nature. The main reason of such influence is it etching by the hydrogen halogenide according to Equation (1).

4. KINETICS OF AlN FILM CRYSTALLIZATION

Growth kinetics of aluminum nitride films by CVD as a function of the initial compound nature, substrate material and crystallization temperature has been studied.

Higher rate of the aluminum nitride films growth rate is noticed (Figure 1) in case of bromide system in some range of conditions. Deposition rate of aluminum nitride for both systems under investigation at increasing temperature pass through a maximum at temperature about 1173 K.

The activation energy (E_a) was determined through Arrhenius plot of kinetics data (Figure 1). For chloride and bromide systems calculated E_a were equal to 90 ± 4 kJ/mole and 92 ± 4 kJ/mole, respectively. The measured E_a for chloride system is comparable with known values [1,2]; the E_a for bromide was measured in this work for the first time.

5. AlN FILM CHARACTERIZATIONS

5.1 X-Ray Diffraction

X-ray study of AlN thin film specimen with about 5 μm thickness deposited by CVD on the sapphire substrates and quartz glass was performed. In optimal growth conditions highly textured (0001) normal to substrate films on quartz glass and single crystalline (heteroepitaxial) films on Al_2O_3 and 6H-SiC was obtained. Crystallite size varied from 14 to 110 nm depending on crystallization conditions. The higher numbers are characteristic usually a higher crystallization temperature.

5.2 Reflection High Energy Electron Diffraction

Reflection high energy electron diffraction (RHEED) demonstrates that single crystalline (heteroepitaxial) films grown at 1223–1273 K on single crystal Al_2O_3 and 6H-SiC were obtained.

5.3 Optical Properties

IR-spectroscopy has demonstrated the influence of temperature on structure and chemical content of the films. Maximal structural perfectness of the films at 1273 K was established. Substrate nature (diamond, sapphire, silicon, silicon carbide, glassy carbon, quartz) seriously influenced morphology and quality of the films.

The band gap was determined by UV-spectra, is equal to 6.1 eV (according to literature data $E_{optical}$ is 6.0 to 6.2 eV).

5.4 Electrical Conductivity

Figure 2 demonstrates a rather high conductivity of CVD AlN films, although comparable to literature data [4].

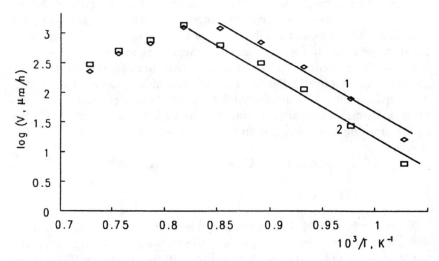

Figure 1. Temperature dependence of deposition rate for AlN films on quartz glass substrate: 1—chloride, 2–bromide systems.

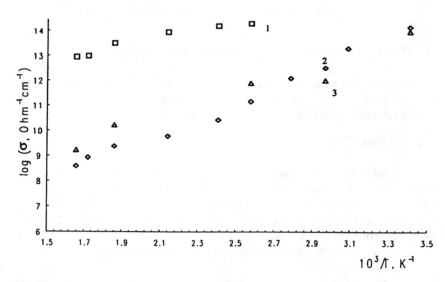

Figure 2. Electrical conductivity of AlN films. 1—[2]; 2—⊥ conductivity; 3—|| conductivity.

5.5 Stoichiometry

Despite the fact that the aluminum to nitrogen ratio is close to stoichiometric, as measured by chemical analysis, specific deviations for bromide and chloride systems have been observed. Aluminum nitride free standing polycrystalline films were prepared by two versions of CVD-chloride and bromide process at temperature of about 1273 K. For the analysis standard Kjeldal procedure [3] have been used. Average contents of the films in chloride and bromide process was determined and are equal to $AlN_{1.03}$ and $AlN_{0.92}$, respectively, with uncertainly equal to ± 0.5% [2].

6. CONCLUSIONS

A comparative study of AlN CVD by NH_4AlCl_4 and NH_4AlBr_4 decomposition was performed. The thermodynamics of several reactions involved AlNs have been calculated. At difference substrates polycrystalline (on quartz glass, glassy carbon, polycrystalline diamond) and heteroepitaxial (on Al_2O_3 and 6H-SiC single crystals) AlN films was regrown at temperature about 1273 K. The kinetics of AlN CVD growth at 973 to 1373 K have been studied. The activation energy for chloride and bromide systems in the 973 to 1173 K range are equal to 90 ± 4 kJ/mole and 92 ± 4 kJ/mole, respectively. The structure, chemical content and some properties (optical, electrical) of synthesized poly- and single crystalline AlN films was investigated.

7. ACKNOWLEDGEMENT

The authors are appreciated very much V. B. Kalinin for X-ray diffraction. S. V. Bantsekov for estimating electrical conductivity, A. E. Gorodetsky for carrying out RHEED studies. For providing of SiC substrates we are grateful to our colleagues from Prof. Yu. A. Vodakov's Laboratory.

The work was performed with a partial support of Sandier National Laboratories under the contract AN-8800.

REFERENCES

1. V. A. Dobrynin, Ph. D. Thesis. Moscow Electron Technique Institute (1986).

2. V. P. Stoyan, Master Thesis. Moscow State University (1996).

3. G. V. Samsonov, O. P. Kulik, V. S. Politchuk. 1978. Preparation and Methods of Nitrides Analysis. Kiev, pp. 254–257.

4. D. W. Lewis. 1970. "Properties of Aluminum Nitride Derived from AlCl₃NH₃," *J. Electrochem. Soc.*, 117(7):978–982.

Growth of AlN Films and Diamond/AlN Layer System Application in Acoustoelectronics

A. F. BELYANIN*

Central Research Technological Institute "TECHNOMASH"
4 Ivan Franko str.
121355 Moscow
Russia

A. N. BLAUT-BLACHEV, L. L. BOUILOV AND B. V. SPITSYN
Institute of Physical Chemistry
Russian Academy of Sciences
31 Leninsky Prospekt
117915 Moscow, Russia

ABSTRACT: The effect of growth conditions on the structure of AlN films manufactured by RF magnetron sputtering in Ar-N$_2$ gas mixture has been considered. Multilayer structure based on polycrystalline diamond and <001> textured AlN film has been used to produce surface acoustic wave filters and delay lines operating up to 1.52 GHz.

KEY WORDS: AlN films, RF magnetron sputtering, diamond CVD, acoustoelectronics.

1. INTRODUCTION

AlN films are typically manufactured by chemical deposition from vapor phase, by reactive Al evaporation and sputtering in nitrogen-con-

---

*Author to whom correspondence should be addressed.

taining gas, ion implantation of nitrogen into Al etc. The techniques for obtaining AlN films are considered in detail, for instance in References [1,2].

For practical application of AlN film growth process, it should be compatible with other operations of electronic devices manufacturing, which dictate a growth temperature in the range of 200–400°C. To such low-temperature techniques for growing AlN films with regular structure relate those using plasma ion sputtering. Among the latter, magnetron sputtering is a promising one, yielding AlN films with structure close to that of single crystal.

2. EXPERIMENTAL

AlN films were grown by reactive RF magnetron sputtering of Al in $Ar + N_2$ gas mixture. Planar and cylindrical magnetron sputtering systems were used [3]. Strong textured along <001> (crystallites misorientation of 0.5–3°) AlN films with thickness up to 5 μm were grown on single crystal, polycrystalline and amorphous substrates. The process parameters varied in a wide range: gas mixture pressure of $3 \cdot 10^{-2} - 0.5$ Pa; N_2 concentration 30–100%; substrate temperatures 100–400°C; target to substrate distance 40–120 mm. The films grown contained both crystalline and amorphous phases. The crystalline phase content was 15–100%. In some cases of employing a planar magnetron of special construction [4], the increased pressure in the gas mixture allowed the formation of <100> textured AlN films. The crystallinity decreased in this case.

AlN films were grown on the substrates which were fixed relative to the sputtered Al target. It was found that the growth rate and the structure of AlN films (crystallinity, grain misorientation, texture axis orientation relative to the substrate, crystallite size and lattice parameters) depend essentially on the direction of film-forming particle flow.

Nonequilibrium crystallization conditions are characteristic for magnetron sputtering. In the case of nonorienting (amorphous or polycrystalline) substrates, crystallization occurs by normal (nontangential) mechanism. Magnetron sputtering allows to grow AlN films with regular structure on the crystalline and amorphous substrates. In this case orientation of the crystalline phase in AlN film is not related, for the wide range of growing parameters, to the orientation of single crystal substrates, but is determined by the direction of the flow of particles forming the film. For this reason the films possess the fiber interior structure. The axis of the texture of AlN films coincides with the fiber orientation. The AlN (space group $P6_3mc$) film texturing can occur along the screw symmetry axes $6_3 - <001>$ or $2_1 - <110>$. Epitaxy reveals for AlN films growing on single crystal substrates only at temperatures exceeding 300°C.

3. POTENTIAL FOR COMBINING AlN AND DIAMOND LAYERS FOR APPLICATIONS

AlN is a strong piezoelectric which possesses high speed ($V = 6$ km/s) of surface acoustic waves (SAW) propagation. To create the devices based on microwave SAW, the diamond/AlN pair is the most promising, since diamond has the maximal SAW propagation speed ($V = 10$ km/s) among the existing materials and the SAW speed for AlN is the most close to that for diamond. Using it as a sound duct in the acoustoelectronic devices enables the upper limit of frequency range of the signal processing to be raised. To produce SAW-based filters, the Si/diamond/AlN/IDT (IDT—interdigital transducer) and W/diamond/AlN/IDT layer structures (Figure 1) were used. In these structures Si and W are basic substrates, diamond is a sound duct and AlN—AlN is piezoelectric which generates SAW and converts the latter by means of IDT into the electrical signal.

Figure 2 shows the advantage of application for manufacturing the acoustoelectronic microwave devices (>1.5 GHz) of multilayer diamond/AlN structure with the highest SAW speed ($V = 9$–10 km/s) as compared to the single crystal piezoelectrics (e.g., LiNbO$_3$, $V = 3.8$ km/s) and layer structures based on sound ducts with low SAW speed (fused

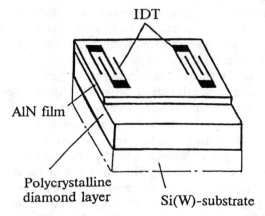

Figure 1. Construction of SAW filter based on layer diamond/AlN system.

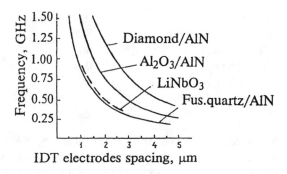

Figure 2. SAW filter operation frequency as a function of IDT electrode spacing for different sound ducts.

quartz/AlN, $V = 3.6$ km/s; Al_2O_3/AlN, $V = 6$ km/s). This is due to the frequency (f) proportionality to the SAW speed: $f = V/2d$, where d is IDT electrode spacing.

4. MANUFACTURING OF SAW DEVICES BASED ON DIAMOND/AlN LAYER STRUCTURE

We used 12 to 36 μm thick polycrystalline diamond films grown on the 6×8 and 9×11 mm Si and W substrates by means of industrial setup "ALMAZ" [5] provided with arc discharge [6]. Diamond films were deposited from the $H_2 + CH_4$ (1–10%) gas mixture at a pressure of $(1-20) \cdot 10^3$ Pa within the temperature range of 900–1200°C, at a rate of 5–10 μm/h. The diamond layers grown contained both crystalline (90–100%) and X-ray amorphous phases, with the crystalline phase formed by the grains textured axially along <111> and <110>. Diamond grains were stretched in the growth direction, having cross size of 2–5 μm and lamellar structure with the thickness of lamellae (coherent scattering regions) of 90–100 nm. On manufacturing SAW devices, unpolished (arithmetic mean deviation for growth surface profile was $R_a = 0.2085$–0.4280 μm) 12–14 μm thick diamond layers, as well as mechanically polished (the relief height difference $R_z < 0.04$ μm) 30–36 μm diamond films were used.

Piezoelectric AlN films were grown on diamond surface in a modified setup UVN-62P-3 [3], with reactive RF magnetron sputtering at the following conditions:

Preliminary vacuum	$7 \cdot 10^{-4}$ Pa
Gas composition	$Ar + N_2$ (40:60)
Gas pressure	0.6 Pa

Target	Al (99.99%)
Substrate temperature	300°C
Target-substrate distance	90 mm
RF power	0.6–1.5 kW
RF voltage	300 V
Bias on substrate holder	+25 to +40 V
Growth rate	2.5 μm/h

AlN films grown for the purposes of manufacturing SAW filters consisted of crystalline (30–60%) and amorphous phases. The crystalline phase was <001> textured, with orientation independent of that for the polycrystalline diamond grains at the above growth parameters. The texture axis of AlN films used for constructing SAW devices was inclined at <3° with respect to the normal to the substrate surface, and the grain misorientation relative to the texture axis was 1–2°. Adhesion of AlN film to the diamond layer exceeded 0.5 kg/mm². To increase adhesion, transition ~0.1 μm thick amorphous AlN layers were deposited when growing piezoelectric AlN films on polished diamond layers.

Si(W)/diamond/AlN layer structures were used to manufacture SAW delay lines and filters operating at 0.8149 and 1.52 GHz. The length of Al electrodes in IDT of delay lines was 3 mm; the IDT step was 11.25 μm; the distance between IDT groups was 2.8 mm. An IDT was formed lithographically using the typical contact photocopying technique. 814.9 MHz SAW delay lines were 36 μm thick diamond layer and 3.2 μm AlN layer were manufactured (SAW speed—9.17 km/s, electromechanical coupling coefficient—0.09, insertion loss for mismatch conditions—46 dB).

On manufacturing SAW filters, an IDT was formed by optical photolithography in a projection exposure setup. Electron-beam lithography was used to make a photomask for IDT exposing. Two IDT groups with 10 pairs of electrodes in each group (200 μm in length, spacing 1.5 μm, step 6 μm) were used in filter construction. The distance between IDT groups was 100 μm. Filter central frequency was 1.52 GHz; calculated SAW speed for the Rayleigh mode was 9.2 km/s.

5. CONCLUSIONS

1. By means of RF magnetron Al reactive sputtering in the Ar+N₂ gas mixture, the <001> textured AlN films have been grown. AlN film growth occurs via normal (nontangential) mechanism. Under certain growth conditions, the <110> and <100> texturing and also the epitaxial crystallization of AlN film are possible.
2. On the base of Si(W)/diamond/AlN/IDT multilayer structures using

polycrystalline diamond and with <001> textured AlN film as sound duct, the 0.8149 and 1.52 GHz SAW delay lines and filters have been produced.

ACKNOWLEDGEMENTS

The work has been performed under the partial support of Sandia (USA) AN-8800 Contract.

REFERENCES

1. Davis, R. F. 1991. "III-V Nitrides for Electronic and Optoelectronic Applications," *Proceedings of the IEEE*, 79–5:702.
2. Belyanin, A. F. 1996. "AlN Films Manufacturing (Review)," *Materialy VII Mezhdunarodnogo Simposiuma "Tonkie plyonki v electronike"* [in Russian], Russia, Yoshkar-Ola, pp. 167–213.
3. Belyanin, A. F., N. A. Bulyonkov, A. B. Bogomolov and V. G. Balakirev. 1990. "Composition and Applications of Thin AlN Films Obtained by RF Magnetron Sputtering" [in Russian], *Tehnika sredstv svyazi, ser. TPO*, vyp. 3, pp. 4–24.
4. Belyanin, A. F., P. V. Pastchenko and A. P. Semenov. 1991. RF Magnetron Sputtering Setup for Growing Thin Films" [in Russian], *Pribory i tehnika experimenta*, N3, pp. 220–222.
5. Alexenko, A. E., A. F. Belyanin, A. A. Botev, L. L. Bouilov, P. V. Pastchenko and B. V. Spitsyn. 1995. "The Setup Constructions for Polycrystalline Diamond Film Growing by Means of Arc Discharge," *Materialy VI Mezhdunarodnogo Simposiuma "Tonkie plyonki v electronike,"* [in Russian], Ukraine, Herson, pp. 17–26.
6. Bouilov, L. L., A. E. Alexenko, A. A. Botev and B. V. Spitsyn. 1986. "Certain Characteristics of the Growth of Diamond Layers from an Active Gas Phase," *Soviet Physical Chemistry*, pp. 302–306.

AlN Thin Film Deposition by Ion Beam Sputtering

A. F. BELYANIN*
Central Research Technological Institute
"TECHNOMASH"
121355 Moscow
Ivan Franko str. 4, Russia

A. P. SEMENOV AND V. M. HALTANOVA
Buryat Institute of Natural Sciences Siberian Division of Russian
Academy of Sciences
670047 Ulan-Ude, Sakhjanova str. 6, Russia

1. INTRODUCTION

Our efforts have been devoted to the deposition of AlN protective films onto resistive heating elements of thin thermoprinting matrixes. AlN possesses sufficient hardness and is characterized by high thermal stability, good insulating properties and a high thermal conductivity. A promising means for growing AlN films at low temperatures is reactive ion beam sputtering [1] in which a hard target is sputtered by a beam of accelerated nitrogen ions. Phase formation occurs on separate arrival of the reaction components to the substrate.

2. EXPERIMENTAL DETAIL AND RESULTS

Metallic Al disks 30–50 mm in diameter were used as targets. The experiments were performed in a modernized setup reported earlier [2] with a plasma ion source [3]. The N_2-Ar gas mixture was cleaned by the

---

*Author to whom correspondence should be addressed.

successive use of a pyrogallol solution A ($C_{12}H_{12}O_6$), copper at 773 K, strong H_2SO_4, $CaCl_2$ and ascarite. The cleaning system lowered the oxygen and water vapor concentrations in the gas mixture to the minimum acceptable values. The optimal process parameters for deposition are shown in Table 1.

The films were studied with an X-ray diffractometer ($Cu_{k\alpha}$). The X-ray studies showed that AlN films grown at a substrate temperature lower than 673 K were amorphous. In order to achieve crystallinity it is necessary to use higher substrate temperatures [4]. Accordingly, our textured AlN films were grown on fused quartz substrates at 673–723 K. The diffractogram (Figure 1) shows a peak corresponding to the (0002) AlN reflection. The film is textured with the texture axis perpendicular to the substrate surface. The grain disorientation and the tilt of the texture axis relative to the normal to the substrate surface were measured with rocking curve diffractograms [Figure 1(b)]. The disorientation of the structures was $\sigma = 3.9°$. The deposition temperature, 673–723 K, exceeded however, the permissible heating temperature of the resistive thermoelement. Therefore, the synthesis of AlN films was limited to temperatures around 423 K.

The composition of the films was investigated with X-ray electron spectroscopy using an ESCA-5 spectrometer. Figure 2 presents X-ray electron spectra of AlN thin films grown at ~423 K. A satellite maximum at 403 eV is observed next to the N1s line of AlN, which corresponds to aluminum oxynitride. The 01s line corresponds to Al_2O_3. Aluminum nitride and alumina appear in AlN films due to the $1.33 \cdot 10^{-3}$ Pa residual pressure in the

Table 1. Deposition process parameters.

Parameter	Value
Accelerating voltage, kV	5–10
Ion beam current, mA	4–6
Ion incident angle, degree	45–60
Residual pressure, Pa	$1.33 \cdot 10^{-3}$
Working pressure, Pa	$(2.66–3.99) \cdot 10^{-3}$
Working gas composition ratio (argon:nitrogen)	1:20
Substrate temperature, K	423–723
Growth rate, μm/h	0.14–0.21
Deposition time, h	5–7
Ion beam diameter at the target, mm	20–30
Substrate type	fused quartz, film structure with resistive thermoelement

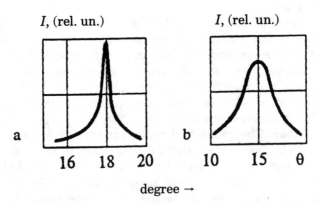

Figure 1. A partial diffractogram of an AlN film grown on a fused quartz substrate (a) and a rocking curve diffractogram (b).

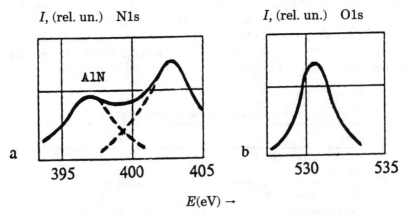

Figure 2. X-ray spectra: N1s (a) and O1s (b) lines in AlN films.

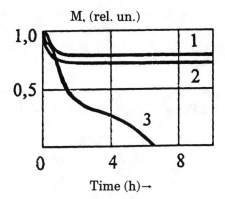

Figure 3. Mass loss vs. time dependence for AlN films: 1—boiling water; 2—sulphuric acid; 3—NaOH aqueous solution.

working chamber. The partial pressure of the oxygen stays comparatively high, while the growth rates of the films are low.

3. FILM CHARACTERISTICS

The solubility of the films in water, acids and alkalis have been studied. Figure 3 shows the time dependence of AlN film solubility in boiling water, sulphuric acid and aqueous caustic soda solutions. AlN films dissolve comparatively fast in NaOH aqueous solutions, but are resistive to boiling water and H_2SO_4 solutions. The high solubility of AlN in NaOH is consistent with the well-known active interaction between aluminum nitride and an aqueous solution of caustic soda.

In the course of AlN film deposition, the affect of growing film upon the structure of the resistive thermoelements was examined. Measuring the electrical resistance in situ, showed that the protective AlN films exert no influence on the electrical characteristics of the film resistors. A pilot batch of thermoprinting matrixes coated with protective AlN films was tested in an actual thermoprinting device. The service life of the matrix with AlN coating is 2–3 times higher than that for continuous work of the matrix with typical protection based on SiO_2 and Al_2O_3.

REFERENCES

1. Semenov, A. P., A. F. Belyanin and V. M. Haltanova. 1988. *Composition Peculiarities of Thin Films Formed by Ion Beam Sputtering.* Tehnika Sredstv Svyazi, ser. TPO, 1:25–31.

2. Semenov, A. P. 1986. "Ion Beam Sputtering Device in Universal Exhaust Unit (VUP-4)," *Pribory I Tehnika Experimenta,* 2:220–221.

3. Semenov, A. P. 1993. "Generation of Strong Current Ion Beams in Ion Sources Based on Cold Cathode Discharges," *Pribory I Tehnika Experimenta,* 5:128–133.

4. Morosanu, C. E. 1980. "The Preparation, Characterization and Application of Silicon Nitride Thin Films," *Thin Solid Films,* 65(2):171–208.

Optical and Electrical
Properties of AlN Films

G. A SOKOLINA, A. N. BLAUT-BLACHEV,
L. L. BOUILOV, T. A. KARPUKHINA
AND E. I. KOCHETKOVA
Institute of Physical Chemistry
Russian Academy of Sciences
31, Leninsky Prospekt
117915 Moscow, Russia

A. F. BELJANIN
Stock Co Central Science Research and Technology Institute
"Technomash"
121355 Moscow, Iv. Franko str. 4, Russia

The wide bandgap semiconductor AlN possesses a number of interesting properties that can find applications in various areas of electronics. They are: a large bandgap, the piezoelectric effect, high resistivity and insulating characteristics and an appreciable thermoconductivity. As a result both growth of AlN films by different methods and studies of their properties have been carried out for an appreciable length of time.

In our work we have investigated optical and electrical properties of AlN films obtained by magnetron RF sputtering. A pure aluminum target was sputtered in Ar + N atmosphere under the action of an RF discharge (13.6 MHz). The working pressure in a chamber was about 1 Pa and the ratio $Ar:N_2 = 35:65$. The substrate temperature was held between 20–300°C.

A UV spectrometer, SPECORD M-40, recorded the optical absorption spectra within the wave range 200–800 nm for the films deposited on quartz and sapphire substrates. A two beam scheme was employed with the initial substrate as a reference, enabling calculation of the absorption caused by the AlN film alone. By means of a computer, the spectral parameters were calculated and the absorption coefficient dependence on photon energy E was plotted.

Figure 1 shows the spectral dependence of the absorption coefficient on energy for an AlN film at room temperature. From these data a bandgap, E_g, can be estimated. The problem in E_g estimation, as well as that of determining the type of valence-conduction band electron transition (direct or indirect), reduces to the analysis of optical absorption over the region of energies close to the gap (absorption edge). For all samples studied the absorption coefficient exceeded 10^4 cm^{-1} at $E > 5.5$ eV. Such a large value can be an indication of a direct transition in the films tested,

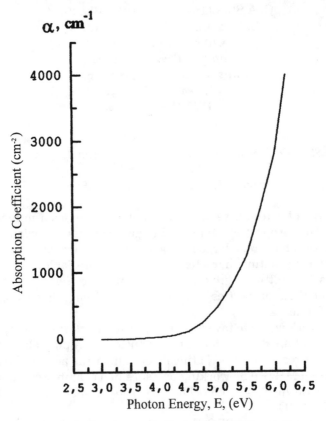

Figure 1. Dependence of the absorption coefficient upon photon energy near the fundamental edge of an AlN film at room temperature.

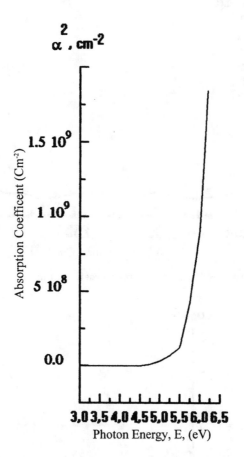

Figure 2. Dependence of the absorption coefficient (squared) vs photon energy.

since the typical absorption coefficient estimate for indirect transitions lies in the range of $10–10^3$ cm^{-1}. Supporting evidence for the direct band character of AlN films is the fact that absorption coefficient is proportional to $(E–E_g)^{0.5}$ for all test films, as can be seen from the Figure 2, which presents the absorption coefficient squared as a function of photon energy. Such dependence is known to take place for allowed direct transitions.

The bandgap values were estimated from Figure 2 by extrapolation of the absorption coefficient to zero. For the majority of samples studied the gap was $E_g = 6.1$ eV, which is close to the values obtained elsewhere for the most perfect of specimen.

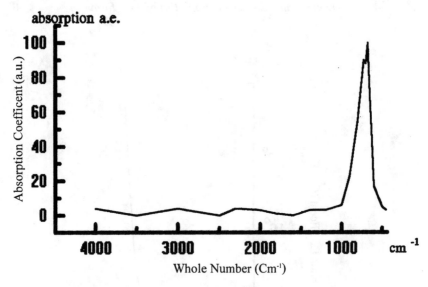

Figure 3. Value of the absorption coefficient in the IR region.

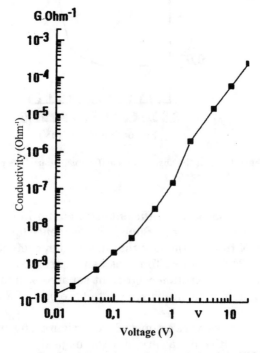

Figure 4. Conductivity-voltage characteristic of an AlN film.

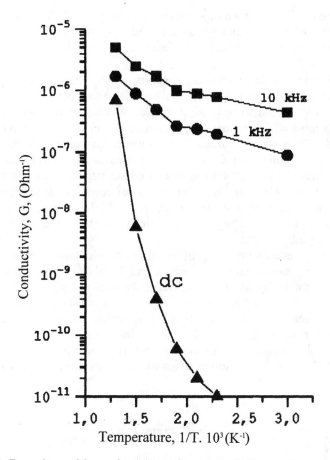

Figure 5. Dependence of the conductivity on the reciprocal of temperature measured for d.c. and 1 and 10 kHz a.c.

Figure 3 shows an I. R. absorption spectrum for an AlN film on a Si substrate. A peak at 675 cm^{-1} is characteristic for this spectrum, which corresponds to the frequency of transversal optical (TO) phonon of AlN. An additional sharp peak at large wave numbers is related to Si substrate absorption.

Temperature and frequency dependencies of static G_{dc} and dynamic G_{ac} electrical conductivity of AlN films were measured. High insulating characteristics of the AlN films demand that the conductivity measurements be carried out at high temperatures and under conditions which exclude leakage currents. Therefore, the electrical resistance of the films was measured in a special vacuum set-up with optical heater. This enabled one to eliminate leakage currents through the adsorption surface layers and to preclude an ingress of impurities from the heater onto the test specimen. $G_{dc}(T)$ and $G_{ac}(T)$ were measured in the temperature range 150–800 K.

AlN films were deposited on tungsten, silicon, AlN ceramic and sapphire substrates. The films thickness varied from 0.8 to 5 micrometer. Aquadag electrodes were used to measure conductivity with the two-electrode method. To measure the G_{dc}, an electrometer with current sensitivity of 10^{-14} A was employed. Previous current-voltage characteristics were taken for the test specimens (one of them is shown in Figure 4), and the conductivity was measured in the linear region of current-voltage characteristics. For the films deposited on conducting substrates the transverse conduction was measured, from which the conductivity of the film was estimated. In the case of the dielectric substrates the longitudinal conduction was measured.

Figure 5 presents typical graphs of the conductivity dependence on temperature, measured for d.c. and 1 and 10 kHz. a.c. The value of the electrical resistivity at room temperature differs by several orders of magnitude for d. c. and a. c., which may be due to the considerable relaxation processes occurring in the films. The activation energy of G_{dc} in the high temperature region is 0.75 eV, which indicates the presence of impurities or defect complexes responsible for electrical properties.

Figure 6 shows the conductivity and the capacitance of AlN film as a

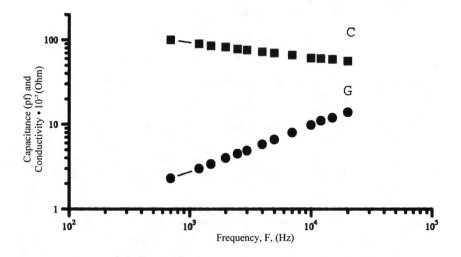

Figure 6. Dependencies of the conductivity (G) and the capacitance (C) of AlN films on frequency.

function of frequency. The change of G_{ac} (f) can be written as G_{ac}-f^s with s = 0.6. Polycrystalline CVD diamond films showed analogous frequency-temperature dependencies. The activation energy values and the character of G_{ac} (f) plots were explained by the presence of impurities and energy barriers at crystallite boundaries.

The experimental data obtained demonstrate that RF magnetron sputtering presents a means of manufacturing at relatively low temperatures AlN films of a few micron in thickness, which possess both good optical and electric properties.

This work is supported partially by the Sandia National Laboratories, USA under Contract AN-8800.

Some Features of Initial Stages of the GaN Epigrowth by Low Pressure Metalorganic Chemical Vapor Deposition Method

W. V. Lundin,* B. V. Pushnyi, A. S. Usikov,
M. E. Gaevski and M. V. Baidakova

A. F. Ioffe Physical-Technical Institute
26, Polytechnicheskaya
St. Petersburg, Russia

ABSTRACT: The growth of the buffer GaN layer and its influence on perfection and PL properties of GaN epilayers grown by LP-MOCVD method on the $(00 \cdot 1)$ and $(10 \cdot 2)$ sapphire substrates have been studied. At different gas flow conditions we have found the presence of some sort of "delay time" when GaN deposition does not occur on the sapphire substrate at 500°C. The substrate nitridation has a strong effect on crystallographics properties of partly amorphous buffer layer and, following annealing, produce the strain relaxation and the relatively undefective region formation. The optimal buffer layer thickness gives rise to relatively an unstrained epilayer as thin as 1 μm. The initial epigrowth step depends on the Ga-metalorganic flow. The buffer and epilayer thickness as a function of substate position on the graphite susceptor is also presented.

KEY WORDS: GaN growth, metalorganic chemical vapor deposition (MOCVD), buffer layer, sapphire surface nitridation.

INTRODUCTION

There is considerable interest in III-N semiconductors recently, resulting from the possibility of producing light sources and photo-

---

*Author to whom correspondence should be addressed.

detectors in the blue and UV spectral range, giving rise to the necessity of comprehensive studies of the process of its growth. These studies were performed on the layers grown by metalorganic chemical vapor deposition (MOCVD) method on sapphire substrates [1].

PROCESS TECHNOLOGY

The GaN films were grown in a horizontal flow, inductively heated reactor operated at reduced pressure 200 mbar using H_2 as a carrier gas, trimethylgallium (TMG) and ammonia (NH_3) as a component precursors. Optical grade polished sapphire of $(00 \cdot 1)$ or $(10 \cdot 2)$ orientations was used as the substrate. Before loading into the reactor chamber the substrates were etched with a solution of HF + HNO_3 (1:3). After that, on special occasions, its surfaces were treated by H^+ ions.

The typical growth sequence is given in Figure 1. Prior to the GaN

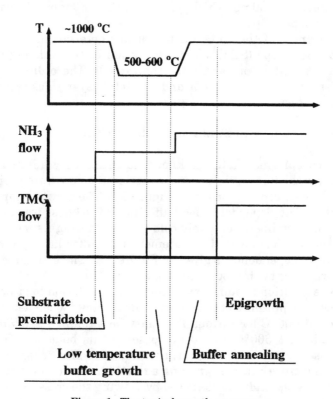

Figure 1. The typical growth sequence.

epitaxial deposition the substrates were thermally treated at 990°C for 10 min to remove the traces of possible surface contamination and to begin nitridation process. Then, the growth process involves the growth of buffer GaN layer at low temperature (about 500°C) followed by annealing and GaN epilayer growth at higher temperature (990°C). The GaN buffer layer was deposited using 24 μmol/min trimethylgallium (TMG), 1.5 standard liter per minute (SLM) NH$_3$, and 4.5 SLM H$_2$. During the GaN epilayer deposition, the flow rates of TMG, NH$_3$ and H$_2$ were maintained at 36 μmol/min, 2.5 SLM and 4.5 SLM respectively.

The thicknesses of the buffer layers were determined by direct measurements of heights of ridged structures using a profilometer. The ridge structure with vertical walls was created by a photolithography process using special mask followed by Ar-ions sputtering of uncovered place. The sputtering was controlled by mass spectrometer and stopped when the signal from gallium disappeared. The ridge width was about 300 microns. For a given temperature the buffer layer thickness variation in range 10–150 nm which was accomplished mainly by variating its growth time.

The thickness of the epitaxial layers was determined by scanning electron microscope (SEM) or by estimation of the difference in the substrate weight before and after the growth. The ordinary growth rates were about 0.8–2.5 μm/h and the total layer thickness ranged from 0.5 to 2 μm.

RESULTS AND DISCUSSION

The GaN epilayer quality is known to be strongly affected by the buffer layer thickness [2]. Its dependence on the substrate position on the graphite susceptor is shown in Figure 2(a). The growth temperature was 500°C. The dependence for buffer layer thickness we suppose to arise from warming of relatively cold gas flow during its passage over the susceptor. To support this statement the buffer layer growth rate temperature dependence is given in Figure 2(c). The estimated activation energy of growth process in this case is 270 meV.

The investigation of buffer layer thickness in relation to growth time has shown the presence of some sort of "delay time" when deposition is not occur at 500°C for various TMG flow (less than 25 μmol/min) [see Figure 2(b)]. At 600°C we do not observe this behavior. Thus, the widely used procedure of determination of buffer layer thickness through the growth rate by growing a thick film for a long time at buffer layer growth conditions followed by viewing the cross section using SEM is not correct procedure [2].

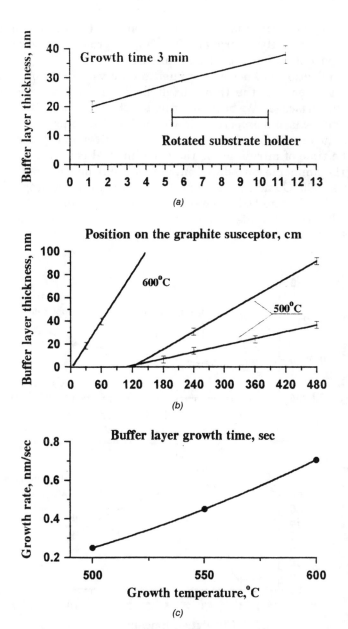

Figure 2. The dependence of buffer layer thickness on the substrate position on (a) the susceptor, (b) growth time and (c) growth temperature.

According to X-ray diffraction measurements (XRD), the as grown buffer layer is partly amorphous. The higher growth temperature the smaller part of buffer layer is amorphous.

The nitridation is known to influence on crystallinity of epilayers. During this process the thin AlN film is supposed to form on the substrate surface [3]. We have investigated the influence of the nitridation on buffer layers properties grown on $(00 \cdot 1)$ sapphire substrates at $500°C$. The results for grown buffer layers are given in Figures 3(a) and 3(b). The time of nitridation is the time interval prior the growth between the ammonia introduced into reactor during thermally treated process at $990°C$ and the moment when temperature is dropped down to $700°C$ (see Figure 1). For lower temperatures nitridation does not occur [4]. The typical buffer layers thickness was 25–30 nm and did not depend on the nitridation time. The lattice constant C in relation to the

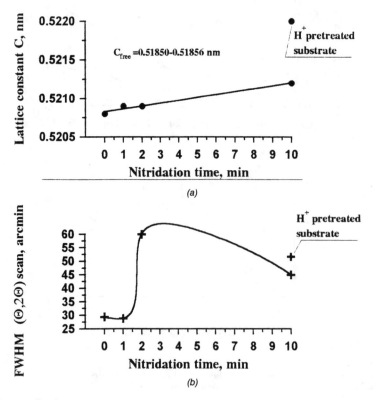

Figure 3. The dependence of lattice constant C (a) and crystallinity (b) of buffer layers on the nitridation time.

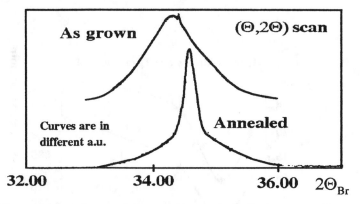

Figure 4. The influence of the annealing on the buffer layer crystallinity.

nitridation time exhibit the linear dependence, whereas the full width at high maximum (FWHM) of X-ray rocking curve ($\theta,2\theta$ scan.) displays considerable change in buffer layer crystallinity. The longer the nitridation time the thicker is the AlN sub-layer, the stronger GaN buffer layer connects with substrate, and the higher is the buffer layer strain. It was found that, other conditions being equal, the epilayers with the best crystallinity were grown on the buffer layers with the worst crystallinity (i.e., with highest FWHM of $\Theta,2\Theta$ scan). In our growth conditions it corresponds to nitridation time about 2 minutes which is in good agreement with previously published results [3]. High defect density of these buffer layers we suppose helped to accommodate the high lattice mismatch between GaN and sapphire. Utilizing the sapphire substrates treated by H^+ ions results in poor crystallographic perfection as well as in decreasing growth rate of the epilayer. The H^+ ions treatment we believe creates additional unsaturated bonds on the sapphire surface which results in stronger bonding of the buffer layer to the substrate.

After the buffer layer deposition it was exposed to heat treatment at the temperature where epitaxial growth takes place. The influence of the annealing on the shape of the XRD curve is given in Figure 4.

Noticeable is the shifting of the rocking curve maximum corresponding to the crystal relaxation and the narrowing of its FWHM after annealing at 990 °C for 5 minutes. The XRD curve, after annealing, can be regard as consisting of two parts. One has extremely wide FWHM and low intensity and can be attributed to extremely defective regions in the buffer layer. The other has a narrow FWHM and a higher intensity that can be attributed to relatively undefective regions. The

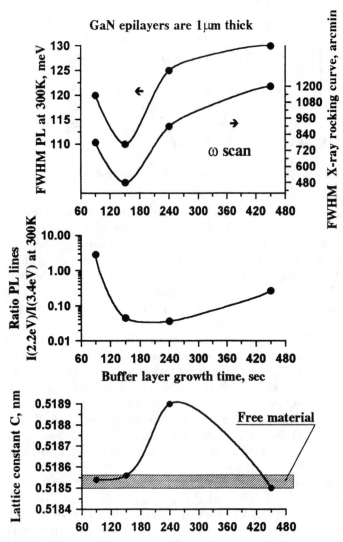

Figure 5. The influence of buffer layer growth time (i.e., buffer layer thickness) on the properties of 1 μm thick GaN epilayer.

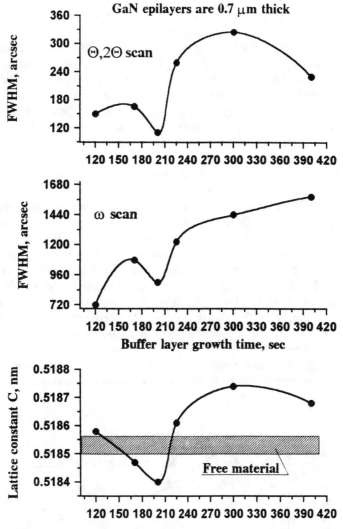

Figure 6. The influence of buffer layer growth time (i.e., buffer layer thickness) on the properties of 0.7 μm thick GaN epilayer.

previous investigations of the buffer layer by atomic force microscopy have established the formation of several large islands during annealing [5]. These islands we suppose to be these undefective regions and act as nucleation sites during the epigrowth. Thus, the annealing of the buffer layer substantially improve its crystallinity.

The influence of buffer layer growth time (i.e., buffer layer thickness) on the properties of epitaxial GaN films for two sets of samples is shown in Figures 5 and 6. These sets are distinguished in buffer layer growth conditions and epilayer thickness. For both sets the optimal thickness of the buffer layer which results in the best surface morphology, PL and crystalline properties is about 30 nm. For this optimal thickness the intensity of PL line at 2.2 eV is more than 30 times than the one at 3.4 eV at 300 K (He-Cd laser excitation with density of 1 W/cm²). The PL line at 2.2 eV is supposed to be determined by defects in epilayer.

The buffer layer thickness was found to influence on the epilayer strain. The optimal buffer layer thickness approaches relatively unstrain epilayers as thin as 1 μm (see Figure 5). To compare the epilayers grown on AlN buffer layer are strained up to 4 μm and more [6]. This clearly shows the advantage of the GaN buffer layer in comparison with AlN one. In the case of thin epilayer the dependence of lattice constant C on the buffer thickness exhibits the strong minimum (Figure 6). Change of the strain sign in this dependence is surprising and to the best of our knowledge was not observed earlier in GaN/-sapphire system.

The epilayer thickness as a function of substrate position on the graphite susceptor given in Figure 7 has the opposite slope relative to

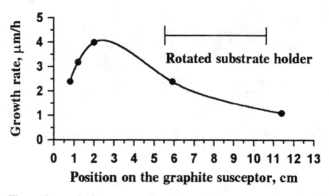

Figure 7. The epilayer thickness as a function of substrate position on the graphite susceptor.

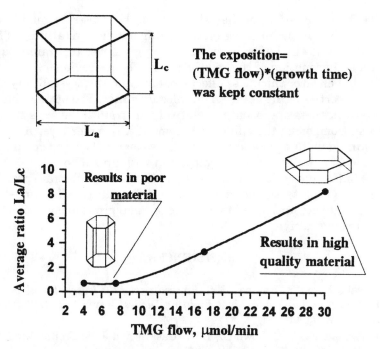

Figure 8. The initial stage epigrowth surface morphology as a function of the TMG flow.

the one for buffer layer mainly on account of parasitic deposition from the vapor phase on the susceptor.

The SEM surface morphology study has shown the formation of separate growth crystallites at the initial stages of epigrowth. The higher the TMG flow during the epigrowth the higher the growth crystallites density, the higher L_a/L_c ratio (Figure 8), and the higher the thick epilayer quality. After several minutes of epigrowth under high TMG flow conditions the growth crystallites connect to form single epilayer surface.

For the $(10 \cdot 2)$ sapphire substrate we have found that when the buffer layer thickness is less than 25 nm the $(11 \cdot 0)$ GaN layer can be grown on the $(10 \cdot 2)$ sapphire substrates.

SUMMARY

1. The method of direct buffer layer thickness measurement has shown a strong temperature dependence and the presence of some sort of "delay time" when deposition does not occur at 500 C.

2. The XRD investigation has shown that the substrate nitridation has a remarkable influence on the strain and the crystallinity of the buffer layer. The buffer layer annealing gives rise to strain relaxation and formation of the relatively undefective region.

3. The buffer layer thickness affects, not only on the crystallinity and PL properties of the epilayer, but also its strain. The optimal buffer layer thickness approaches a relatively unstrain epilayer as thin as 1 μm. Comparing the epilayers grown on AlN buffer layer requires straining up to 4 μm and more [6]. This clearly show the advantage of the GaN buffer layer in comparison with an AlN one.

4. The buffer and epilayer thickness as a function of substrate position on the graphite susceptor are quite different. This fact must be taken into account for adjustment of optimal growth conditions and the reactor design.

REFERENCES

1. Akasaki, I. and H. Amano. 1996. *Proceedings of the International Symposium on Blue Laser and Light Emitting Diodes,* March 5–7, 1996, Chiba University, Japan.

2. Doverspike, K., L. B. Rowland, D. K. Gaskill and J. A. Freitas, Jr. 1995. *Journal of Electronic Materials,* 24(4).

3. Uchida, K., A. Watanabe, F. Yano, M. Kouguchi, T. Tanaka and S. Minagawa. 1996. *Proceedings of the International Symposium on Blue Laser and Light Emitting Diodes,* March 5–7, 1996, Chiba University, Japan.

4. Yamamoto, A., M. Tsujina, M. Ohkubo and A. Hashimoto. 1994. *J. Cryst. Growth,* 137:415.

5. George, T., W. T. Pike, M. A. Khan, J. N. Kuznia and P. Chang-Chien. 1995. *Journal of Electronic Materials,* 24:4.

6. Hiramatsu, K., T. Detchprohm and I. Akasaki. 1993. *Jpn. J. Appl. Phys.,* 32:1528.

Epitaxial and Heteroepitaxial Growth of Silicon Carbide on SiC and AlN/Al$_2$O$_3$ Substrates

A. N. Kuznetsov,* A. A. Lebedev, M. G. Rastegaeeva,
E. V. Bogdanova and E. I. Terukov

A. F. Ioffe Physicotechnical Institute
194021, Politechnicheskaya 26
St. Petersburg
Russia

ABSTRACT: SiC films have been grown on SiC and AlN/Al$_2$O$_3$ substrates by reactive magnetron sputtering of a silicon carbide target in an Ar/CH$_4$/SiH$_4$ mixed plasma. Crystal perfection was investigated as a function of substrate temperature in a temperature range 700–1400 K. The SiC epilayers surface were made into Schottky diodes by deposition of molybdenums films. Current-voltage and capacitance-voltage measurements showed an n-type conductivity with uncompensated donor concentration of $5 \cdot 10^{18}$ cm^{-3}. The barrier high was 1.0 eV.

KEY WORDS: SiC, MESFET, AlN, heteroepitaxy, magnetron sputtering.

INTRODUCTION

The rapid development of silicon carbide technology in recent years has brought us to the verge of developing rf devices based on this material. One such device is the field-effect transistor with a Schottky-diode gate or MESFET. High-quality operation of a MESFET requires that there be no shunting of the nonconducting channel. It thus becomes necessary to use substrates with a high resistivity (semi-insulating substrates) [1]. However, it is difficult to make progress in the production of

---

*Author to whom correspondence should be addressed.

homosubstrates even for such well-studied semiconductors as GaAS. There are at present no semi-insulating SiC substrates.

One way to solve this problem is to carry out a heteroepitaxial growth of SiC on high-resistivity substrates of other semiconductors. If the potential of SiC is to be fully realised, this substrate must (first) match SiC in terms of lattice constant and coefficient of linear expansion. Second, it must be at least as thermally stable as silicon carbide. These requirements are met by the AlN/Al_2O_3 system. This system has been used previously for growth of SiC films by chemical vapour deposition [2] and molecular beam epitaxy [3]. In this paper we are reporting the production of SiC epitaxial films by reactive magnetron sputtering on Al_2O_3 substrates with an AlN layer.

SETUP FOR GROWTH OF EPITAXIAL SiC FILMS BY REACTIVE MAGNETRON SPUTTERING

Methods for growing layers of metals, semiconductors, and isolators which involve sputtering of the corresponding materials by ion bombardment are finding progressively wider use. The possibility of using gases and solid targets as sources of the components of a growing layer is one of the main advantages of magnetron reactive sputtering. This method makes it possible to vary the energy of the particles and their flow rate into a growing surface over wide ranges and thereby to model different growth conditions. In the presented work we have used a planar magnetron operating at low discharge powers to ensure the low sputtering rates needed to meet the conditions for epitaxial layer growth as the substrate temperature is lowered. The cathode of the magnetron consisted of a cylindrical electromagnet with a flat holder for a 55 mm diameter target on its end. An annular gap in the magnetic circuit under the target allowed the magnetic field to enter the reactor vessel. The magnitude of the magnetic field was regulated by changing the current in the magnetic winding and could be varied from 0 to 60 mT. The capability of varying the magnetic field strength in the magnetron made it possible to control independently the relationship between the discharge voltage and current and, thereby, the rate of sputtering and the energy of the sputtered particles. A voltage of 300 to 500 V could be applied to an annular anode mounted in the cathode plane. The choice of sputtering voltage was determined by the need to sputter the target at the atomic level with sputtering yields of less than unity, which condition is satisfied when voltage is below 400 V. The need to limit the sputtering power, on the one hand, and to satisfy the conditions for maintaining the discharge, on the other, restricted the choice of discharge current to 15–30 mA.

The magnetron was placed in standard vacuum vessel pumped by rotary and diffusion pumps. The base pressure was 10^{-4} Pa. The apparatus included a two-channel mixing and gas feed system for maintaining a gas pressure of from 10^{-1} to 10 Pa in the magnetron reactor. The pressure of the gas mixture was typically 1 Pa.

Slabs of boron-doped single-crystal Si and of pyrolitic graphite were used as targets, along with a target assembled from single crystals of SiC of diameter 5–7 mm and thickness 0.5 mm. The gas mixtures included argon, silane and methane.

The substrate holder (susceptor) was positioned 8 cm above magnetron target and had a heater for keeping the substrate temperature at 350–1500 K. The electrical potential of the holder could be chosen arbitrarily, biased positively or negatively with respect to the plasma potential by means of an external source. This technology was chosen by us for growth of the SiC epilayer in our experiments.

ELECTRICAL PROPERTIES OF THE SiC FILMS GROWN ON THE SiC SUBSTRATE

The first group of samples were grown on slabs of single-crystal silicon carbide containing an uncompensated nitrogen impurity of about $3 \cdot 10^{18}$ cm^{-3} that had been mechanically worked. The structure of the samples was evaluated by x-ray diffraction and RHEED. It was found that growth of crystalline layers of SiC required that the substrate temperature exceed 1100 K, and the deposition rate was about 1 nm/min. These parameters were typical for all types of targets. The main factor determining whether growth was polycrystalline or monocrystalline was the preparation of the substrate surface. Studies of samples grown on silicon carbide substrates also indicate that either polycrystalline or single-crystal layers with the hexagonal structure can be grown, with most probably a modification of 6H.

Matrices of Schottky barriers with molybdenum contacts were produced on layers grown on SiC substrates. The I-V and C-V characteristics of these barriers were investigated. The typical I-V characteristics exhibit a distinct rectifying effect and n-type conductivity of the layers. The barrier height was 0.8–1.0 eV. The C-V characteristics of these barriers in a plot of V versus C^{-2} form good straight lines (see Figure 1) and can be used to determine the uncompensated donor density of the N_d–N_a impurity, $6 \cdot 10^{18} \, cm^{-3}$.

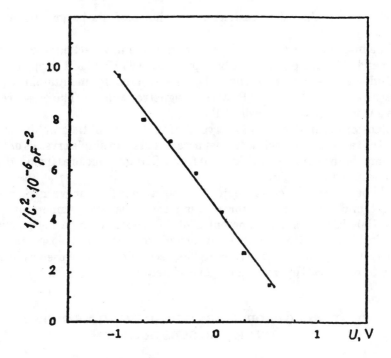

Figure 1. A typical capacitance-voltage characteristic of Mo-SiC barrier.

ELECTRICAL PROPERTIES OF THE SiC FILMS, GROWN ON THE AlN/Al₂O₃ SUBSTRATE

The resulting epitaxial films adhered well to the AlN surface. Figure 2 shows the resistivity of the SiC films at 300 K versus the substrate temperature. We see that the films grown at low temperatures (the "low temperature films") have a resistivity seven orders of magnitude higher than those of the films grown at high temperatures (the "high temperature films").

The conductivity of the high temperature films increased as the sample was heated, with an activation energy of 0.025 eV (Figure 3). This behaviour indicates a semiconductor conductivity. The ionisation energy of the primary shallow donor (nitrogen) in heavily doped 6H-SiC is [4] 0.1 eV. In the heavily doped SiC (N_d–N_a > 10^{18} cm³), on the other hand, some other donor levels have been observed. They are evidently a defect of nature and have an ionisation energy of 0.03–0.04 eV [5].

At room temperature, the low temperature films are essentially insulators. Their resistance remains constant up to 500°C and then decreases

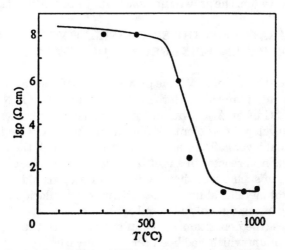

Figure 2. Resistivity of obtained SiC layers versus substrate temperature (AlN).

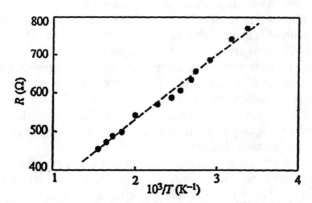

Figure 3. Temperature dependence of sputtered film resistivity.

425

irreversibly approximately as shown in Figure 2. This effect may be due to a change in the crystal structure of the resulting films. These evidently occurred both when the temperature of the substrate during the growth was raised and when the grown films were subjected to heat treatment.

PREPARATION OF THE SEMI-INSULATING SUBSTRATE FOR HETEROEPIXY GROWTH OF THE SiC FILMS

The epitaxial layers of AlN have proved successful as layers for matching the substrate material with the layer. The use of an AlN sublayer in the epitaxy of GaN on Al_2O_3 was one of the decisive factors which made it possible to develop an efficient LED emitting in the UV region.

The epitaxial layers of AlN have been grown through the use of a flow-through horizontal reactor in a multizone resistance furnace. The reactor was designed in such a way that the following steps could be carried out: chlorination of metallic aluminium with hydrogen chloride (in the source zone), mixing of gaseous chlorides and ammonia, and deposition of AlN on the substrates (the deposition zone). In contrast with the chlorination of molten aluminium which had been carried out earlier [6,7], the chlorination was carried out at a temperature below the melting point. This change made it possible to resolve the problem of a container for the aluminium and also to substantially increase the productivity of the process. This productivity increase was achieved by arranging conditions for the formation of gaseous aluminium trichloride, in contrast with the aluminium monochloride produced during high temperature chlorination [8].

It had been shown previously that AlN particles form in the gas phase in a chloride-hydride system, in addition to the heterogeneous growth of a layer of aluminium nitride [9,10]. This effect substantially degrades the uniformity of the layers and also their structure. A detailed study of the effect of the basic technological parameters on the rate at which AlN particles form in the gas phase made it possible to develop regimes which completely suppressed this process [11].

Carrying out the growth in these regimes made it possible to produce (0001) AlN layer on (0001) α-Al_2O_3 substrates productively. During growth under optimum conditions, the surface of the layer had a mirror finish; the height of the irregularities did not exceed 0.2 μm at a layer thickness of 5 μm. Analysis of the resulting layers by x-ray diffraction revealed a block structure, with individual blocks 1000–2000 Å in size. There was a stain Δ d/d ~ 4.1 · 10^{-4}–1.1 · 10^{-3}. Study of the AlN/Al_2O_3 samples by Auger spectroscopy revealed an oxygen impurity. Its concentration in the interior of the material varied from sample to sample over the range 0.3–5 at%.

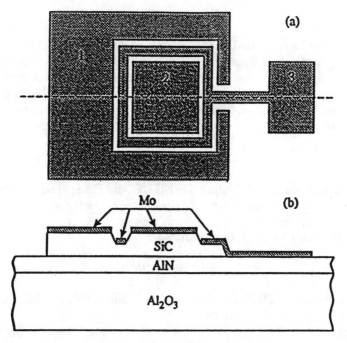

Figure 4. General view of MESFET on a semi-insulating substrate. (a) Plain view: 1—source, 2—drain, 3—gate. (b) Mesa structure cross section.

On a base of "high temperature" films MESFET-like device structures were prepared (Figure 4). Unfortunately, attempts to produce a barrier contact on the surface of the films were unsuccessful. The deposition of various metals (Al, Au, and Mo) and the attachment of a metal probe resulted in non-linear ohmic contacts. A similar situation may arise in the case of heavily doped SiC ($\sim 10^{20}$ cm^{-3}).

In general our results are close to those which have been found on epitaxial SiC films grown under similar technological conditions in ceramic substrates [12]. Nagai et al. [12] concluded that the SiC films had a polycrystalline structure and that the conductivity of the films was governed by the conductivity of boundary regions between single-crystal grains. They also mentioned the possibility of using silicon carbide films of this sort as thermoresistors. We believe that the SiC films which were grown have a basically polycrystalline nature, and that their low resistivity is a consequence of a high concentration of defects which forms shallow donor levels. If this is the case, an optimization of the growth of some SiC films with a relatively light doping (N_d-$N_a \sim 10^{16}$–10^{17} cm^3), would be suitable for device structures.

CONCLUSION

This study has resulted in the growth of SiC films on AlN/Al_2O_3 substrates by magnetron sputtering. The films adhere well to the substrate surface and have a low resistivity. Further refinements of the growth parameters (and of the substrate material) may make it possible to produce high-quality SiC films on high-resistance substrates by reactive magnetron sputtering.

ACKNOWLEDGEMENT

We wish to thank Yu V. Melnik (CREE EED St. Petersburg) for furnishing the AlN/Al_2O_3 substrates which they grew at the State Electrotechnic Institute in St. Petersburg. This work was supported in part by Arizona University (USA) and INTAS grant 93-2651.

REFERENCES

1. DiLorenzo, J. and D. Khandelwal, eds. 1982. *GaAs FET Principles and Technology*. Norwood, MA: Artech Hse.

2. Nishino, S., K. Takahashi, H. Tanaka, and J. Sarade. 1994. *Inst. Phys. Conf.,* Ser. No. 137, p. 63.

3. Misawa, S., S. Yoshida, and S. Gonda. 1983. Extens. Abstracts (The 44th Autumn Meeting), *Jpn. Soc. Appl. Phys.,* 27:4-A-11.

4. Pensl, G. and W. J. Choyke. 1993. *Physica B,* 185:264.

5. Euwage, A. O., S. R. Smith, and N. Mitchel. 1994. *J. Appl. Phys.,* 75:3472.

6. Yim, W. M., E. J. Zanzuechi, J. I. Pankowe, M. Ettenberg, and S. L. Gilbert. 1973. *J. Appl. Phys.,* 44: 296.

7. Callaghan, M. P., E. Patterson, B. P. Richards, and C. A. Wallace. 1974. *J. Cryst. Growth,* 22:85.

8. Pichugin, I. G. and A. M. Tsaregorodtsev. 1983. *Izv. LETI,* 302:3.

9. Lebedev, A. O., Yu. V. Melnik, and A. M. Tsaregorodtsev. 1990. *Abstracts First Int. Conf. on Ep. Growth,* Budapest, April 1–7, p. 116.

10. Efimov, A. N., A. O. Lebedev, Yu V. Melnik, and A. M. Tsaregorodtsev. 1992. *Abstracts, Eighth All-Unition Conference on Crystal Growth,* Kharkov, p. 153.

11. Lebedev, A. O., Yu V. Melnik, and A. M. Tsaregorodtsev. 1994. *Inst. Conf. Ser. No. 137,* p. 405.

12. Nagai, T., K. Yamamoto, and I. Kobayashi. 1985. *Thin Solid Films,* 125:355.

Room-Temperature Photoluminescence of Amorphous Hydrogenated Silicon Carbide Doped with Erbium

E. I. TERUKOV,* V. KH. KUDOYAROVA AND A. N. KUZNETSOV
A. F. Ioffe Physico-Technical Institute
194021 St. Petersburg
Politechnicheskaya 26
Russia

W. FUHS
Hahn-Meitner Institut
Rudower Chausse 5
D-12489, Berlin
Germany

G. WEISER
Fachbereich Physik Philipps-Universitat Marburg
Renthof 5
D-35032 Marburg an der Lahn
Germany

ABSTRACT: For the first time strong room temperature photoluminescence of Er ions at 1.54 μm corresponding to $^4I_{13/2} - {}^4I_{15/2}$ transition in the ion f-shell was observed in erbium-doped hydrogenated amorphous silicon carbide films (a-Si$_{1-x}$C$_x$:H<Er>). Films of a-Si$_{1-x}$C$_x$:H<Er> have been prepared by cosputtering of graphite and Er targets applying the magnetron-assisted silane-decomposition (MASD) technique with mixtures of Ar and silane. The composition of films (x) was varied in the range 0–0.29. The concentration of incorpo-

*Author to whom correspondence should be addressed.

rated Er-ions was 6.10^{19} cm^{-3}. The weak temperature dependence of PL in amorphous hydrogenated silicon carbide doped with erbium is discussed within the framework of defect-related Auger excitation (DRAE) model.

KEYWORDS: photoluminescence, room temperature, silicon carbide, erbium doped.

1. INTRODUCTION

Recently the luminescent properties of erbium-doped crystal-line silicon (c-Si:Er) have attracted much attention. The reason for this interest originates from the idea to fabricate LEDs which are integrable into silicon electronic devices and emit at a wavelength of 1.537μm where the absorption of silica glass optical fibers is the lowest.

However, the existing attempts to obtain efficient photoluminescence (PL) on the basis of erbium-doped crystalline silicon (c-Si) have met serious difficulties connected with a strong temperature quenching of erbium luminescence in this material and segregation of rare earth impurities limiting the highest concentration of erbium optical centers by the value 10^{18} cm^{-3}. It should be noted that in order to achieve high emission intensity crystalline silicon c-Si<Er> the samples should be co-implanted with erbium and oxygen and optimization need extended high-temperature annealing [1].

These difficulties can be avoided if amorphous but not crystalline semiconductor is doped with erbium impurity. In the case of amorphous matrix the disorder in the environment of the rare earth ion favors the mixing of the internal f-states of the rare earth ion which leads to an increase of probability of radiative transition between these states and shortens the response time of photoluminescence. Besides, the activation energy of defects involved in excitation of rare earth is greater in amorphous as compared to crystalline material and it is believed that this fact is responsible for the absence of strong temperature quenching of erbium luminescence in amorphous silicon (a-Si:H<Er>) [2].

It has been shown that the intensity of the Er emission strongly depends on the band gap energy of the host semiconductor, mainly for the room temperature emission. To obtain an intense room temperature emission, a wide-gap semiconductor must be used [3,4].

The optical band gap of amorphous hydrogenated silicon carbide (a-Si$_{1-x}$C$_x$:H) dependence on the composition (x) and it is larger than in amorphous hydrogenated (a-Si:H) and crystalline silicon (c-Si) [5].

In this paper, the results obtained on erbium-doped amorphous hydrogenated silicon (a-Si:H:Er) [2] are extended to erbium-doped amorphous hydrogenated silicon carbide (a-Si$_{1-x}$C$_x$:H:Er) to increase the band gap energy limit. For the first time it is shown that a-Si$_{1-x}$C$_x$:H:Er exhibits strong room-temperature PL at 1.54 μm, which is assigned to the internal 4f-shell transition in Er ions.

2. EXPERIMENT

We have used magnetron assisted silane decomposition technique (MASD) for obtaining a-Si$_{1-x}$C$_x$:H<Er> films. The reactor system used for MASD was described in Reference [6] in detail. The magnetron target was used of C and Er of 99.999% purity. The Si/C ratio was varied by changing SiH$_4$ percentage in the gas mixture of Ar + SiH$_4$. The substrate temperature was 300°C.

The use of novel technology (MASD) permits insertion into the semiconductor matrix concentrations of Er ions by two orders of magnitude higher than those in the case of crystalline matrix. The composition of films (x) and the presence of Er in the films have been monitored by Rutherford back scattering (RBS), using RBS of ^{4}He^{2+} ions. The energy of ^{4}H^{2+} was 3.7 MeV, the angle of backscattering was 170° and the detector resolution energy was 50 keV. The typical RBS spectra for the film with $x = 0.29$ and for the crystalline silicon carbide c-6H SiC are shown in Figure 1. The

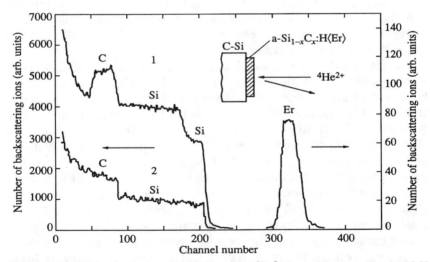

Figure 1. Rutherford backscattering (RBS) spectra ^{4}He^{2+} ions with energy E = 3.7 MeV backscattered at Si, C and Er as function of the energy (channel number) for: 1—a-Si$_{0.71}$C$_{0.29}$:H<Er>; 2—c-6H-SiC (crystalline silicon carbide). Concentration of incorporated Er ions was 6.10^{19} cm^{-3}.

crystalline carbide c-6H SiC was used as calibration sample. It has been shown that Si/C ratio was estimated as 1.022 accurate for c-6H SiC [7]. The content of C (x) was varied in the range 0–0.29. The concentration of incorporated Er ions was -6.10^{19} cm^{-3}. The content of hydrogen was estimated at 9–12 at % from the integrated IR absorption due to the Si-H wagging band (630 cm^{-1}) and C-H stretching band (2900 cm^{-1}). The optical band gap (E_g^{opt}) was determined by constant photocurrent measurement (CPM) as an energy where the absorption coefficient is 10^3 cm^{-1} varied from 1.59 (x = 0) to 1.82 eV (x = 0.29).

The photoluminescence (PL) spectra were recorded with excitation by a krypton laser with λ = 649.6 nm in the temperature range 77–400 K. The PL spectra were measured by means of germanium detector cooled to the temperature of liquid nitrogen and a SPEX-1403 double monochromator with a resolution of 20 A. The PL spectra were detected in the linear regime.

3. RESULTS AND DISCUSSION

Figure 2 shows the PL spectra of Er doped amorphous hydrogenated silicon carbide a-Si$_{0.71}$C$_{0.29}$:H<Er> and amorphous hydrogenated silicon a-Si:H<Er> taken at 77 K. It should be noted that PL-spectra of a-Si$_{0.71}$C$_{0.29}$:H<Er> and a-Si:H<Er> measured at 77 K exhibit emission from both the Er^{+3} centers and the amorphous Si and Si-C matrix. The later is intrinsic band and is seen as a broad band at 1.2–1.4 eV. The position of this band dependent on x.

Usually, the PL data on a-Si$_{1-x}$C$_x$:H alloys (intrinsic band) have been discussed in terms of the models put forward for a-Si:H. The 1.2–1.4 eV band is attributed to transition between localized band tail states, radiative recombination occurring by tunneling [5]. The spectrum in Figure 2 is dominated by strong Er-induced emisssion at 0.8 eV. It should be noted that intensity Er-induced emission in amorphous hydrogenated silicon carbide is smaller than in amorphous hydrogenated silicon. The reason for such behavior of the intensity PL may be the decreasing the number of radiative Er centers due to the absorption constant (at the energy of the excitation PL 1.92 eV) in amorphous hydrogenated silicon carbide is smaller than that in amorphous hydrogenated silicon. At 300 K the intrinsic emission from amorphous silicon carbide matrix (1.2–1.4 eV) is suppressed by temperature quenching and the remaining signal is due to Er^{3+} ions (Figure 3) on the whole. As in the case of a-Si:H, the dangling bonds act as non radiative centers, quenching of PL (intrinsic band) occurring by the tunneling to defect. The temperature quenching of lumi-

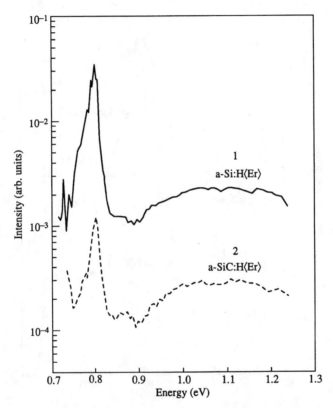

Figure 2. Photolumiscence spectra of Er doped: 1—amorphous hydrogenated silicon a-Si:H<Er>; 2—amorphous hydrogenated silicon carbide a-Si$_{0.71}$C$_{0.29}$:H<Er> (concentration of incorporated Er ions was 6.10^{19} cm^{-3}) taken at 77 K excited with 649.6 nm (1.92 eV) at $I = 20$ mW/cm^2.

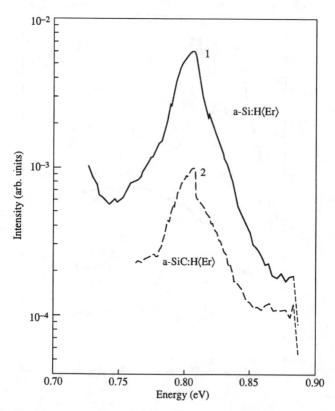

Figure 3. Photolumiscence spectra of Er doped: 1—amorphous hydrogenated silicon a-Si:H<Er>; 2—amorphous hydrogenated silicon carbide a-Si$_{0.78}$C$_{0.22}$:H<Er> (concentration of incorporated Er ions was 6.10^{19} cm^{-3}) taken at 300 K excited with 649.6 nm (1.92 eV) at I = 20 mW/cm^2.

434

nescence above 50 K is explained by the increased mobility of carriers which diffuse to the recombination center [5].

Temperature dependencies of intensity of PL for different optical transitions are displayed in Figure 4 (curve a—PL of erbium ions in a-Si:H<Er>, b—PL of erbium ions in a-Si$_{1-x}$C$_x$:H<Er>, c—PL of intrinsic emission of the amorphous silicon matrix in a-Si:H<Er> and d—PL of erbium ions in erbium doped crystalline silicon c-Si<Er,O>). We used the samples of a-Si:H<Er> and a-Si<Er,O> in which are observed the maximum intensity PL for comparison. At low temperature (T < 100 K) the Er-induced emission is large by 1.5–2 orders magnitude than that in optimized c-Si<Er,O> (the concentration of Er ions and O are 10^{17} and 10^{18} cm^{-3} accordingly). As it is seen in Figure 4 the onset of temperature quenching of PL in the case of a-Si$_{1-x}$C$_x$:H<Er> is observed at higher temperatures than for a-Si:H<Er> and c-Si<Er,O> and the quenching is less pronounced.

The weak temperature dependence of PL in a-Si$_{1-x}$C$_x$:H<Er> is discussed in the terms of the model previously proposed by us for a-Si:H<Er> [8]. In this model the mechanism of electronic excitation of Er ions is based on defect-related Auger excitation (DRAE).

Our studies of PL and EL [8] in a-Si:H<Er> in a wide temperature range show that dangling-bond defects play an important role in the exci-

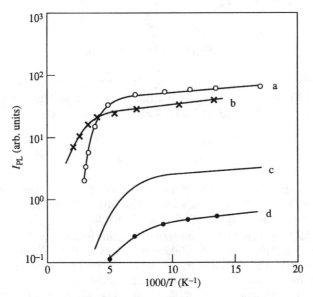

Figure 4. Temperature dependence intensity photoluminescence for: a—a-Si:H<Er> (N$_{Er}$ = 1 10^{20} cm^{-3}); b—a-Si$_{0.71}$C$_{0.29}$:H<Er> (N$_{Er}$ = 6 10^{19} cm^{-3}); c—a-Si:H intinsic band; d—c-Si<Er,O> (N$_{Er}$ = 10^{17} cm^{-3}, N$_{O}$ = 10^{18} cm^{-3}).

tation of erbium ions. Non-radiative transition $D^0 + e - D^-$ accompanied by excitation of erbium ion due to Coulomb interaction between the electron being captured by D^0-center and f-electron of Er^{+3}. The DRAE process is especially effective because of its nearly resonant character since the energy difference of D^0 and D^- states is close to the energy difference of $^4I_{13/2}$ and $^4I_{15/2}$ states of the f-shell. The excess energy of the transition is transferred to local phonons. In the work [8] the probability of DRAE process is calculated and is shown to be significantly larger than that of similar radiative transition. A correlation between the intensities of defect and erbium luminescence supports this mechanism.

4. CONCLUSIONS

In conclusion, we have studied photoluminescence of erbium ions in erbium doped amorphous hydrogenated silicon carbide (a-$Si_{1-x}C_x$:H<Er>). For the first time strong room temperature photoluminescence of Er ions at 1.54 μm corresponding to $^4I_{13/2} - {}^4I_{15/2}$ transition in the ion f-shell was observed. The weak temperature dependence of PL in amorphous hydrogenated silicon carbide doped with erbium is discussed within the framework of the defect-related Auger excitation (DRAE) model previously proposed for a-Si:H<Er>.

ACKNOWLEDGEMENTS

This work was partly supported by the Arizona University (USA), by Volkwagen-Stiftung, by Russian Foundation of Basic Research through grants 96-02-16931-a and by INTAS grant no. 93-1916.

REFERENCES

1. Coffa S., Franzo G., Priolo F., Polman A., Serna R. 1994. "Temperature Dependence and Quenching Processes of the Intra-4f Luminescence of Er in Crystalline Si," *Phys. Rev.* B49(23):16313–16320.
2. Bresler M. S., Gusev O. B., Kudoyarova V. Kh., Kuznetsov A. N., Pak P. E., Terukov E. I., Yassievich I. N., Zakharchenya B. P., Fuhs W., Sturm A. 1995. "Room-Temperature Photoluminescence of Erbium-Doped Hydrogenated Amorphous Silicon," *Appl. Phys. Lett.* 67(24):3599–3601.
3. Favennec P. N., Haridon H. L., Salvi M., Mountonnet D., Guillou Y. LE. 1989. "Luminescence of Erbium Implanted in Various Semiconductors: 1V.111-V and 11-V1 Materials," *Electronic Letters,* 25(11):718–719.

4. Choyke W. J., Clemen L. L., Yoganathan M., Pensl G., Hassler Ch. 1994. "Intense Erbium 1.54 μm Photoluminescence from 2 to 525 K in Ion-Implanted 4H, 6H, 15R and 3C SiC," *Appl. Phys. Lett.* 65(13):1668–1670.

5. Bullot J., Schmidt M. P. 1987. "Physics of Amorphous Silicon-Carbon Alloys," *Phys. Stat. Sol. (b)*, 143:345–418.

6. Marachonov V., Rogachev N., Ishkalov J., Marachonov J., Terukov E. I., Chelnokov V. 1991. "Electrical and Optical Properties of a-Si:H, a-SiGe:H and a-SnSi:H Deposited by Magnetron Assisted Silane Decomposition," *J. Non-cryst. Solids*, 137/138:817–820.

7. Sorokin N. D., Tairov Yu. M., Tsvetkov V. F., Chernov M. A. 1983. "Investigation of the Crystall-Chemical Properties of the Silicon Carbide Polytypes," *Kristalographia*, 28:910–914.

8. Yassievich I. N., Gusev O. B., Bresler M. S., Fuhs W., Kuznetsov A. N., Masterov V. F., Terukov E. I., Zakharchenya B. P. 1996. "Photo- and Electroluminescence Study of Excitation Mechanism of Er Luminescence in a-Si:H<Er>," Spring Meeting, April 8–12, San Francisco, *Proc. Mat. Res. Soc.*, D7.3:73.

Optical and TEM Study of Copper-born Clusters in DLC

V. I. Ivanov-Omskii,* S. G. Yastrebov,
A. A. Suvorova and A. A. Sitnikova
A. F. Ioffe Physical-Technical Institute
St. Petersburg, 194021
Russia

A. V. Tolmatchev
St. Petersburg State Tech. University
St. Petersburg, 195251
Russia

ABSTRACT: Transmission Electron Microscopy (TEM) (100 keV), Selected Area Electron Diffraction (SAED) and optical absorption data (0.6-5 eV) of copper-doped Diamond-like Carbon (DLC) films are presented. The size distribution functions (SDF) of copper-born clusters embedded in DLC were obtained both with direct observation of TEM images of clusters and from numerical analysis of optical data. Validity of optical SDF is discussed. TEM SDF is analyzed in the frames of fluctuation mechanism of early stage of cluster formation.

KEY WORDS: nanocluster, size-distribution function, copper, diamond-like carbon.

INTRODUCTION

Interest in the conductive clusters physics is stimulated partially by its possible applications to nano- and micro-electronics [1]. Modification of DLC properties by introduction of metallic nanoclusters is of much interest to extend our knowledge of DLC structure [2]. Using

---

*Author to whom correspondence should be addressed.

metal such as copper, which is not capable of forming a carbide, different kinds of conductive clusters distributed within DLC matrix can be formed [3]. The main goal of the present investigation is to process SDF of embedded in DLC copper-born clusters both from the optical spectra and directly from TEM images to compare them to judge a validity of optical methods of SDF evaluation.

EXPERIMENTAL

DLC films were produced by ion (DC magnetron) co-sputtering of graphite and copper targets in argon-hydrogen (80% Ar and 20% H_2) plasma onto KBr and amorphous quartz substrates. The copper concentration was checked with secondary ion mass spectroscopy. The substrate temperature was close to 500 K. The average magnetron power was varied in the range from 0.35 to 0.45 kW. It was possible to grow films having thickness from 0.1 to 1.5 μm.

The optical absorption spectra (1-6 eV) of DLC films deposited onto fused silica substrates were measured with Hitachi-U3410 spectrophotometer to characterize the microstructure of grown films by an analysis of processed optical data.

Ellipsometrical measurements were performed with LEF-3m ellipsometer (1.96 eV) to find out both geometrical characteristics of grown films and their optical constants. A Philips-400 TEM operating at 100 keV was used for direct observation of microstructure and electron diffraction analysis of copper-doped DLC films. For this purpose, freestanding films approximately 100 nm thick were prepared by dissolving KBr substrates in water.

OPTICAL STUDIES

Averaged optical transmission spectra of copper-doped DLC are presented in Figure 1. Shape of optical spectra shows that the dielectrical function is a result of superposition of four Lorentz oscillators as follows [3]:

1. A low-frequency oscillator associated with the free-carrier absorption
2. A mid-frequency oscillator associated with surface plasmon excitations in insulated spherical conductive clusters
3. A high-frequency oscillator associated with surface plasmon excitations in non-spherical conductive clusters
4. A high-frequency oscillator associated with the fundamental absorption of the DLC matrix

The algorithm described in Reference [3] was used to evaluate imagi-

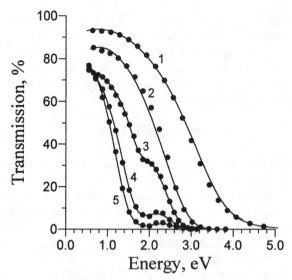

Figure 1. Smoothed optical transmission spectra of DLC doped with copper with different copper concentrations: 1—0%, 2—6%, 3—18%, 4—20%, 5—23%. The points are experimental values. The solid curve is the result of fitting procedure.

nary and real parts of the dielectrical function. To reach this goal a functional containing experimental optical transmission and ellipsometrical data included their theoretical expressions was minimized to process relaxation times and resonance frequencies of Lorentz oscillators to convert experimental data into imaginary and real parts of the dielectrical function. The success of the outlined procedure despite its multiparameter nature derives from the fact that the spectral ranges where the type (1), (2) and (3) oscillators are of first importance are noticeably resolved, as it is seen in Figure 1. Therefore, conditions of negligible overlap of different oscillator contributions may be found to determine the oscillators' parameters properly. SDFs of conductive copper-containing clusters embedded in DLC matrix were neglected in the course of calculations. In this paper we estimate SDFs of oscillators 2 and 3 by simple averaging (integration) of contribution of oscillators with their SDF weights. Because phase-transition behavior of physical properties appearing in most of composites with increase of the content of metallic phase [4], it is reasonable to choose the SDF shape in accordance with the fluctuation theory as follows [4,5]:

$$P(D) = \frac{A}{\sqrt{2\pi} \cdot bD_{av}} \ \exp\left(-\frac{(D - D_{av})^2}{2(bD_{av})^2}\right) \tag{1}$$

Here D_{av} being the average size of the conductive cluster is a correlation length of the fluctuation, b is a parameter. In accordance with the fluctuation theory $b = 1$ [5]; A is a constant calculated by means of the renormalization of $P(D)$. To perform this procedure, it is necessary to integrate $P(D)$ over all possible clusters sizes. Integration limits have been chosen from about 0.3 nm for lower limit (size of single copper atom) up to infinity for upper one. Normalization condition requires the equivalence of the integral to the unity. In other words, A is a function of correlation length only. In the course of calculation it was also assumed that the relation between the relaxation time of each of the oscillator and their geometrical characteristics corresponds to a size-confined free-electrons gas model [3]. As regards to the algorithm described in Reference [3], introduction of SDF in the model brings no additional parameters to the calculation scheme. In the course of calculations it was found that chosen distribution function fits the experimental data (Figure 1). As a result of fitting, we were able to find the spectral dependencies of optical constants which are shown in Figure 2(a and b). The SDF for 2 types of oscillator is depicted in Figure 3(a), and for type 3 is shown in Figure 3(b). The experimental points for correlation length of conductive clusters associated with oscillators 2 and 3 are shown in Figure 4 vs the copper content. The correlation length fulfills the relation [4,5]

$$D_{av} = L_0 \left(\frac{x_c - x}{x_c} \right)^{-v} \tag{2}$$

where x is a fraction of conductive phase. Expression (2) fits experimental data as it is also shown in Figure 4. Fitted values are: $L_0 = 10$ Å is an extrapolated value of conductive cluster at zero concentration of copper, this value is close to the average size of graphite-like constituent of DLC matrix [6], $x_c \approx 18$ at.% is a percolation threshold, $v = 0.81$ is a critical index of the correlation length. It should be emphasized that such the dependence correlates with DC conductivity data previously discussed in References [7,8] . Formation of different types of clusters at the percolation threshold, which is evident from Figures 3 and 4, could appear as a result of precipitation process of copper-born conductive inclusions into clusters with higher density than for ones formed below x_c.

TEM STUDIES

A dark-field TEM image of copper-born clusters for sample with 9% of copper is shown in Figure 5. This figure shows clusters of 2.88 nm aver-

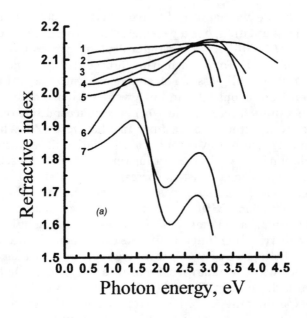

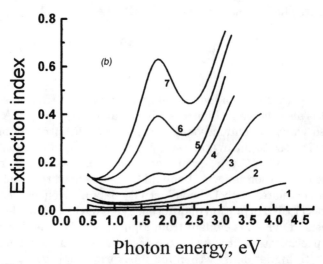

Figure 2. The spectral dependencies of refractive index (a) and extinction index (b) vs copper concentration for Cu-doped DLC. 1—0%; 2—3%; 3—6%; 4—14%; 5—18%; 6—20%; 7—24%.

442

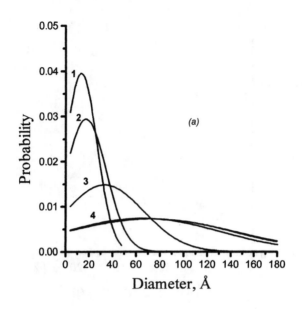

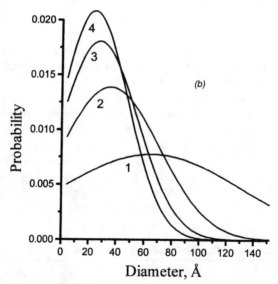

Figure 3. (a) SDF obtained from optical data (Figure 1) for non-spherical conductive copper-containing clusters embedded in DLC. 1—3%; 2—6%; 3—14%; 4—18%, 20%, 24%. (b) SDF obtained from optical data (Figure 1) for spherical conductive copper-containing clusters embedded in DLC. 1—14%; 2—18%; 3—20%; 4—24%.

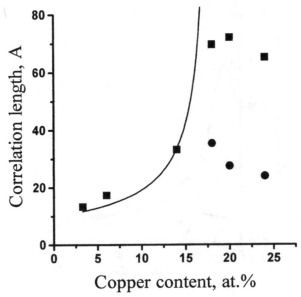

Figure 4. The correlation length of copper-containing clusters vs copper concentration. Full squares stand for non-spherical clusters; full circles stand for spherical clusters; line is fit according to Equation (2).

age diameter distributed in DLC matrix. The SAED (Figure 5, Inset) shows diffuse rings and the lattice spacing corresponding to each ring provides a reasonable match to the copper (Table 1).

The size variation of copper-born nanoclusters was analyzed from TEM micrographs directly and the SDF is plotted in Figure 6, full squares stand for experimental points. Full line presents the best fit to the Gaussian law, Equation (1) with $b = 0.63$, $D_{av} = 2.6$ nm. Dotted curve demonstrates SDF for 9% Cu-doped DLC obtained from Equations (2) and (1) using fitted optical data ($b = 1$). There is seen a reasonable agreement between optical and TEM data. Rather good agreement between optical and TEM SDF can be ascribed in terms of early (nucleation) stage

Table 1. SAED data for copper-born clusters (from Figure 5 SAED pattern).

Ring	Lattice Spacing d (nm)			
	1	2	3	Error
Experimental value	0.252	0.186	0.156	±0.02
Standard (ASTM) value for Cu	0.208	0.1798	0.1271	
(hkl) Plane	(111)	(200)	(220)	

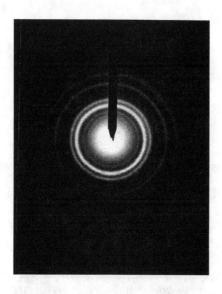

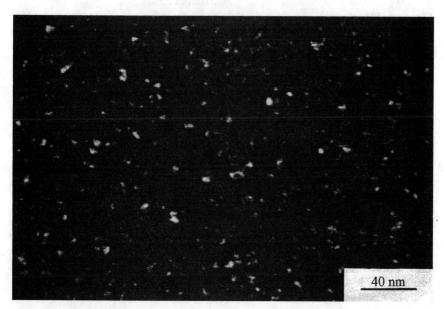

Figure 5. TEM micrograph of copper-born clusters for sample with 9% of copper. The average diameter of copper-born clusters is 2.88 nm. Inset is the SAED pattern.

445

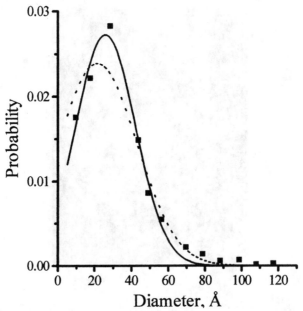

Figure 6. SDF obtained from TEM micrograph shown in Figure 5. Full squares stand for experimental points, full line is a fitted Gaussian curve. Dotted curve is SDF obtained from optical data (Figure 1).

of copper-based clusters formation process. In this case average (critical) size of the cluster is a function of interface free energy per unit area σ and bulk free energy ΔG decrease per unit volume [5,9]:

$$D_{av}/2 = 2\sigma/\Delta G \qquad (3)$$

It is worth noting that at copper concentrations exceeding percolation threshold an analysis of TEM images is fairly complicated because of difficulties in the spatial resolution of different geometrical forms of copper-born clusters and, in this case, a special analysis technique should be used.

CONCLUSIONS

1. An optical method of analysis of specific optical absorption bands in copper-doped DLC can be used to find sizes and shapes of conductive copper-containing clusters and to estimate their SDFs.

2. The Gaussian curve reasonably fits experimental data of TEM SDF at copper concentration of about 9%. This indicates nucleation stage of conductive clusters formation for the samples with copper concentrations lying about the percolation threshold.
3. TEM SDF at 9% copper agree well with optical one.

ACKNOWLEDGEMENT

This work was supported in part by the Arizona University grant.

REFERENCES

1. Li, J., T. E. Seidel and J. W. Mayer. 1994. "Copper-based Metallization in ULSI Structures," *MRS Bulletin XIX*, (8):15–18.
2. Ivanov-Omskii, V. I. and G. S. Frolova. 1995. "Activation of Raman Frequencies in the IR Absorption Spectra of Amorphous Hydrogenated Carbon Doped with Copper," *Tech. Phys.*, 40(9):966–967.
3. Ivanov-Omskii, V. I., A. V. Tolmatchev and S. G. Yastrebov. 1996. "Optical Absorption of Amorphous Carbon Doped with Copper," *Philos. Mag. B*, 73:715–722.
4. Efros, A. L. and B. I. Shklovskii. 1976. "Critical Behaviour of Conductivity and Dielectric Constant near the Metal-Non-Metal Transition Threshold," *Phys. Stat. Sol. B*, 76:475–485.
5. Landau, L. D. and E. M. Lifshitz. 1981. *Statistical Physics, Course of Theoretical Physics*, Pergamon, New York.
6. Robertson, J. 1986. "Amorphous Carbon," *Adv. Phys.*, 35:317–374.
7. Ivanov-Omskii, V. I., A. B. Lodygin and S. G. Yastrebov. 1995. "Characteristics of Conducting Structures in Copper-Doped Diamond-Like Carbon," *Phys. Solid State*, 37(6):920–922.
8. Ivanov-Omskii, V. I., M. I. Abaev and S. G. Yastrebov. 1994. "Electric and Optic Properties of Copper-Doped Amorphous Carbon," *Tech. Phys. Lett.*, 20:917–919.
9. Landau, L. D. and E. M. Lifshitz. 1981. *Physical Kinetics, Course of Theoretical Physics*, New York: Pergamon.

UHV Deposition Cell for Carbon Atoms Containing No Carbon Clusters

N. R. GALL, E. V. RUT'KOV,
AND A. YA. TONTEGODE
A. F. Ioffe Physico-Technical Institute RAS
Polytechnicheskaia str. 26
St. Petersburg, 194021, Russia

P. B. KUZNETSOV AND R. N. GALL*
CADIX (R), Ltd.
P. O. Box 576
Russia

1. INTRODUCTION

A large variety of carbon chemical forms is known in nature and from technical sources, such as graphite, diamond and diamond films, carbides of boron and the most recent, newly discovered and already widely studied forms, the fullerenes.

Fullerenes stand out in their very high reactivity: they can exist only in the form of a flux of carbon atoms in ultra high vacuum. Atomic carbon is an ideal building material for all other forms of carbon. During surface deposition, it is potentially able to continue the crystal lattice of any of the materials named above with complete crystalline and geometric order. Until lately atomic carbon in its pure form, i.e., in the form of a flux of carbon atoms, free of carbon clusters (C_2, C_3, C_4 ...) and of noncarbon admixtures, has not been obtained. During graphite sublimation clusters in the

*Author to whom correspondence should be addressed.

flux amount to about 60% [1], during ion sputtering of carbon targets this value is 10–40% [2]. When carbon is deposited onto a surface from a plasma (e.g., by CVD methods) noncarbon admixtures as well as some amount of carbon clusters are present in the absorbing flux.

2. SOURCE OF CARBON ATOMS

To produce a flux of carbon atoms free of carbon clusters and noncarbon admixtures in UHV, we have developed, designed, produced and tested a novel source. The source is a separate unit, designed to be built into UHV scientific and technological installations. It was tested in a high-resolution of Auger-electron spectrometer, described in Reference [3]. The source showed parameters, presented below:

1. Carbon atom beam density, at/s·sq. cm $\qquad$ 10^{11}–$5 \cdot 10^{14}$
2. Contents of admixtures in the beam (%)
   foreign atoms $\qquad$ 10^{-2}
   carbon clusters $\qquad$ 10^{-2}
3. The number of carbon atoms engaged in the process
   up to the new regeneration $\qquad$ 10^{18}
4. The number of regenerations (without vacuum system depressurization) $\qquad$ 100
5. Time of regeneration (min) $\qquad$ 10
6. Nonhomogeneity of the beam (%) (exposed area $0.5 \cdot 3$ cm^2, distance 20 mm) $\qquad$ 10
7. Vacuum during carbon deposition (Torr) $\qquad$ 10^{-10}

Structurally the source consists of a getter of carbon atoms, a regeneration unit, a control unit intended for controlling the regeneration process and a getter movement mechanism. Carbon atoms are stacked in the getter during the regeneration cycle which is performed without depressurization and even without degradation of vacuum conditions. In a regeneration regime the directly heated getter is positioned near the regeneration unit, which is a Knudsen deposition cell in which a flux of carbon containing molecules comes to the heated surface of the getter. The molecules break up on the surface and carbon penetrates into the bulk of the getter. The type of molecules is specially chosen so that foreign particles do not enter the vacuum chamber.

The regeneration process is controlled by registration of the thermoelectric current emitted by the getter. The current is monitored by a control unit which is a collector with antidynatron grid, placed onto a screen. During regeneration, a small increase (2–3 times) is observed and at the final stage it increases 100 times. In our opinion such behavior can be explained from the knowledge that until finishing the regeneration the work function of the surface is constant and the small increase of the thermoelectric current is a result of an increase of the getter electroresistivity (and, therefore, its temperature) during carbon accumulation in a getter bulk. At the final stage of regeneration a two-dimensional graphite film forms on the surface which drastically diminishes the work function. This leads to a strong increase in the thermoelectric current from the surface. Carbon containing molecules do not break up on graphite films, and the regeneration process is finished.

After finishing the regeneration cycle, the heaters of the deposition cell and the getter are switched off and the getter is put inside the vacuum chamber. Under UHV conditions it can be stored for months without any loss of carbon. To produce the flux of carbon atoms the getter is heated to a temperature larger than that used for regeneration ($T >$ 2100 K). At this temperature a flux of carbon atoms free from carbon clusters is emitted from the getter surface. The purity of the flux was tested by mass-spectrometry. Mechanically the source is mounted on a 63 mm "Conflat" flange. The length from the flange to the surface of the getter is 240 mm and can be easily varied, by utilization of the appropriate size bars, from 80 to 300 mm.

3. APPLICATIONS OF A CARBON ATOM FLUX

A flux of carbon atoms free of carbon clusters and noncarbon admixtures as designed by us, appears to be a very effective tool in solving a number of scientific tasks. Using it we have determined the role of carbon clusters in the carbon adlayer on the (111) surface of Pt [4,5], in low temperature ($T = 400$ K) growth of SiC films on metal substrate.

The application of a flux of carbon atoms free from carbon clusters to the growth of diamond films appears very promising. Carbon clusters present in the flux have a structural difference from that of the growing diamond film. They can be centers of nucleation of the parasite graphite phase. The absence of carbon clusters must minimize the amount of such centers and probably make possible the growth of diamond films using MBE technology.

Let us evaluate a possible growth rate of a diamond film, assuming 100% utilization of the carbon atoms that fall onto the surface. If one monolayer of diamond contains 1×10^{15} at/cm^2 one would need two seconds to grow it. The thickness of a carbon monolayer is 1.3 Å, so during 10 hours of carbon deposition the thickness would be 36,000 × 0.5 × 1.3 = 2.5 · 10^4 Å = 2.5 μm, which is rather large.

4. SUMMARY

A source of carbon atoms free from carbon clusters and noncarbon admixtures was developed, designed, produced and tested by us for the first time. Atomic carbon seems to be very promising for epitaxial growth of diamond films.

ACKNOWLEDGEMENT

Three of us (N. R. Gall, E. V. Rut'kov and A. Ya. Tontegode) acknowledge support from the Russian Research Program, "Fullerenes and Atomic Clusters," Contract No. 96134.

REFERENCES

1. Nesmeianov, N. A. 1961. "Vapor Pressure of Chemical Elements," Moscow: Academic Press.
2. Werner, H. W. 1978. In "Electron and Ion Spectroscopy of Solids," eds., L. Fierman, J. Vennik, and W. Dekeyser, London: Plenum Press, pp. 361–420.
3. Gall, N. R., S. N. Mikhailov, E. V. Rut'kov, and A. Ya. Tontegode. 1987. *Surf. Sci.*, 191:185–202.
4. Smirnov, M. Yu., N. R. Gall, A. R. Cholach, V. V. Gorodetskii, A. Ya. Tontegode, E. V. Rut'kov, and D. Yu. Zemlianinov. 1991. *Catalysis Letters,* 8:101–106.
5. Smirnov, M. Yu., A. R. Cholach, V. V. Gorodetskii, N. R. Gall, E. V. Rut'kov, and A. Ya. Tontegode. 1991. *Surface and Coating Technology,* 47:224–232.

Optical Studies of UV-Stimulated Mid-Range Order Transform of DLC

A. V. CHERNYSHEV,* I. N. TRAPEZNIKOVA AND S. G. YASTREBOV
A. F. Ioffe Physical-Technical Inst.
St. Petersburg, 194021
Russia

A. V. TOLMATCHEV
St. Petersburg State Technical University
St. Petersburg, 195251
Russia

ABSTRACT: Studies have been made of ellipsometrical (1.96 eV) and interferometrical (1.4 eV) UV-induced photostimulated transform of DLC films. The films were grown onto amorphous quartz substrates by magnetron sputtering using a graphite target in an Ar-H_2 (4:1) plasma. Results obtained in the course of the UV treatment of these DLC films are discussed in the framework of the effective medium approximation model (EMA) in order to study contributions of UV generated voids and structural reordering to the dielectric properties of DLC.

KEY WORDS: DLC, photoinduced changes, mid-range order.

1. INTRODUCTION

The observance of UV-stimulated photoinduced changes in ion-sputtered DLC films irradiated with light at energies greater than the gap of fundamental optical absorption, E_g, [1] gives sufficient promise to investigate DLC as a new photosensitive material. The main goal of

---

*Author to whom correspondence should be addressed.

the present investigation was to study the dependence of E_g as a function of irradiation time, and to reveal the dependence of the average size of the graphite-like constituents on treatment time.

2. EXPERIMENT

DLC films were deposited onto amorphous quartz and silicon substrates by magnetron sputtering of a graphite target in Ar-H_2 (4:1) plasmas at pres-

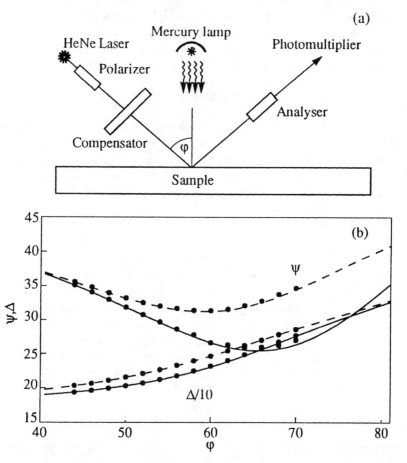

Figure 1. (a) Experimental set-up for *in situ* ellipsometric monitoring of the surface state. (b) Ellipsometric angles Ψ, Δ vs. incident angle for one hour irradiated (dotted curve) and non-irradiated (solid curve) films deposited on a silicon substrate. Probe light wavelength is 632.8 nm [1].

sures in the growth chamber of the order of 8.5 m Torr and at a magnetron potential of 400 V. The substrate temperature was close to 500 K. Optical absorption spectra were measured with an SF-4 spectrophotometer (1.4 eV). The geometric characteristics of films deposited onto silicon substrates and their optical constants at 1.96 eV were controlled ellipsometrically with an LEF-3m ellipsometer operating in the reflection mode at various angles of incidence. Typical film thicknesses were found to be about 0.3 micrometers. The films were irradiated with a 200 W mercury lamp at room temperature under atmospheric conditions and tested optically after treatment.

A schematic of the set-up for ellipsometric monitoring of UV-stimulated changes of DLC films is shown in Figure 1(a). The range of measured angles is shown in Figure 1(b) [1].

After UV treatment, we performed measurements of the optical absorption spectra on films deposited onto quartz substrates.

3. RESULTS AND DISCUSSION

It was found that the frequency dependence of the absorption coefficient of the fabricated films followed the Tauc rule [2]. The spectral dependence of the optical constants was obtained from the absorption spectra using the Kramers-Kronig technique [3] taking into account the ellipsometric data of the dielectric function at the 1.96 eV wavelength, and the films' thickness.

According to Moss [3]:

$$n(\varepsilon_0) = 1 + \frac{2}{\pi} \int_0^\infty \frac{k(\varepsilon)d\varepsilon}{\varepsilon^2 - \varepsilon_0^2} \tag{1}$$

in which the first term (one) was changed to k, a constant obtained from optical absorption spectra using the extended Buger's law [2]. In the course of processing the experimental data we neglected the k value and the dispersion of the refractive index for the preexponential factor. Tauc's law was taken into account in extrapolating $k(\varepsilon)$ over the high-frequency region,

$$(\varepsilon\alpha)^{1/2} = G(\varepsilon - E_g) \tag{2}$$

where ε is the photon's energy, G is a constant and $\alpha = (4\pi k/\lambda)$ so that we obtain,

$$k(\varepsilon) = \frac{G^2}{\varepsilon} \frac{\lambda}{4\pi} (\varepsilon - E_g) \tag{3}$$

G and E_g are experimental values. In the attempts to satisfy the equality conditions between the calculated refractive index and the experimental values from ellipsometric measurements we did use the appropriate constant in Equation (1).

The dependence of the refractive index on time and wavelength is presented in Figure 2(a). The inset of Figure 2 demonstrates the dependence of the refractive index of DLC on irradiation time for the 1.96 eV wavelength. For a correlation between the microstructural changes of DLC matrix and its dielectric properties resulting from UV treatment we have made use of the dependence of the average size (diameter) D of graphite-like constituents and the Tauc's E_g, which is really a parameter of the mid-range order [4], which gives,

$$D = d(2b)^2 / E_g^2 \qquad (4)$$

in which $b = 1.4$ eV is a constant. The size of each graphite ring d, a cluster constituent, was taken to be 2.8 Å.

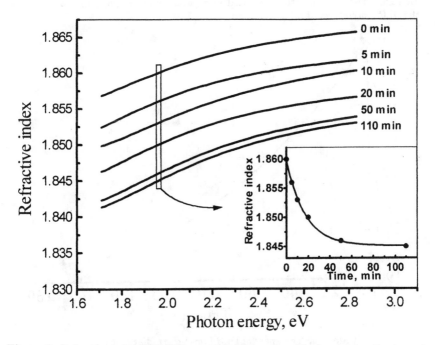

Figure 2. Refractive index of DLC films vs. irradiation time. The inset shows the dependence of the refractive index of DLC on irradiation time for 1.96 eV.

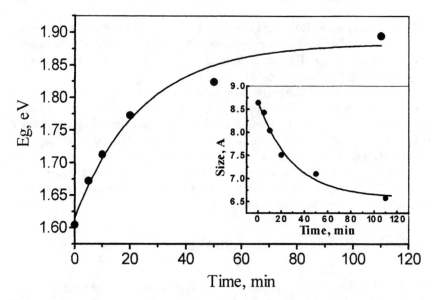

Figure 3. The dependence of the value of the optical energy gap (E_g) of DLC films on irradiation time. The inset gives the calculated average size of the graphite-like constituents in the DLC matrix, according to Equation (4).

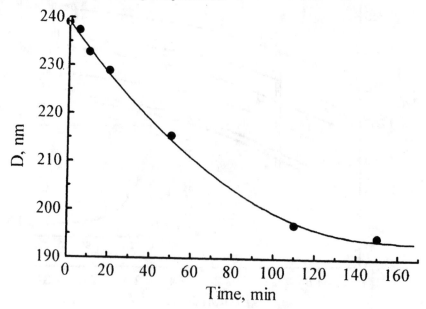

Figure 4. Thickness of DLC films vs. irradiation time.

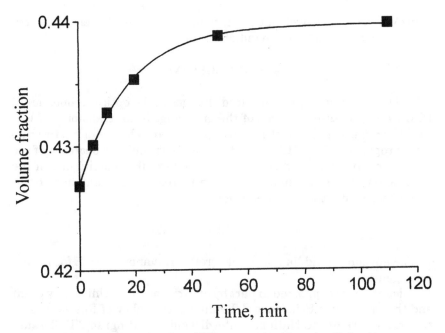

Figure 5. Volume fraction of UV-generated voids vs. irradiation time.

It was found (Figure 3) that E_g increases dramatically as a function of irradiation time and saturates after about one hour. This dependence indicates a decrease in the quantity of the six-membered rings of graphite-like constituents of the DLC matrix in the course of the irradiation (Figure 3, inset). The refractive index [Figure 2 (inset)] and the films thickness (Figure 4) show a decreasing trend and saturation. The behavior of the refractive index may be explained in terms of the appearance of UV-generated voids resulting from the treatment. According to a simple EMA model [2] the dielectric function of DLC could be written as,

$$\varepsilon = v + (1 - v)\varepsilon_{d,g} \tag{5}$$

where ε is a dielectric function of DLC, v is the volume fraction of the voids and $\varepsilon_{d,g}$ is the dielectric function of the dense constituent of DLC (the index d stands for diamond and g stands for graphite). For further estimations it was assumed that $\varepsilon_{d,g} = 4.44$, which is the same for one of the refraction indices of the graphite [5]. Calculations were performed for the

1.96 eV irradiation wavelength. The volume fraction of voids is depicted in Figure 5 as a function of irradiation time.

4. CONCLUSIONS

The UV-stimulated photoinduced changes of the optical properties of DLC were discussed in terms of the mid-range order transform of UV-irradiated films. This effect may be ascribed to a change in microstructural properties of DLC films, that is to a UV-stimulated decrease of the number of six-membered rings of the graphite-like constituents of the DLC matrix. These changes lead to an increase in the number of UV-stimulated voids in the treated films.

ACKNOWLEDGEMENTS

The authors would like to thank Prof. V. I. Ivanov-Omskii for his encouragement.

This work was supported in part by an Arizona State University grant and the Ministry of Science and Technological Policy of Russia, within the framework of the National Interdisciplinary Program "Fullerenes and Atomic Clusters" (Grant No. 94007) and the Russian Foundation for Basic Research (Grant No. 96-02-16851-a).

REFERENCES

1. Yastrebov, S. G., A. P. Skvortsov, A. B. Lodygin, V. F. Masterov and O. V. Prichodko. 1994. "Threshold Nature of Photostimulated Changes in Thin Amorphous Carbon," *Mol Mat.*, 4:233.
2. See e.g., Ivanov-Omskii, V. I., A. V. Tolmatchev and S. G.Yastrebov. 1996. "Optical Absorption of Amorphous Carbon Doped with Copper," *Philos. Mag. B.*, 73:715–722.
3. Moss, T. S. 1959. *Optical Properties of Semi-Conductors,* London, Butterworth: Scientific Publications.
4. Robertson, J. and E. P. O'Reilly. 1987. "Electronic and Atomic Structure of Amorphous Carbon," *Phys. Rev. B.*, 35:2946–2957.
5. Dolginov, A. Z. and V. I. Siklitsky. 1992. "Stokes Parameters of Radiation Propagating through an Aligned Gaseous-Dust Medium," *Mon. Not. R. Astr. Soc.*, 254:369–3824.

Nitrogen-Doped DLC Films: Correlation between Optical and Mechanical Properties

N. I. KLYUI,* B. N. ROMANYUK, V. G. LITOVCHENKO,
B. N. SHKARBAN AND V. A. MITUS
Institute of Semiconductor Physics
Nat. Acad. Sci. of Ukraine
45 Prospect Nauki
252028, Kiev, Ukraine

V. A. SEMENOVICH AND S. N. DUB
Institute for Superhard Materials
Nat. Acad. Sci.
2 Avtozavodskaya str.
254074, Kiev, Ukraine

ABSTRACT: Optical and mechanical properties of nitrogen-containing diamond-like carbon (NC-DLC) films deposited by rf plasma decomposition of a CH_4:H_2:N_2 gas mixture have been investigated. Nitrogen was incorporated into DLC films both during the film growth and after the deposition of the film by implantation of nitrogen ions. It could be shown that both optical and mechanical properties of the films strongly depend on their nitrogen content. Films deposited in low N_2 concentration gas mixtures have a high refractive index (up to 2.4), a hardness of up to 10 GPa and a low optical gap (less than 2.0 eV). An increase in the nitrogen content of the gas mixture results in a lowering of the hardness and refractive index values because the incorporated nitrogen causes more polymer-like films. It could also be shown that intrinsic stresses are substantially reduced by nitrogen incorporation in the films. A significant increase in microhardness and Young's modulus values (up to 3 times) of above mentioned films was observed after nitrogen ion implantation. The mechanisms of the effects mentioned are also discussed.

---

*Author to whom correspondence should be addressed.

Our results have permitted the development of efficient, durable antireflective and protective coatings for silicon, germanium and especially ZnSe optics.

1. INTRODUCTION

Practical applications of diamond-like carbon films are likely to be extended by nitrogen-containing modifications. Moreover, Liu and Cohen [1] have predicted a crystalline carbon nitride, β-C_3N_4, material with possible unique mechanical properties. A number of researchers have attempted to synthesize this material and have investigated the properties of nitrogen containing DLC films [2,3]. Various authors have reported conflicting film properties with an increase of the nitrogen content in the films [2–8]. Further investigations of NC-DLC properties can therefore be useful to find out the peculiarities of film formation. In the present work we report the results of optical (index of refraction, n, extinction coefficient, k), electronic (bandgap E_g) and mechanical (hardness and Young's modulus) property measurements of NC-DLC films enriched by nitrogen during both film growth and after deposition by nitrogen ion implantation.

2. EXPERIMENTAL DETAILS

The NC-DLC films used in this study were deposited in a CH_4:H_2:N_2 gas mixture plasma, in a parallel plate reactor. Silicon substrates were put directly on the 200 mm diameter water cooled cathode which was capacitively connected to a 13.56 MHz rf generator. The total pressure in the reaction chamber varied between 0.1 to 0.8 torr. During the plasma decomposition experiments the rf basis voltage was about 1900 volts. The thickness of the films varied between 100 and 10,000 nm.

The optical constants of NC-DLC films were measured in the spectral energy range 1.5–4.5 eV by ellipsometry with a rotating analyzer. The optical constants (n,k) and film thicknesses were also determined by laser ellipsometry at a fixed wavelength ($\lambda = 632.8$ nm). The optical bandgaps of NC-DLC films were calculated using Tauc's equation. The mechanical properties of NC-DLC films were derived from microprobe indentation measurements (Nano Instrument Inc., Knoxville, TN, USA). The measurements were made at a fixed indentation depth of 150 mm. All measurements were carried out at room temperature. Auger Electron Spectroscopy (AES) was used to estimate the nitrogen content in the doped a-C:H films.

3. RESULTS AND DISCUSSION

We find a clear dependence of the n and k values on nitrogen content in the gas mixture. Increasing the nitrogen content leads to an n and k maximum (Figures 1 and 2) between about 10 and 20 at.% of nitrogen in the gas mixture. It should be noted that particularly the extinction coefficient decreases greatly with an increase of the N_2 content in $CH_4:H_2:N_2$ gas mixtures. This implies that the films with nitrogen incorporated are more transparent, and this effect is connected with an increase of the optical gap [7] and the decrease of the near edge state density values [5,7]. Indeed, we found that nitrogen doping of the a-C:H films in gas mixtures up to 20 at.% gives rise to a narrowing of the optical gap. On the other hand, the nitrogen doping of the a-C:H films in the range of 20–45 at.% gives rise to widening of the optical gap (Figure 3). These narrowing and widening effects in nitrogen doping a-C:H films are caused mainly by the nitrogen-hydrogen attached to carbon.

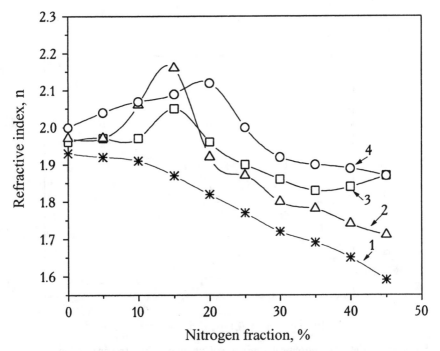

Figure 1. Refractive index as a function of the nitrogen fraction in $CH_4:H_2:N_2$ gas mixtures during deposition at different total gas pressures: 1 = 0.8 torr; 2 = 0.6 torr; 3 = 0.4 torr; 4 = 0.2 torr.

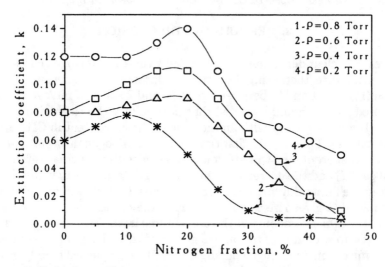

Figure 2. Extinction coefficient as a function of nitrogen fraction in CH_4:H_2:N_2 gas mixtures during deposition at different total gas pressure: 1 = 0.8 torr; 2 = 0.6 torr; 3 = 0.4 torr; 4 = 0.2 torr.

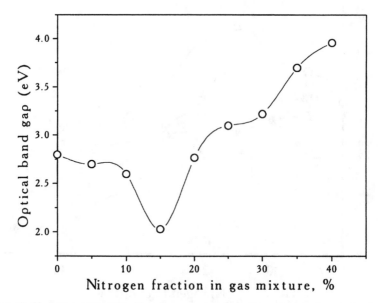

Figure 3. Variation of the optical band gap as a function of the nitrogen fraction in the gas-mixture for as-prepared a-C:H:N films (total gas pressure, 0.8 torr).

Earlier studies of ion-implanted DLC films have shown that ion implantation leads to the degradation of their optical, mechanical and other properties [6,9]. It is important to note that most of these observations were made at implantation doses right up to 10^{16} ions/cm^2. A further dose increase does not result in any changes [9]. Figure 4 shows the behavior of the Young's modulus as a function of carbon ion dose implantation in DLC films after Jiang et al. [9] compared to some of our results. As can be seen, the Young's modulus for DLC films [9] monotonically decreases for doses between 10^{14} up to 10^{16} ions/cm^2 with a saturation above 10^{16} ions/cm^2. In contrast to this, our experiments ("soft" films) led to an increase of the Young's modulus for DLC films in the same range. This could be explained by different mechanisms occurring at ion bombardment of "hard" [9] and "soft" films (our experiments). We assume that during ion bombardment of "soft" DLC films destruction of C-H bonds takes place which favours a partial reconstruction of sp^3 bonds. It is probable that new defect centers are produced that lead to saturation at high doses of ion implantation.

We also found that the hardness dependence on nitrogen concentration correlates with that of the extinction coefficient k (Figure 2). The upper curves in Figure 5 present the data for NC-DLC films implanted with nitrogen and carbon ions of intermediate energy (150 keV) at 10^{16} ions/cm^2. Implantation leads to substantial increases in film hardness, with the increase slightly dependent on the nitrogen fraction in the gas mixture. This is, in our opinion, connected with the formation of more dense and disordered films when the nitrogen fraction in the gas mixture is about 15%. The ion implantation, in this case, slightly modifies the film structure. A further increase in the nitrogen fraction results in the formation of more ordered (polymer-like) and lower films. In this case, ion implantation leads to a disordering of the film and an increase in film density and hardness (Figure 4). These conclusions are confirmed by the results of n and k measurements for initial (Figures 1 and 2) and implanted films (not shown). The increase of film hardness after nitrogen and carbon implantation may, however, not be connected only with film disordering. For example, the hardness of the film grown at a nitrogen fraction of 20% and a total gas pressure of 0.2 torr (upper curves in Figure 1) increased from 8.1 GPa for the unimplanted film to 15.5 GPa for the film implanted by nitrogen ions ($E = 150$ keV, $D = 10^{16}$ ions/cm^2). After further implantation up to a total dose of 2.5 x 10^{16} ions/cm^2, the film hardness decreased to the initial value of 8.1 GPa. The increase in film hardness after the first stage of implantation, may therefore result from not only film disordering but from the formation of carbon-nitrogen bonds as well.

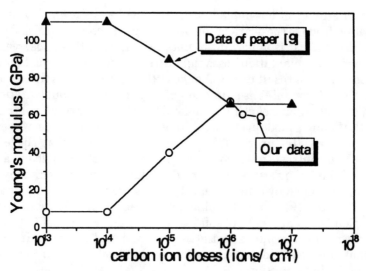

Figure 4. Young's modulus versus carbon ion implantation dose for "hard" a-C:H films [9] (triangles) and "soft" a-C:H films (circles).

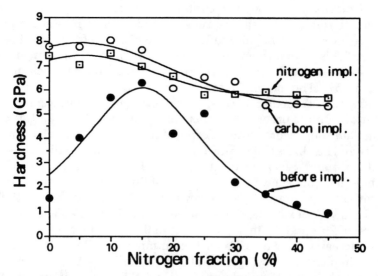

Figure 5. Variation of hardness as a function of the nitrogen fraction in the gas mixture for as-prepared a-C:H:N films (filled circles) and carbon implanted (open circles) and of nitrogen implanted (squares) films. The solid lines represent Lorentz fits to the data.

4. CONCLUSIONS

It could be shown that the optical and mechanical properties of nitrogen containing diamond-like carbon films strongly depend on the nitrogen fraction in the feed gas mixture. Films deposited at high nitrogen concentrations (30 at.%) display a high transparency and could be used as a protective coating for solar cells. Significant increases in film hardness after nitrogen and carbon ion implantation were also observed.

REFERENCES

1. Liu, A. Y. and M. L. Cohen. 1990. *Phys. Rev. B,* 41:10727.
2. Feng, J., C. Long, Y. Zheng, F. Zhang and Y. Fan. 1995. *J. of Crystal Growth,* 147:333.
3. Seth, J., R. Padiyath and S. V. Babu. 1994. *Diamond and Relat. Mater.,* 3:210.
4. Wang, X., P. J. Martin and T. J. Kinder. 1995. *Thin Solid Films,* 256:148.
5. Mansour, A. and D. Ugolini. 1993. *Phys. Rev. B.,* 47:10201.
6. Kalish, R., O. Amir, R. Brener, R. A. Spits and T. E. Derry. 1991. *Appl. Phys. A.,* 52:48.
7. He-Xiang Han and B. J. Feldman. 1988. *Solid State Comm.,* 65:921.
8. Kaufman, J. H., S. Metin and D. D. Saperstein. 1989. *Phys. Rev. B.,* 39:13053.
9. Jiang, X., J. W. Zou, K. Reichelt and P. Grunberg. 1989. *J. Appl. Phys.,* 66:4729.

Compositional Modulated DLC Films for Improvement of Solar Cell Efficiency and Radiation Stability

V. A. SEMENOVICH

Institute for Superhard Materials
National Acadamy Science
2 Avtozavodslaya str.
254074, Kiev, Ukraine

N. I. KLYUI,* V. P. KOSTYLYOV,
V. G. LITOVCHENKO AND V. V. CHERNENKO

Institute of Semiconductor Physics
National Acadamy Science of Ukraine
45 prospect Nauki
252028, Kiev, Ukraine

ABSTRACT: Parameters of silicon-based solar cells (SC) covered with compositionally modulated diamond-like carbon (DLC) films before and after proton and ultraviolet (UV) irradiations have been investigated. The significant improvement of efficiency was observed for SC with thin (~50–100 nm) antireflecting DLC films. Thick (~1300–1500 nm) compositional modulated DLC films deposited on a working side of the SC increases its stability from the effect of proton and UV irradiation. So, one-stage (E = 50 keV) or multi-energy (50 + 100 keV) implantation of proton into protected SC (the SC + DLC films system) did not practically influence SC parameters. Only after the third-stage of implantation (E = 150 keV) did the SC parameters of the protected SC deteriorate slightly. At the same time the unprotected SC drastically deteriorated after treatment.

*Author to whom correspondence should be addressed.

1. INTRODUCTION

A **great deal** of effort has recently been directed to the development of opto-electronic devices using carbon films as both antireflecting and protective coatings [1–4]. Silicon-based SC are extensively used to convert solar energy into electrical energy in space. Bare Si has a solar reflectance of approximately 35%. A single-layer coating can reduce this loss to 5–10%. Additional coatings can generate additional improvements. However, the conversion efficiency of Si solar cells decreases rapidly with increasing damage caused by proton and UV irradiation [5]. Therefore, two types of coatings are very important for the Si-based SC: antireflecting coatings directly on the SC and a protective coating. At the same time, there is one problem with the properties of protective coatings which are used for Si-based SC. Namely, the absorption increases as the thicknesses of protective film increase in the region of 400–700 nm, where both a thick film and a large optical gap are required for the cosmic space SC.

It has been previously reported [3,4] that the photovoltaic properties of Si-based SC can be increased by using i-C and a-C films. However, any information concerning the effects of proton and UV treatments on the basic properties of SC + DLC systems has not been reported. Meanwhile, we have made efforts to protect degradation of the basic characteristics of SC under proton and UV irradiation treatments.

2. EXPERIMENTAL DETAILS

Silicon solar cells with shallow (~1000 nm) combined inversion-diffusion junction were used. To increase photocurrent the isotype barrier was created on the back side of the elements. The front side of some elements was textured by etching in KOH solution. As a result, random pyramidal relief was formed. The current-voltage (I-V) relationships for the SC both with and without DLC coating were measured under AMO illumination conditions.

The DLC (a-C:H) films were deposited by r.f. glow discharge in a parallel plate reactor. The diameter of the parallel plate electrodes was 200 mm and they were 200 mm apart. The r.f. power (13.56 MHz) was applied to the lower electrode, the upper electrode was grounded. The gases (CH_4, H_2 and N_2) were introduced through the upper electrode in a shower head flow configuration. The substrates for deposition were put directly on the lower electrode which was cooled by water. The stainless steel chamber was evacuated to a base vacuum of about 10^{-4} Torr before each deposition run. The total pressure of the reaction chamber was varied from 0.1 to 0.8 Torr. During the plasma decomposition experiment r.f. bias value was varied from 500 to 2000 volts under fixation of other deposition parameters.

Optical constants (refractive index n and extinction coefficient k) of DLC films were measured by spectroellipsometry method [6] within spectral range from 1.5 to 5.6 eV and laser ellipsometry at fixed wavelength ($\lambda = 632.8$ nm). Using dispersion dependencies of n and k the reflection spectra for SC + DLC film systems were calculated. The optical band gap of DLC films was determined from Tauc equation. SC with and without DLC coating were irradiated by proton of intermediate energies (50–150 keV) at doses up to 10^{16} cm^{-2}. The ion beam current density did not exceed 1 μA/cm^2. The UV irradiation treatments of SC were made by xenon (8.4 eV) and krypton (10 eV) lamps.

3. RESULTS AND DISCUSSION

Using CH_4-H_2 as carbon sources, we have fabricated a-C:H films and have conducted a series of experimental investigations of optical properties in these films. It has been found through these investigations that the a-C:H films are almost constructed with tetrahedrally coordinated carbons attached with a hydrogen (Figure 1). Hydrogen has levelled one fragments sizes of the network structure. At that network structure of the films will be determined predominantly by the strongly cross-linked carbon-carbon bonds that is supported by infrared active absorption features near 2925 cm^{-1}. At the same time the 2850 cm^{-1} peak in the "hard" films becomes less strong than in the "soft" films.

The refractive index (n) of these a-C:H films increases monotonically (near-linear relationship) with increasing r.f. bias values. At the same time the optical band gap (E_g) and extinction coefficient (k) of these films depend on the r.f. bias values in opposing ways. It is interesting in this case, that narrowing and widening effects of E_g are caused by the hydrogen attached to carbon and variations in the proportion of the bonds SP^2/SP^3.

Special attention has been paid to the "hard" ($n \approx 2.0$) and "soft" ($n \approx 1.6$) a-C:H films. As seen from Figure 2 the refractive index for "hard" a-C:H films remains roughly constant within spectral range from 1.5 to 4.5 eV ($n \approx 2.0$) and satisfies the condition for optimal antireflecting effect. So, experimental reflection spectra of SC with antireflecting coating show significant decreasing of reflection coefficient, which can attain ~1% in the minimum of the reflection band (~600 mm). The optimiza-

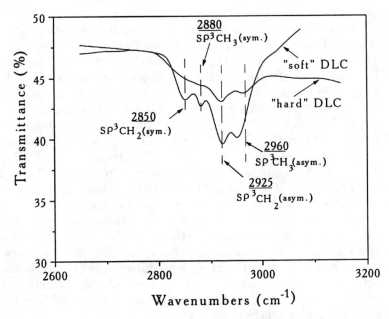

Figure 1. Infrared transmission spectra of the "hard" and "soft" DLC films on silicon.

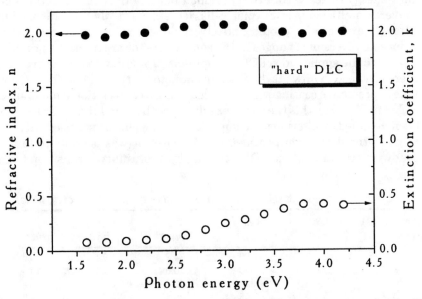

Figure 2. Refractive index (n) and extinction coefficient (k) vs photon energy for "hard" DLC film.

tion of the deposition parameters for achieving the maximum figure of merit for SC applications has been performed and in Table 1 we summarize properties of silicon SC with and without DLC antireflecting films.

The results indicate that increasing of short circuit current J_{sc} attains 25–40%, while the open circuit voltage V_{oc} does not change, but in some cases increases. At the same time the fill factor FF increases as well. This results in increasing of SC efficiency 1.3 times. The parameters for a typical silicon SC with (sample 1a) and without (sample 1) DLC antireflecting film are presented in Table 1. It should be noted that DLC films used as antireflecting coating demonstrated a high thermal stability. Thermal annealing at 350°C for 30 min neither the optical constants of DLC films nor the parameters of SC with antireflecting coating (see sample 1b in Table 1) was affected. In Table 1 the parameters of textured silicon SC without (sample 2) and with (sample 2a) DLC antireflecting coating are presented. Taking into account the obtained results (see Table 1) and the fact that the texturing procedure allows to improve the SC efficiency up to 1.1–1.15 times. It may be concluded that due to deposition of antireflecting DLC coatings on textured surface of SC the reflection losses could be practically excluded.

It is a well known fact [5] that radiation-induced lattice defects do produce energy states within the forbidden band of Si semiconductor. The conversion efficiency of Si solar cells decreases rapidly with increasing damage produced by proton and UV irradiation. Light and radiation-induced irreversible changes in the semiconductor are a serious problem related to the application of DLC in opto-electronic devices and therefore, there is a need to prevent the Si semiconductor surface. Therefore, a new type of protective coating was developed using a compositional modulated a-C:H films. To do this the CH_4 was diluted with H_2 and the C:H ratio has been modified by altering the composition of the plasma gas used with respect to the deposition parameters. Moreover, it was possible to find a sandwich structure of bulk DLC coating by a continuous transition from

Table 1. Parameters of silicon SC with and without DLC antireflecting films.

Sample Number	DLC Film	V_{oc} mV	J_{sc} mA/cm^2	FF	η, %
1	–	580	34.2	0.72	10.6
1a	+	590	40.8	0.75	13.4
1b*	+	590	40.7	0.75	13.4
2**	–	555	28.5	0.78	9.0
2a**	+	575	34.0	0.784	11.3

*The sample 1b is the sample 1a after thermal annealing at 350°C for 30 min in N_2.
**The samples 2 and 2a have texture front surface.

high (2.4) index ($\lambda/4$ thickness) to low (1.6) index ($\lambda/4$ thickness). In addition, such coatings show significant decreasing of reflection coefficient and are not particularly sensitive to the exact shape of the refractive index profile.

Results of proton implantation effect on Si sollar cell parameters are presented in Table 2. To protect the SC the thick (1300–1500 nm) compositional modulated DLC films described above were deposited onto SC working surface. As seen from Table 2 one-stage ($E = 50$ keV) or multi-energy (50 + 100 keV) implantation of proton into protected SC practically does not influence its parameters (sample 3a and 3b in Table 2, respectively). And, only after the third stage of implantation ($E = 150$ keV) are the SC parameters slightly deteriorated (sample 3c), while the unprotected SC (sample 4) drastically degraded after this treatment (sample 4a). Some decrease in the parameters of protected SC is due to penetration of implanted ions throughout top DLC film into SC that results in disordering of silicon subsurface layer. This conclusion is confirmed by results of Monte-Carlo simulation of ion depth distribution as well as by experimental results obtained for SC covered by thicker compositional modulated DLC film ($d = 1500$ nm, sample 5) and exposed to implantation of greater ion dose (sample 5a).

Indeed, as seen from Table 2 deterioration of sample 5a parameters is substantially less than for sample 3c in spite of higher ion dose. The unprotected SC (sample 6) exposed to this high dose implantation dramatically degraded (sample 6a). Irradiation of SC covered by protective compositional modulated DLC films ($d \sim 1300$–1500 nm) by using UV

Table 2. Parameters of silicon SC with and without compositional modulated DLC protective films.

Sample Number	Thickness of DLC Film (nm)	Proton Implantation		V_{oc} mV	J_{sc} mA/cm^2	FF	η, %
		E, keV	D, 10^{15} cm^{-2}				
3	1300	—	—	608	30.8	0.78	10.8
3a	1300	50	3.3	607	30.7	0.78	10.8
3b	1300	50/100	3.3/3.3	606	30.5	0.78	10.7
3c	1300	50/100/150	3.3/3.3/3.3	572	27.2	0.71	8.1
4	—	—	—	524	20.2	0.78	6.1
4a	—	150	3.3	412	15.0	0.57	2.6
5	—	—	—	580	29.8	0.7	8.9
5a	1500	150	10	565	26.2	0.7	7.6
6	—	—	—	595	37.5	0.66	10.8
6a	—	150	10	350	31.4	0.44	3.5

lamps (xenon-8.4 eV or krypton-10 eV) did not result in any marked changes in both the SC parameters and optical constants of films. Moreover, improvement of parameters of SC with protective compositional modulated DLC films and previously implanted by proton was observed after UV treatment. At the same time, the UV treatment of unprotected and implanted SC led to further degradation of its parameters.

4. CONCLUSIONS

Optical properties of a-C:H films can be controlled by manipulation of the composition of gas mixture and value of r.f. bias voltage. We have shown that compositional modulated DLC films play a major role in the significant improvements of SC efficiency and radiation stability. The results obtained have opened up a wide variety of potential applications of compositional modulated DLC films in the fields of electronics and optoelectronics.

REFERENCES

1. Enke, Knut. 1985. *Applied Optics*, 24:509.
2. Lu, F. X., B. X. Yang, D. G. Cheng, R. Z. Ye and J. B. Sun. 1992. *Thin Solid Films*, 212:220.
3. Moravec, T. J. 1982. *J. Vac. Sci. Technol.*, 20:338.
4. Alaluf, M., J. Appelbaum, L. Klibanov, D. Brinker, D. Scheiman and N. Croitoru. 1995. *Thin Solid Films*, 256:1.
5. Yang, L., L. Chen and A. Catalano. 1991. *Appl. Phys. Lett.*, 59:840.
6. Litovchenko, V. G., S. I. Frolov and N. I. Klyui. 1993. *Proc. SPIE*, 2113:65.

Spectroscopy of
DLC Films
Deposited with a
Hydrogen-Free Technique

I. V. GLADYSHEV,* V. I. SHISHKIN,
A. S. SIGOV AND V. I. SVITOV
*Moscow State Institute of Radioengineering
Electronics & Automation
Vernadsky Av. 78
117454 Moscow, Russia*

1. INTRODUCTION

DLC films were deposited from hydrocarbon vapors with an ECR-assisted MW plasma technique. The pressure in the vacuum chamber (0.03 to 1 Pa) and the substrate stage–ECR zone distance, were varied during the deposition process. In-situ temperature control indicated that the substrate spontaneous heating by plasma action did not exceed 400 K. The reactants were acetone C_3H_6O, propanol-2 C_3H_8O and halogen-containing hydrocarbons. The main feature of the technique is that it avoids gaseous hydrogen application. The films as-deposited can be subdivided into two types: "transparent" films, deposited from acetone and propanol-2 and "black" films, obtained with chlorine-containing hydrocarbons vapors. All films appear to be amorphous. The hardness of the deposited DLC films is higher than 25 Gpa and their mass densities differ considerably: 1.3–1.5 g/cm^3 for "transparent" films and about 2.9 g/cm^3 for "black" films.

The films deposited onto silicon and visually transparent polymers, were investigated with photo- and cathodeluminescence (CL), Raman

---

*Author to whom correspondence should be addressed.

473

scattering, Fourier-transform infrared (FTIR) spectroscopy, and optical spectroscopy in the UV and visible regions.

2. FILM PROPERTIES

"Transparent" films appear to show significant photo- and cathodeluminescence which is not typical for diamond-like carbon. CL-spectra show the characteristic peaks with the main peak at $\lambda \approx 530$ nm. A rather broad peak at $\lambda \approx 476$ nm is also present accompanied by a low intensity peak at $\lambda \approx 600$ nm for some samples (see Figure 1). The CL-peak intensities depend on the deposition pressure and on the substrate stage–ECR zone distance. Figure 2 shows that the main peak($\lambda \approx 530$) intensity decreases when the above distance increases and, according to Figure 3, it rises with the deposition pressure. At the same distance no film is detected using deposition pressures higher than 1 Pa. Deposition parameters giving the highest CL intensity were chosen to produce the coatings on transparent polymer and glass for an investigation of the DLC optical properties. The data are presented by Figure 4.

The UV, visible and near IR spectra of "transparent" DLC have several characteristic sections. The optical energy band gap is estimated as 2.3 eV. The refractive index, for the entire transparency range is ~1.75 and for the anomalous dispersion range it changes between 1.46 to 1.8. There are no Raman scattering data because the strong fluorescence excited by the analyzer laser beam ($\lambda \approx 657$ nm) exceeds that typical for DLC by 6 orders of magnitude.

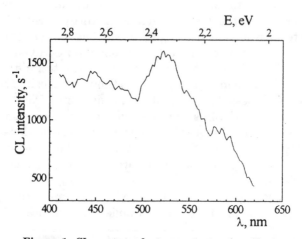

Figure 1. CL-spectrum for transparent carbon films.

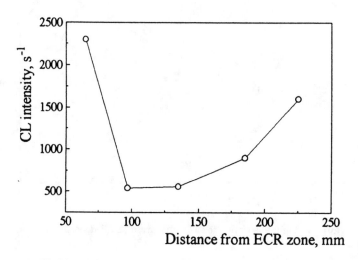

Figure 2. CL dependence on substrate–ECR zone distance for "transparent" films.

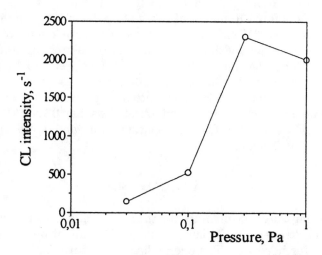

Figure 3. CL dependence on the deposition pressure for a "transparent" film. Stage–ECR distance = 65 mm.

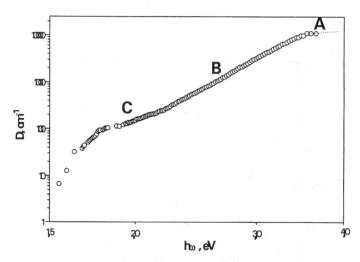

Figure 4. Light absorption for a "transparent" ECR-deposited carbon.

The films we call "black" show neither photo- nor cathodelumines-cence and fully absorb UV and visible light. They do, however, exhibit, however, well-defined Raman scattering at 1580 cm⁻¹, typical for DLC, and a wide band at 1317 cm⁻¹ normally considered as belonging to the dia-mond phase. The halfwidth for the 1317 cm⁻¹ peak (about 50 cm⁻¹) is, however, too large for high quality diamond, so we prefer to designate it sp³-bonded carbon instead, although the diamond phase may in reality be present as nanometer particles. The content of sp³-bonded carbon for our "black" films approaches 50 at.% and contain less than 0.1 at.% hydrogen as shown in FTIR spectra. The "transparent" films and typical DLC con-tain up to 30 at.% hydrogen.

3. DATA ANALYSES

Considering the experimental data, the development of an energy band diagram for "transparent" diamond-like carbon might be considered a subject for further work. The energy band structure analysis assumes identity of the film properties independent of the substrate material along with a relationship between the optical spectroscopy data and those of CL-spectroscopy.

As noted earlier, the absorption spectra of these films show several characteristic sections (Figure 4). The dependence $D \sim (1/\omega)(\omega - \omega_0)^2$ governs the section B, and allows estimation of the energy band gap from the absorption edge ($E_g = 2.3$ eV). Such a dependence was reported recently for DLC [1]. A dependence of the type $D \sim (1/\omega)(\omega - \omega_0)^{1/2}$ is observed for section A with $E_g \approx 3.1$ eV. Section C corresponds to the Urbach absorption edge typical for amorphous materials [2]. Section C can be characterized by the dependence $D \sim \exp(\Gamma\hbar\omega)$ with $\Gamma \approx 5$. The deviation observed may be connected with the traps in the band gap. The results have been compared with the CL-spectra. Thermal glow of the peaks at $\lambda_{max} \approx 600$ and 476 nm and stability of the main $\lambda \approx 530$ nm peak, were taken into account and are considered to be caused by the decrease of the localized states area. A possible energy band diagram is shown in Figure 5. E_C–E_B and E_V–E_A are the areas of localized states in the conduction band and the valence band respectively. E_X and E_Y are the "impurity" levels. Considering this $E \approx 3.1$ eV corresponds to the width of the carrier mobility band gap (the absorption approaches 10^4 cm^{-1}) and $E_g \approx 2.3$ eV is the energy band gap.

The CL-spectral behaviour is due to plasma-chemical process dynamics: the reactions of atomic oxygen and hydrogen the concentration of which is strongly dependent on the deposition pressure and the substrate stage–ECR zone distance. This reflects on the chemical composition and structure of the deposited carbon material.

ECR-plasma deposited carbon has specific mechanical, optical and electric properties that merits consideration for many applications of

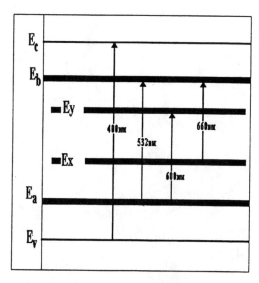

Figure 5. Energy bands structure of ECR-deposited carbon.

these films. Further investigations of their physical properties, would promote the development of optimal deposition processes.

REFERENCES

1. Sleptsov, V. V., V. M. Elinson and S. S. Gerasimovitch et al. 1991. 136:(1–2): 53–59.
2. Mott, N. F. and E. A. Davis. 1979. *Electron Processes in Non-Crystalline Materials,* Oxford: Clarendon Press.

The Development of C KVV X-Ray-Excited Auger Spectroscopy for Estimation of the sp2:sp3 Ratio in DLC Films

M. N. Petukhov* and A. P. Dementjev

IRTM, RRC "Kurchatov Institute"
Kurchatov sq.
Moscow 123182
Russian Federation

ABSTRACT: The experimental X-ray-excited Auger lineshapes for graphite, diamond and diamond-like carbon (DLC) film have been obtained. The comparison of N(E) C KVV spectral features of graphite, diamond and DLC film has also been carried out. The C KVV Auger lineshape of DLC film has been represented as a synthesis of the diamond and graphite Auger spectra. A sp2:sp3 ratio has been determined on the basis of the synthesis. The first-derivative X-ray-excited Auger spectra for various mixtures of sp2 and sp3 states have been compared.

KEY WORDS: sp2:sp3 ratio, diamond-like carbon, diamond, graphite, electron spectroscopy.

1. INTRODUCTION

Raman spectroscopy is widely used method rendering information about different bonds of carbon atoms, however, this method is applicable for bulk samples. Auger Spectroscopy provides information on the depth in the order of a few atomic layers and is capable of yielding information concerning the chemical state of atoms in the near-surface region of a solid [1]. The C KVV Auger lineshapes for graphite (sp2 hybridization) and

---

*Author to whom correspondence should be addressed.

diamond (sp3 hybridization) have different characteristic features and N'(E) C KVV spectra of diamond and graphite are widely used to identify sp2 and sp3 states of carbon atoms in solids [2]. Obviously, AES is the preferential method for investigation of a carbon nucleation at an early stage of carbon films growth and chemical interaction on the surface of carbon films. Nevertheless, in spite of obvious advantages the power of this method is not fully used for DLC films studies. This is due to poor interpretation of the N'(E) C KVV experimental data and the absence of accepted standardization for the diamond spectrum [2].

The X-ray-excited AES (XAES) allows a decrease of the influence of a high-energy electron beam and, in conjunction with the XPS data, places the Auger spectrum precisely on the energy scale. In the previous study we obtained the high-quality X-ray-excited C KVV spectra for diamond, polyethylene and graphite [2].

In this work we have considered the N(E) C KVV spectra for graphite, diamond and DLC film and compared characteristic features. The Auger processes in the mixture of sp2 and sp3 states in DLC films are regarded as a sum of independent Auger processes for sp2 and sp3 hybridized atoms. The N(E) C KVV spectrum of DLC film has been represented as a synthesis of the diamond and graphite spectra. The derivative N'(E) C KVV spectra for these substrates have also been exhibited and treated.

2. EXPERIMENTAL

The XAES and XPS data were obtained using a MK II VG Scientific spectrometer. The photoelectron and Auger processes were excited using an Al Kα source with a photon energy of 1486.6 eV and the vacuum in the analysis chamber was maintained at 5×10^{-10} Torr. Spectra were collected in the analyzer constant energy mode, with a pass energy $Ep = 20$ eV for XPS, and $Ep = 50$ eV for XAES analysis. The C 1s XPS spectra were acquired with a 0.1 eV step size; a 0.3 eV step size was used for the C KVV XAES spectra. The Auger spectra were obtained with an overall collecting time tc = 20 ÷ 30 min and a signal-to-noise ratio of 500.

Samples of graphite, diamond and DLC film were investigated. A calibration of C KVV Auger spectra was made by taking into account the charging effect, and according to the C 1s energy position; Eb(C 1s) = 284.4 eV for diamond [2], Eb(C 1s) = 284.6 eV for graphite [2] and Eb(C 1s) = 284.5 eV was used for DLC film. Background subtraction was made on the base of appearing in Reference [3] method. Derivative Auger spectra were obtained by numerical differentiation of the raw N(E) Auger spectra.

The graphite spectrum was obtained from the plane of highly oriented pyrolytic graphite (HOPG). A surface of an artificial diamond specimen 3 × 3 × 1 mm³ was prepared by polishing with a diamond paste and ultrasonically washing in acetone immediately before insertion into the spectrometer [2]. A sample of DLC film was prepared by CVD method.

3. RESULTS AND DISCUSSION

The N(E) C KVV Auger spectra of graphite, diamond and DLC film after correction for charging effect and background subtraction are shown in Figure 1.

All spectra have been normalized and have the same area. The comparison analysis of characteristic features for diamond and graphite Auger spectra was performed in detail in Reference [2]. There was shown that the right shoulder of the Auger spectrum for graphite (Ek > 277 eV) contains π-bond folds [2]. Consequently, the greater C KVV spectrum intensity at Ek > 277 eV, the greater concentration of sp2 states.

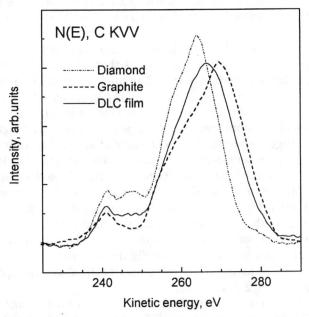

Figure 1. The N(E) Auger lineshapes of diamond, graphite and DLC film after background subtraction.

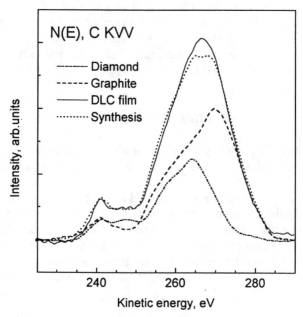

Figure 2. The experimental Auger spectrum for DLC film and a least-squares-fit synthesis of DLC film spectrum using the diamond and graphite spectra as component signals.

The Auger line of DLC film is intermediate between the diamond and graphite lines. This clearly demonstrates that DLC film is a mixture of sp2 and sp3 states of carbon atoms. The initial excitation for the Auger transition is the ionization of an inner atomic level. Since the created hole is strongly localized at the atom the Auger process is limited in spatial extent to the ionized atom and its nearest and next nearest neighbors [1]. Hence, the Auger processes in the mixture of sp2 and sp3 states like in DLC films may be considered as a sum of the independent Auger processes for sp2 and sp3 hybridized atoms. In Figure 2 the N(E) Auger spectrum of DLC film is treated as a superposition of the diamond and graphite spectra. The difference between the experimental C KVV spectrum and the synthesized spectrum may occur due to changes in many-particle processes for mixed and pure states. The ratio of areas for diamond and graphite spectra in Figure 2 gives an estimation of sp2:sp3 ratio in the near-surface region of DLC film. The sp2:sp3 ratio for the given in Figure 2 film equals 2.

In order to compare our data with an electron excited AES data the raw N(E) CKVV spectra have been differentiated. The N'(E) CKVV spectra of diamond and graphite are shown in Figure 3. The derivative spec-

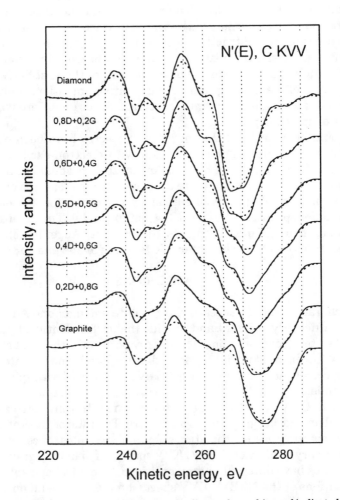

Figure 3. The first-derivative XAES spectra for diamond, graphite and indicated combinations of the raw N(E) Auger spectra for diamond (D) and graphite (G) obtained by numerical differentiation (solid line). Demonstration of losses of characteristic features and changes of energy positions due to modulation in electron-excited Auger spectroscopy (dashed line).

tra for intermediate states have been obtained by using the differentiation of indicated in Figure 3 linear combinations of the raw diamond and graphite spectra. The raw diamond and graphite spectra have previously been normalized by taking into account equality of areas for the background subtracted spectra. Those N'(E) C KVV spectra in Figure 3 characterize the mixed states and may be used for estimation of sp2:sp3 ratio in carbon films. The comparison of the first-derivative C KVV spectra of diamond, graphite and their combinations shows that the energy of the main minimum of spectrum varies for different concentrations of sp2 and sp3 states.

It should be mentioned that our Auger spectra have been obtained with 0.3 eV step size. For comparison with electron-excited spectra modulated, for example by 1.5 eV, the initial spectra should be averaged by 5 points (dashed lines in Figure 3). Obviously the features of the left shoulder of C KVV spectra significantly smoothed for the electron-excited Auger spectra (References [3,4] for example). The greater value of modulation in AES measurements, the smoother Auger line shape and the greater losses of the characteristic features of sp2 and sp3 states.

4. CONCLUSIONS

The characteristics features of the graphite and diamond Auger spectra are sufficiently different and provide the useful identification of sp2 and sp3 states in carbon solids. The energy position of the right shoulder of C KVV Auger spectrum is very sensitive to π-bonds and the intensity of Auger spectrum at kinetic energy >277 eV is associated with sp2 states in carbon solids.

The N(E) C KVV spectrum of a carbon film may be considered as a synthesis of diamond and graphite Auger spectra. The ratio of areas of the synthesizing spectra gives an estimation of sp2 and sp3 states concentrations.

The primary minima of N'(E) C KVV spectra of diamond, graphite and mixed states have different energy positions. The lineshape and the energy positions of the N'(E) C KVV spectral features depend on value of modulation.

The energy positions of the extreme left features of the diamond and graphite spectra (Ek = 240 eV, see Figure 1) may be used as an energy reference for electron-excited Auger spectra.

ACKNOWLEDGEMENT

The authors are grateful to V. V. Khvostov from Moscow State University for his program of background subtraction.

REFERENCES

1. Weissman, R. and K. Muller. 1981. "Auger Electron Spectroscopy—A Local Probe for Solid Surfaces," *Surf. Sci. Rep.*, 105:251–309.

2. Dementjev, A. P. and M. N. Petukhov. 1996. "Comparison of X-ray-excited Auger Lineshapes of Graphite, Polyethylene and Diamond," *Surf. Interface Anal.*, 24:517.

3. Burrel, M. C. and N. R. Armstrong. 1983. "A Sequential Method for Removing the Inelastic Loss Contribution from Auger Electron Spectroscopic Data," *Appl. Surf. Sci.*, 17:55; Khvostov, V. V., V. G. Babaev and M. B. Guseva. 1985. "Auger Spectroscopy of Amorphous Carbon Films," *Solid State Physics*, 27:887–891 (in Russian).

4. Skytt, V., E. Johansson, N. Wassdahl, T. Wiell, J.-H. Guo, J. O. Calsson and J. Nordgren. 1993. "A Soft X-ray Emission Study of HFCVD Diamond Films," *Diamond Relat. Mater.* 3:1–6; Hoffman A., 1994. "Fine Structure in the Secondary Electron Emission Spectrum as a Spectroscopic Tool for Carbon Surface Characterization," *Diamond Relat. Mater.*, 3:691–695.

Size-Distribution Analysis of Diamond- and Graphite-Like DLC Constituents

V. I. IVANOV-OMSKII,* A. B. LODYGIN, A. A. SITNIKOVA,
A. A. SUVOROVA AND S. G. YASTREBOV

A. F. Ioffe Physical-Technical Institute
St Petersburg, 194021, Russia

A. V. TOLMATCHEV

St. Petersburg State Tech. University
St. Petersburg, 195251, Russia

ABSTRACT: Nucleation of diamond nanocrystals through self-assembling of carbon atoms within the surface of growing DLC films during ion sputtering on graphite has been studied. TEM (100 keV), Selected Area Electron Diffraction (SAED) and optical absorption (1–6 eV) studies of the prepared DLC films were performed. Observation of diamond nanocrystals and graphite nanoclusters and their size-distribution functions are reported. The average sizes of sp^2-bonded clusters obtained by analyzing TEM images are compared with those determined with optical data. The results are discussed in the framework of the fluctuation mechanism of nucleation of graphite and diamond phases.

KEY WORDS: diamond, carbon, graphite, nanoclusters, optical properties, size-distribution function.

INTRODUCTION

Close attention given to the mechanism of self-assembling in diamond nucleation could result in a new approach to the synthesis of dia-

---

*Author to whom correspondence should be addressed.

mond. The basic idea of this approach is to study conditions of diamond nucleation and to create them locally inside elementary units of a growing structure by controlling the flow of carbon atoms/ions or other precursors to the surface of the growing structure. Local fluctuations of growth parameters, such as temperature and pressure within the surface of growing film in the course of ion bombardment of the film, should be considered as a possible driving force of diamond phase formation. In this case the mean pressure in the growth chamber remains low, which classifies the method with low pressure chemical techniques [1] widely accepted by science and industries in recent years. In the context of further development of the local hyperpressure method, nucleation of diamond nanocrystals through self-assembling of carbon atoms within the surface of growing DLC films was studied for the case of ion sputtering of graphite.

EXPERIMENTAL

DLC films were produced by ion (DC magnetron) sputtering of a graphite target in argon-hydrogen (4:1) plasma on KBr and amorphous quartz substrates. The substrate temperature was close to 500 K. The average magnetron power was varied in the range 0.25 to 0.45 kW. Films having thickness from 0.1 to 1.5 micrometers were grown. Optical absorption spectra (1–6 eV) of DLC films deposited onto fused silica substrates were measured with a Hitachi-U3410 spectrophotometer in order to characterize the microstructure of grown films by means of analysis of the optical absorption edge.

A Philips 420 TEM operating at 100 keV and Selected Area Electron Diffraction (SAED) was used for direct microstructural analysis of DLC films grown under various growth chamber gas pressures (2–20 mTorr). For this purpose, free-standing films approximately 100 nm thick were prepared by dissolving KBr substrates in water.

RESULTS AND DISCUSSION

It was found that the frequency dependencies of the absorption coefficient of the fabricated films follow the Tauc rule [2] with optical gaps E_g ranging from 1.6 to 2.2 eV (Figure 1). Being a parameter of short- and medium-range order, the Tauc gap E_g increases with decreasing growth pressure at given magnetron voltage. This corresponds to a decrease of the average number of six-member rings in sp^2-bonded fragments of the disordered matrix occurring when pressure is lowered, as evident from the inset in Figure 2. The inset shows cal-

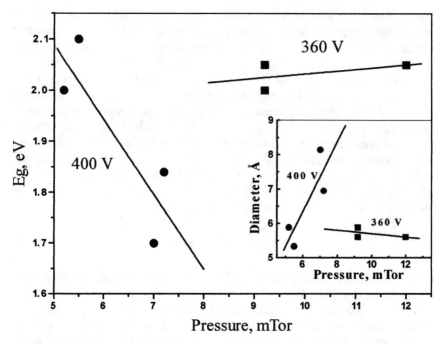

Figure 1. Tauc gap of DLC vs pressure for two magnetron voltages: 360 V and 400 V. The inset shows the average diameter of graphite constituent of DLC as a function of growth potential and pressure according to Equation (1).

culated sizes (diameter of the sp^2-bonded cluster D) of sp^2-bonded clusters according to the relation [3]:

$$D = d(2b)^2 / E_g^2 \qquad (1)$$

$b = 1.4$ eV is a constant, the size of each graphite ring d from which the cluster consists of was supposed to be 2.8 A.

Microstructural peculiarities of DLC films were studied by TEM and SAED. TEM image of the nanocrystalline carbon phase in DLC films grown at 10 mTorr and a magnetron potential of 360 V is shown in Figure 3. This figure shows clusters of average diameter 5.6±1 nm distributed in the DLC matrix. The inset presents SAED pattern which shows diffuse rings, and the lattice spacing corresponding to each ring provides a reasonable match to the spacing for the diamond lattice (Table 1).

The study of the size-distribution function of the diamond nanocrystals provides evidence of the mechanism of their nucleation into the

Table 1. SAED data for diamond nanocrystals (from Figure 2 SAED pattern).

Ring	Lattice Spacing d (nm)				Error, nm
	1	2	3	4	
Experimental value	0.206	0.146	0.132	0.104	±0.01
Standard (ASTM) value	0.206	0.1783	0.126	0.1075	
(hkl) plane	(111)	(200)	(220)	(311)	

DLC matrix. The fluctuation theory of phases nucleation [4,5] gives the Gaussian size-distribution function, which can be represented as:

$$P(D) = \frac{1}{\sqrt{2\pi} \cdot D_{av}} \exp\left(-\frac{(D - D_{av})^2}{2D_{av}^2}\right) \quad (2)$$

where D is a correlation radius of the fluctuation and D_{av} is average correlation radius. The size variation of diamond nanocrystals was analyzed directly from the TEM micrographs shown in Figure 2. The resulting size distribution is plotted in Figure 3. The experimental points are seen to fit reasonably with the distribution Function (2), in which the correlation radius of the fluctuation D substituted for the nanocrystal's diameter, and the average correlation radius of the fluctuation D_{av} substituted for the average diameter of nanocrystal. This means that the nanocrystal sizes are distributed as the fluctuation sizes do, that is, the fluctuation mechanism of diamond nanocrystal nucleation into the DLC matrix does take place. In other words, the Gaussian shape of size-distribution function indicates that the nucleation stage of nanocrystals formation occurs under thermodynamic conditions resulting from the fluctuation mechanism of self-assembling—carbon atoms arriving at the surface are able to create a hyperpressure within the small spatial areas. As a result, the carbon atoms confined within the correlation of the fluctuation at the presence of hydrogen reassemble themselves into the diamond lattice.

In the frames of the theory of a new phase nucleation Landau [4] the average embryo diameter obeys a power law, which can be written as:

$$D_{av} \propto (T(P) - T_c(P))^{-v} \quad (3)$$

where v is the critical index for the diameter D_{av} of the new phase, $T(P)$ is the effective temperature, and $T_c(P)$ corresponds in our case to the equilibrium temperature of the graphite-diamond phase diagram [6]. The dependence $T(P)$ in Equation (3) means that the energy of carbon atoms striking the film surface is crucial for the size of nucleated diamond crystals.

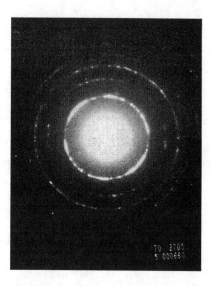

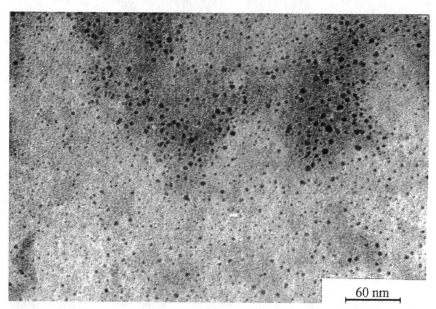

Figure 2. TEM image of diamond nanocrystals in DLC films grown at 10 mTorr and magnetron potential of 360 V. Inset presents SAED pattern.

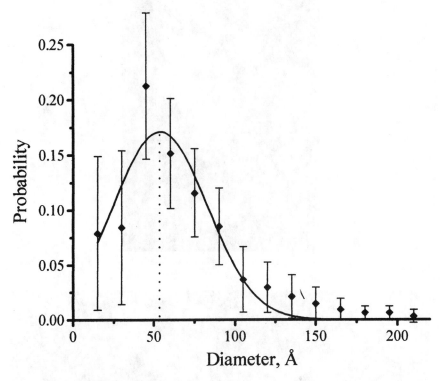

Figure 3. The experimental diamond nanocrystal distribution function obtained from TEM micrograph shown in Figure 2 and its Gaussian approximation using Equation (2). Black grains are diamond nanocrystals.

The nucleation of graphite clusters is demonstrated by a typical TEM picture of DLC films grown at pressures different from 10 mTorr (namely, $P = 6$ mTorr), presented in Figure 3. This figure shows clusters of average diameter 0.66 nm; in this case SAED patterns were diffuse rings with lattice spacing corresponding reasonably to that for graphite (Table 2). The experimental distribution function and its Gaussian approximation are shown in Figure 5. It is seen that the average graphite

Table 2. SAED data for graphite nanoclusters (from Figure 4 SAED pattern).

Ring	Lattice Spacing *d* (nm)				Error, nm
	1	2	3	4	
Experimental value	0.314	0.195	—	0.167	±0.02
Standard (ASTM) value	0.338	0.212	0.202	0.169	
(hkl) plane	(002)	(100)	(101)	(004)	

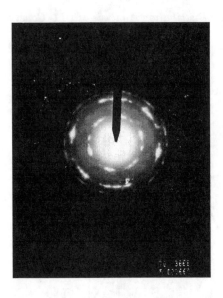

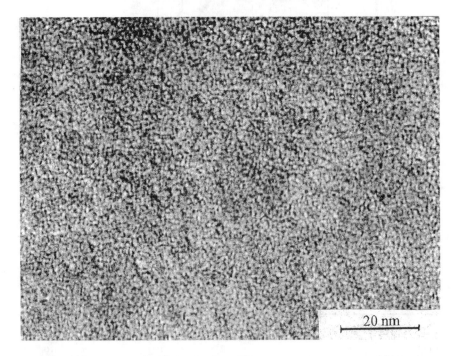

Figure 4. TEM image of DLC films grown at pressure higher than 10 mTorr and magnetron potential of 400 V. Black grains are graphic nanocrystals. Inset presents SAED pattern.

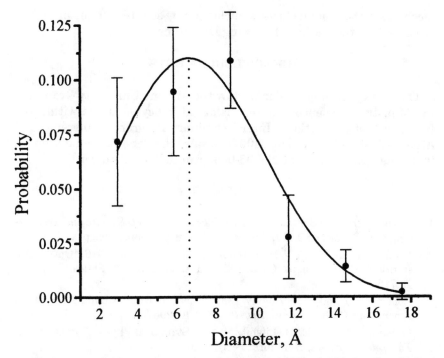

Figure 5. The experimental graphite-like nanocluster size distribution function obtained from TEM micrograph shown in Figure 4 and its Gaussian approximation.

cluster size is in a good agreement with the optical data (Figure 1). A rather good coincidence of the Gaussian curve and experimental data shows that the nucleation stage of graphite cluster formation in DLC films is observed.

CONCLUSION

Nucleation of diamond nanocrystals with average sizes in the range of nanometers is shown to take place at a growth pressure approximately equal to 10 mTorr and a magnetron voltage of about 360 V. These conditions cause formation of DLC with diminished average number of six-member rings in sp^2-bonded fragments of the disordered matrix. Analysis of size-distribution functions for diamond and graphite phases clearly manifests the nucleation stage of carbon cluster formation in the course of ion-sputtered DLC film growth. The existence of a pressure optimum for the formation of the diamond phase is evidence in favor of fluctuation mechanism of diamond clusters nucleation by self-assembling of carbon atoms within the surface of growing DLC films.

ACKNOWLEDGEMENTS

This work was supported in part by the Arizona University Grant and the Ministry of Science and Technological Policy of Russia within the framework of the National Interdisciplinary Program "Fullerenes and Atomic Clusters" (grant No. 94007). We also thank Soros Foundation for individual student's grant No. 95-069 (granted to A. Lodygin).

REFERENCES

1. Angus, J. 1994. In *Synthetic Diamond: Emerging CVD Science and Technology,* K. E. Spear and J. P. Dismukes, eds., Wiley & Sons, Inc., p. 21.
2. Ivanov-Omskii, V. I., A. V. Tolmatchev and S. G. Yastrebov. 1996. "Optical Absorption of Amorphous Carbon Doped with Copper," *Philos. Mag. B.,* 73:715–722.
3. Robertson, J. and E. P. O'Reilly. 1987. "Electronic and Atomic Structure of Amorphous Carbon," *Phys. Rev. B.,* 35:2946–2957.
4. Landau, L. D. and E. M. Lifshitz. 1981. "Statistical Physics," in *Course of Theoretical Physics,* New York: Pergamon.
5. Landau, L. D. and E. M. Lifshitz. 1981. "Physical Kinetics," *Course of Theoretical Physics,* New York: Pergamon.
6. Bundy, F. P. 1979. "The P-T Phase and Reaction Diagram for Elemental Carbon," *J. Geophys. Res.,* 85:6930–6936.

Author Index